# Communications
# in Computer and Information Science  661

*Commenced Publication in 2007*
Founding and Former Series Editors:
Alfredo Cuzzocrea, Dominik Ślęzak, and Xiaokang Yang

## Editorial Board

More information about this series at http://www.springer.com/series/7899

Dmitry I. Ignatov · Mikhail Yu. Khachay
Valeri G. Labunets · Natalia Loukachevitch
Sergey I. Nikolenko · Alexander Panchenko
Andrey V. Savchenko · Konstantin Vorontsov (Eds.)

# Analysis of Images, Social Networks and Texts

5th International Conference, AIST 2016
Yekaterinburg, Russia, April 7–9, 2016
Revised Selected Papers

Springer

*Editors*
Dmitry I. Ignatov (ORCID)
National Research University Higher School
  of Economics
Moscow
Russia

Mikhail Yu. Khachay
Krasovsky Institute of Mathematics and
  Mechanics
Yekaterinburg
Russia

Valeri G. Labunets
Ural Federal University
Yekaterinbug
Russia

Natalia Loukachevitch
Research Computing Center
Lomonosov Moscow State University
Moscow
Russia

Sergey I. Nikolenko
National Research University Higher School
  of Economics
St. Petersburg
Russia

Alexander Panchenko
Technische Universität Darmstadt
Darmstadt
Germany

Andrey V. Savchenko
Laboratory of Algorithms and Technologies
  for Networks Analysis
National Research University Higher School
  of Economics
Nizhny Novgorod
Russia

Konstantin Vorontsov
Dorodnicyn Computing Centre of Russian
  Academy of Sciences
Moscow
Russia

ISSN 1865-0929          ISSN 1865-0937  (electronic)
Communications in Computer and Information Science
ISBN 978-3-319-52919-6          ISBN 978-3-319-52920-2  (eBook)
DOI 10.1007/978-3-319-52920-2

Library of Congress Control Number: 2017930145

Printed on acid-free paper

This Springer imprint is published by Springer Nature
The registered company is Springer International Publishing AG
The registered company address is: Gewerbestrasse 11, 6330 Cham, Switzerland

# Preface

This volume contains proceedings of the 5th International conference on Analysis of Images, Social Networks, and Texts (AIST 2016)[1]. The previous conferences during 2012–2015 attracted a significant number of students, researchers, academics, and engineers working on interdisciplinary data analysis of images, texts, and social networks.

The broad scope of AIST makes it an event where researchers from different domains, such as image and text processing, exploiting various data analysis techniques, can meet and exchange ideas. We strongly believe that this may lead to crossfertilisation of ideas between researchers relying on modern data analysis machinery. Therefore, AIST brings together all kinds of applications of data-mining and machine learning techniques. The conference allows specialists from different fields to meet each other, present their work, and discuss both theoretical and practical aspects of their data analysis problems. Another important aim of the conference is to stimulate scientists and people from industry to benefit from the knowledge exchange and identify possible grounds for fruitful collaboration.

The conference was held during April 7–9, 2016. Following an already established tradition, the conference was organised in Yekaterinburg, a cross-roads between European and Asian parts of Russia, the capital of the Urals region. The key topics of AIST are analysis of images and videos; natural language processing and computational linguistics; social network analysis; pattern recognition, machine learning and data mining; recommender systems and collaborative technologies; Semantic Web, ontologies and their applications.

The Program Committee and the reviewers of the conference included well-known experts in data mining and machine learning, natural language processing, image processing, social network analysis, and related areas from leading institutions of 27 countries including Australia, Bangladesh, Belgium, Brazil, Croatia, Cyprus, Egypt, Estonia, Finland, France, Germany, Greece, India, Ireland, Israel, Italy, Japan, Lithuania, Norway, Portugal, Qatar, Russia, Spain, Switzerland, The Netherlands, UK, Ukraine, and USA.

This year we have received 142 submissions mostly from Russia but also from Australia, Austria, France, Hungary, India, Italy, Mexico, Norway, Sweden, UK, USA, and Vietnam.

Out of 142 submissions only 30 papers were accepted as regular oral papers. Thus, the acceptance rate of this volume was around 21%. In order to encourage young practitioners and researchers, we included three papers by industry speakers to the main volume and 38 papers to the supplementary proceedings. Each submission was reviewed by at least three reviewers, experts in their fields, in order to supply detailed and helpful comments.

---

[1] http://aistconf.org/.

The conference featured several invited talks and an industry session dedicated to current trends and challenges.

The invited talks were:

- Mehdi Kaytoue (Université de Lyon, CNRS, INSA-Lyon, France), "Finding Duplicate Labels in Behavioral Data: An Application for E-Sport Analytics"
- Attila Kertesz-Farkas (Higher School of Economics, Russia), "False Discovery Rate Control for Database Search Methods over Heterogeneous Biological Data"
- Radhakrishnan Delhibabu (Kazan Federal University, Russia), "Analysis of Anisotropic Diffusion in Image Processing"

The business speakers also covered a wide variety of topics[2]. We have included three of these invited talks in the main volume:

- Denis Nikiforov, Alexander Korchagin, and Ruslan Sivakov (Centre of Information Technology, Ekaterinburg, Russia) "An Ontology-Driven Approach to Electronic Document Structure Design"
- Tatyana Prisyach (Speech Technology Center, St. Petersburg, Russia), Valentin Mendelev, and Dmitry Ubskiy (ITMO-University, St. Petersburg, Russia), "Data Augmentation for Training of Noise Robust Acoustic Models"
- Alexander Semenov (Higher School of Economics, Russia) and Peter Romov (Yandex, Russia), "Performance of Machine Learning Algorithms in Predicting Game Outcome from Drafts in Dota"

We would also like to mention the best conference papers selected by the Program Committee within each section:

- "On Complexity of Searching a Subset of Vectors with Shortest Average under a Cardinality Restriction" by Anton Eremeev, Alexander Kelmanov, and Artem Pyatkin (*Machine Learning and Data Mining*)
- "The Problem of The Optimal Packing of The Equal Radius Circles for Non-Euclidean Metric" by Alexander Kazakov, Anna Lempert and Huy Liem Nguyen (*Data Analysis, Social Networks and Complex Data*)
- "Parallel Non-blocking Deterministic Algorithm for Online Topic Modeling" by Oleksandr Frei and Murat Apishev (*Natural Language Processing*)
- "Image Processing Algorithms with Structure Transferring Properties on the Basis of Gamma-Normal Model" by Inessa Gracheva and Andrey Kopylov (*Analysis of Images and Video*)

The final round of the SNA Hackathon 2016[3] co-organised by OK.Ru and the AIST team took place in conjunction with the main conference during April 6–8.

We would like to thank the authors for submitting their papers and the members of the Program Committee for their efforts in providing exhaustive reviews. We would

---

[2] The detailed program of AIST 2016 Business Day can be found at a separate website: http://dataconf.org.

[3] http://snahackathon.org/.

also like to express special gratitude to all the invited speakers and industry representatives.

We deeply thank all the partners and sponsors. Our golden sponsors are Exactpro[4] and OK.Ru[5]. Exactpro, a fully owned subsidiary of the London Stock Exchange Group, specialises in quality assurance for exchanges, investment banks, brokers, and other financial sector organisations. OK.Ru, part of the Mail.Ru Group, is one of the two largest Russian social networking services. Our bronze sponsor is the Centre of Information Technologies (Centre IT)[6]. This is a Russian engineering company that specializes in creating exclusive IT solutions and services for the interdepartmental electronic interaction both on national and international levels.

We would like to acknowledge the Scientific Fund of Higher School of Economics for providing AIST participants with travel grants. Our special thanks goes to Springer for their help, starting from the first conference call to the final version of the proceedings. Last but not least, we are grateful to all organisers, especially to Eugeniya Vlasova, and the volunteers, whose endless energy saved us at the most critical stages of the conference preparation.

Traditionally, we would like to mention the Russian word "aist" is more than just a simple abbreviation (in Cyrillic) – it means a "stork." Since it is a wonderful free bird, a symbol of happiness and peace, this stork brought us the inspiration to organise the AIST conference. So we believe that this young and rapidly growing conference will likewise be bringing inspiration to data scientists around the world!

October 2016

Dmitry Ignatov
Michael Khachay
Valery Labunets
Natalia Loukachevitch
Sergey Nikolenko
Alexander Panchenko
Andrey Savchenko
Konstantin Vorontsov

---

[4] exactprosystems.com.

[5] OK.ru.

[6] centre-it.com.

# Organisation

## Program Committee Chairs

Dmitry I. Ignatov     National Research University Higher School of
Economics, Moscow, Russia

Mikhail Khachay     Krasovsky Institute of Mathematics and Mechanics of
Ural Branch of Russian Academy of Sciences,
Russia

Valery G. Labunets     Ural Federal University, Russia

Natalia Loukachevitch     Computing Centre of Lomonosov Moscow State
University, Russia

Sergey I. Nikolenko     National Research University Higher School of
Economics and Steklov Institute of Mathematics of
Russian Academy of Sciences, St. Petersburg,
Russia

Alexander Panchenko     Technische Universität Darmstadt, Germany and
Université catholique de Louvain, Belgium

Andrey Savchenko     National Research University Higher School of
Economics, Nizhny Novgorod, Russia

Konstantin Vorontsov     Dorodnicyn Computing Centre of Russian Academy of
Sciences, Russia

## Proceedings Chair

Dmitry I. Ignatov     National Research University Higher School of
Economics, Russia

## Organising Committee

Anna Golubtsova     National Research University Higher School of
Economics, Moscow

Irina Dolgaleva     National Research University Higher School of
Economics, Moscow

Eugeniya Vlasova     National Research University Higher School of
Economics, Moscow

Natalia Papulovskaya     Ural Federal University, Yekaterinburg

Ekaterina Borovitina     Chelyabinsk State University, Russia

## Volunteers

| | |
|---|---|
| Maxim Pasynkov | Krasovsky Institute of Mathematics and Mechanics of Ural Branch of Russian Academy of Sciences, Russia, Yekaterinburg |
| Anna Voronova | Yandex, Moscow |
| Eugene Tsymbalov | Webgames and National Research University Higher School of Economics, Moscow |
| Daria Baranetskaya | National Research University Higher School of Economics, Moscow |
| Valeria Bubnova | National Research University Higher School of Economics, Moscow |

## Business Day Chair

| | |
|---|---|
| Evgenia Vlasova | National Research University Higher School of Economics, Moscow |

## Program Committee

| | |
|---|---|
| Mikhail Ageev | Lomonosov Moscow State University, Russia |
| Nickolay Arefyev | Lomonosov Moscow State University and Digital Society Lab, Russia |
| Jaume Baixeries | Universitat Politècnica de Catalunya, Spain |
| Artem Baklanov | Krasovskii Institute of Mathematics and Mechanics of Ural Branch of the Russian Academy of Sciences, Russia and International Institute for Applied Systems Analysis, Austria |
| Pedro Balage | University of São Paulo, Brazil |
| Sergey Bartunov | National Research University Higher School of Economics, Russia and Deep Mind, UK |
| Malay Bhattacharyya | Indian Institute of Engineering Science and Technology, India |
| Elena Bolshakova | Moscow State Lomonosov University, Russia |
| Anastasia Bonch-Osmolovskaya | National Research University Higher School of Economics, Russia |
| Aurélien Bossard | Université Paris 8, France |
| Jean-Leon Bouraoui | Université catholique de Louvain, Belgium |
| Leonid Boytsov | Carnegie Mellon University, USA |
| Pavel Braslavski | Ural Federal University/Kontur Labs, Russia |
| Evgeny Burnaev | Institute for Information Transmission Problems of Russian Academy of Sciences, Russia |
| Aleksey Buzmakov | Inria-LORIA (CNRS-Université de Lorraine), France |
| Artem Chernodub | Institute of Mathematical Machines and Systems Problems of Ukraine National Academy of Science, Ukraine |

| Vladimir Chernov | Institute for Image Processing of Russian Academy of Science, Russia |
| Ekaterina Chernyak | National Research University Higher School of Economics, Russia |
| Marina Chicheva | Samara National Research University, Russia |
| Bonaventura Coppola | IBM Research, USA |
| Hernani Costa | University of Malaga, Spain |
| Boris Dobrov | Lomonosov Moscow State University, Russia |
| Sofia Dokuka | National Research University Higher School of Economics, Russia |
| Florent Domenach | Akita International University, Japan |
| Alexey Drutsa | Lomonosov Moscow State University and Yandex, Russia |
| Julia Efremova | Eindhoven University of Technology, The Netherlands |
| Natalia Efremova | Moscow State Lomonosov University, Russia |
| Maria Eskevich | Radboud University Nijmegen, The Netherlands |
| Myasnikov Evgeny | Samara National Research University, Russia |
| Stefano Faralli | University of Mannheim, Germany |
| Victor Fedoseev | Image Processing Systems Institute of Russian Academy of Sciences and Samara National Research University, Russia |
| Michael Figurnov | Skolkovo Institute of Science and Technology, Russia |
| Elena Filatova | New York City College of Technology, USA |
| Mark Fishel | University of Tartu, Estonia |
| Thomas Francois | Université catholique de Louvain, Belgium |
| Oleksandr Frei | Schlumberger, Norway |
| Binyam Gebrekidan Gebre | Max Planck Institute for Psycholinguistics, The Netherlands |
| Natalia Grabar | STL CNRS Université Lille 3, France |
| Dmitry Granovsky | Yandex, Russia |
| Mena Habib | Maastricht University, The Netherlands |
| Dmitry Ilvovsky | National Research University Higher School of Economics, Russia |
| Vladimir Ivanov | Kazan Federal University, Russia |
| Dmitry Kan | SemanticAnalyzer, Finland |
| Nikolay Karpov | National Research University Higher School of Economics, Russia |
| Egor Kashkin | Vinogradov Russian Language Institute of Russian Academy of Sciences, Russia |
| Yury Katkov | Ecole Polytechnique Fédérale de Lausanne, Switzerland |
| Mehdi Kaytoue | LIRIS, INSA de Lyon, France |
| Alexander Kelmanov | Sobolev Institute of Mathematics of Siberian Branch of Russian Academy of Sciences, Russia |
| Andrey I. Kibzun | Moscow Aviation Institute, Russia |
| Victor Kitov | Lomonosov Moscow State University, Russia |

| | |
|---|---|
| Eduard Klyshinskii | Moscow State Institute of Electronics and Mathematics, Russia |
| Ekaterina Kochmar | University of Cambridge, UK |
| Sergei Koltcov | National Research University Higher School of Economics, Russia |
| Olessia Koltsova | National Research University Higher School of Economics, Russia |
| Alex Konduforov | AltexSoft, Ukraine |
| Natalia Konstantinova | University of Wolverhampton, UK |
| Andrey Kopylov | Tula State University, Russia |
| Kirill Kornyakov | Itseez and Lobachevsky State University of Nizhni Novgorod, Russia |
| Mikhail Korobov | ScrapingHub, Russia |
| Anton Korshunov | Institute for System Programming of Russian Academy of Sciences, Russia |
| Evgeny Kotelnikov | Vyatka State University, Russia |
| Olga Krasotkina | Tula State University, Russia |
| Tomas Krilavicius | Vytautas Magnus University, Lithuania |
| Valentina Kuskova | National Research University Higher School of Economics, Russia |
| Andrey Kutuzov | University of Oslo, Norway |
| Andrey Kuznetsov | Samara National Research University, Russia |
| Alexander Lepskiy | National Research University Higher School of Economics, Russia |
| Vadim Levit | Ariel University, Israel |
| Benjamin Lind | National Research University Higher School of Economics, Russia |
| Natalia Loukachevitch | Lomonosov Moscow State University, Russia |
| Olga Lyashevskaya | National Research University Higher School of Economics & Vinogradov Russian Language Institute of Russian Academy of Sciences, Russia |
| Ilya Markov | University of Amsterdam, The Netherlands |
| Luis Marujo | Carnegie Mellon University, USA and University of Lisbon, Portugal |
| Sérgio Matos | University of Aveiro, Portugal. |
| Yelena Mejova | Qatar Computing Research Institute, Qatar |
| Benjamin Milde | Technische Universität Darmstadt, Germany |
| Olga Mitrofanova | St. Petersburg State University, Russia |
| Andrea Moro | Sapienza, Università di Roma, Italy |
| Hubert Naets | Université catholique de Louvain, Belgium |
| Vassilina Nikoulina | Xerox Research Center Europe, France |
| Damien Nouvel | Institut National des Langues et Civilisations Orientales, France |
| Dimitri Nowicki | Institute of Cybernetics of Ukraine National Academy of Science, Ukraine |

| | |
|---|---|
| Patrick Watrin | Université catholique de Louvain, Belgium |
| Rostislav Yavorskiy | National Research University Higher School of Economics, Russia |
| Seid Muhie Yimam | Technische Universität Darmstadt, Germany |
| Marcos Zampieri | Saarland University, Germany |
| Olga Zvereva | Ural Federal University, Russia |

## Additional Reviewers

Guillaume Bosc
Sujoy Chatterjee
Vladimir M. Chernov
Silvio Ricardo Cordeiro
Samuel Daylis
Anna Denisova
Svyatoslav Elizarov
Denis Fedyanin
Yuri Kan
Benjamin Milde

Alexander Minkin
Andrey Naumov
Paraskevi Raftopoulou
Oleg Slavin
Ivan Sterligov
Dmitry Ustalov
Natali Vaganova
Diliara Valeeva
Lefteris Zervakis

## Sponsors

### Golden sponsors

Exactpro
OK.Ru (Mail.Ru Group)

### Bronze sponsor

IT Centre

# Abstracts of Invited Talks

# Analysis of Anisotropic Diffusion in Image Processing

Radhakrishnan Delhibabu

Kazan Federal University, Kazan, Russia
rdelhibabu@gmail.com

**Abstract.** Anisotropic diffusion is used widely in image processing for edge preserving filtering and image smoothing tasks. One of the important class of such model is by Perona and Malik [1], who used a gradient based diffusion to drive smoothing along edges and not across it. The contrast parameter used in the PM method needs to be carefully chosen to obtain optimal denoising results. Here we consider a local histogram based cumulative distribution approach for selecting this parameter in a data adaptive way so as to avoid manual tuning. We use spatial smoothing based diffusion coefficient along with adaptive contrast parameter estimation for obtaining better edge maps. Moreover, our experimental results indicate that this adaptive scheme performs well for a variety of noisy images. Thus, we obtain better peak signal to noise ratio and structural similarity scores with respect to fixed constant parameter values. The work proposes a modification of fuzzy diffusion coefficient that takes into account local pixel variability for better denoising and selective smoothing of edges.

**Keywords:** Image restoration · Anisotropic diffusion · Fuzzy edge detection · Contrast parameter · Local histogram · Diffusion coefficient · Denoising · Image processing

**Acknowledgment.** The speaker would like to thank Surya Prasath for the provided materials and help during the talk preparation.

# References

1. Perona, P., Malik, J.: Scale-space and edge detection using anisotropic diffusion. IEEE Trans. Pattern Anal. Mach. Intell. **12**(7) (1990) 629–639
2. Canny, J.: A computational approach to edge detection. IEEE Trans. Pattern Anal. Mach. Intell. **8**(6), 679–698 (1986)
3. Prasath, V.B.S.: A well-posed multiscale regularization scheme for digital image denoising. Appl. Math. Comput. Sci. **21**(4), 769–777 (2011)
4. Prasath, V.B.S., Delhibabu, R.: Image restoration with fuzzy coefficient driven anisotropic diffusion. In: Swarm, Evolutionary, and Memetic Computing - 5th International Conference, SEMCCO 2014, Bhubaneswar, India, December 18–20, 2014, 145–155 (2014). Revised Selected Papers

5. Prasath, V.B.S., Delhibabu, R.: Automatic contrast parameter estimation in anisotropic diffusion for image restoration. In: Analysis of Images, Social Networks and Texts - Third International Conference, AIST 2014, Yekaterinburg, Russia, April 10–12, 2014, 198–206 (2014). Revised Selected Papers
6. Prasath, V.B.S., Singh, A.: A hybrid convex variational model for image restoration. Appl. Math. Comput. **215**(10), 3655–3664 (2010)
7. Prasath, V.B.S., Singh, A.: Well-posed inhomogeneous nonlinear diffusion scheme for digital image denoising. J. Appl. Math. **2010**, 763847:1–763847:14 (2010)
8. Prasath, V.B.S., Vorotnikov, D.: On a system of adaptive coupled pdes for image restoration. J. Math. Imaging Vis. **48**(1), 35–52 (2014)

# Finding Duplicate Labels in Behavioral Data: an Application for E-Sport Analytics

Mehdi Kaytoue

Université de Lyon, CNRS, INSA-Lyon, LIRIS UMR5205, 69621,
Villeurbanne, France
mehdi.kaytoue@insa-lyon.fr

**Abstract.** Analyzing behavioral data, by means of data mining, machine learning and visualization, helps answering several industrial challenges and proposing rich applications. When a trace is labeled by the user that generates it, models can be learned to accurately predict the user of an unknown trace in many domains (online security, target marketing, fraud detection) thanks to driving, typing, and mobility patterns. In online systems however, a user may have several virtual identities (avatar aliases): when this mapping is not known, the prediction accuracy drastically drops as the model considers each alias as a different user. In this invited talk, I will present several solutions to tackle this duplicate labels identification problem [1, 2]. We will consider online video games, and eSport in particular, as a use case. I will finish by presenting our recent results in this emerging domain, especially for the mining of strategic patterns from game behavioral data [3–6].

**Keywords:** Sport Analytics · eSport analytics · Behavioral data · Game data science · Video game · DOTA · StarCraft

# References

1. Cavadenti, O., Codocedo, V., Boulicaut, J., Kaytoue, M.: When cyberathletes conceal their game: Clustering confusion matrices to identify avatar aliases. In: IEEE International Conference on Data Science and Advanced Analytics (DSAA) (2015)
2. Labernia, Q., Codocedo, V., Kaytoue, M., Robardet, C.: Découverte de labels dupliqués par l'exploration du treillis des classifieurs binaires. In: 16ème Journées Francophones Extraction et Gestion des Connaissances (EGC). Volume E-30 of RNTI., Hermann-Éditions, 255–266 (2016)
3. Kaytoue, M., Plantevit, M., Zimmermann, A., Bendimerad, A., Robardet, C.: Exceptional contextual subgraph mining. Mach. Learn. (2016)
4. Cavadenti, O., Codocedo, V., Boulicaut, J., Kaytoue, M.: What did i do wrong in my MOBA game? mining patterns discriminating deviant behaviours. In: IEEE International Conference on Data Science and Advanced Analytics (DSAA), 1–10 (2015)
5. Bosc, G., Tan, P., Boulicaut, J.F., Raissi, C., Kaytoue, M.: A pattern mining approach to study strategy balance in RTS games. IEEE Trans. Comput. Intell. AI Game (99) (2015)
6. Low-Kam, C., Raïssi, C., Kaytoue, M., Pei, J.: Mining statistically significant sequential patterns. In: 13th International Conference on Data Mining (ICDM), IEEE Computer Society, pp. 488–497 (2013)

# False Discovery Rate Control for Database Search Methods over Heterogeneous Biological Data

Attila Kertesz-Farkas

National Research University Higher School of Economics, Moscow, Russia
akerteszfarkas@hse.ru

**Abstract.** Database searching methods have become standard methods in identification of biological data, which often involves iteratively matching a query data against large protein sequence databases inexact manner. Accurate assignment is hindered by two problems. First, large number of hypotheses considered; i.e. a high score assigned to a match may not end up being statistically significant after multiple testing correction. Second, the hypothesis space is often combined by various types of data. Application of False Discovery Rate (FDR) control procedures over the combined hypothesis space in a single analysis can be dangerously leading to either overly conservative or overly liberal FDR within any subclasses. For these two problems, we propose a recently published new FDR control procedure, called Cascaded Search.

This method requires two inputs: (1) a user specified statistical confidence threshold, and (2) a series of peptide databases in a user specified order. For instance, such a cascade of databases could include fully tryptic, semitryptic, and nonenzymatic peptides or peptides with increasing numbers of modifications. Cascaded search then gradually searches the databases in a given order, sequestering at each stage query data that is identified with a specified statistical confidence. Our method has been compared to a standard procedure that lumps all of the peptides into a single database, as well as to a previously described Group-FDR procedure that computes the FDR separately within each database. We demonstrate, using simulated and real data, that cascaded search identifies more spectra at a fixed FDR threshold than that with either the ungrouped or grouped approach. Cascaded search thus provides a general method for maximizing the number of identified spectra in a statistically rigorous fashion and it is implemented in the open source Crux mass spectrometry analysis toolkit.

**Keywords:** Heterogeneous Biological Data · Inexact database searching · False Discovery Rate · Computational mass spectrometry

# References

1. Kertesz-Farkas, A., Keich, U., Noble, W.S.: Tandem mass spectrum identification via cascaded search. J. Proteome Res **14**(8), 3027–3038 (2015). PMID: 26084232
2. McIlwain, S., Tamura, K., Kertesz-Farkas, A., Grant, C.E., Diament, B., Frewen, B., Howbert, J.J., Hoopmann, M.R., Kall, L., Eng, J.K., et al.: Crux: rapid open source protein tandem mass spectrometry analysis. J. Proteome Res. **13**(10), 448–4491 (2014)

# Contents

**Natural Language Processing**

**Analysis of Images and Video**

# Industry Talks

# An Ontology-Driven Approach to Electronic Document Structure Design

Denis A. Nikiforov[✉], Alexander B. Korchagin, and Ruslan L. Sivakov

Centre of Information Technology, Ekaterinburg, Russia
{Denis.Nikiforov,Alexander.Korchagin,Ruslan.Sivakov}@centre-it.com

**Abstract.** Over the course of history, humankind used documents as one of the ways of organization of the data. In the recent decades, electronic documentation became increasingly widespread. To make electronic documents exchange possible, standards regulating transmission protocols, representation formats, and rules for document building are necessary. For some protocols (HTTP, SOAP, etc.) and formats (EDI, XML, JSON, etc.), relatively fixed and generally accepted standards are available. As for the electronic document design, there is an abundance of approaches where a leader could hardly be established; all of them have their benefits and drawbacks. This study explores some of these approaches (UN/CEFACT CCTS, WCO DM, ISO 20022, and NIEM). These approaches have different features but from the conceptual perspective they are intended to describe sets of details of some real-world objects. The paper proposes to describe such objects using an ontology and then, based on this ontology, build conceptual structures of electronic documents that can be converted to platform-independent structures of electronic documents in accordance with one of the standards. The introduced approach allows harmonizing the standards under consideration.

**Keywords:** Document engineering · Ontology · Model-driven architecture · Platform-independent model

## 1 Introduction

Electronic Data Interchange (EDI) is among the first standards in electronic data exchange. It regulates protocols, formats and rules for building electronic documents. The Transportation Data Coordinating Committee started to develop this standard in the 1960s and its first version was published in the 1970s. Later, standards like ANSI ASC X12[1], UN/EDIFACT [3], HL7 [1], and many others were developed based on EDI. To this day, those standards are dominant in electronic commerce [2].

In the 1990s–2000s, with Internet expansion, the focus has shifted from protocols and formats of data transmission to structure and semantics of electronic documents. Moreover, structure is described in platform-independent

---

[1] http://www.x12.org.

© Springer International Publishing AG 2017
D.I. Ignatov et al. (Eds.): AIST 2016, CCIS 661, pp. 3–16, 2017.
DOI: 10.1007/978-3-319-52920-2_1

form, for example, in UML language. Documents themselves can be presented in different platform-depended languages (mainly, EDI or XML). Main standards of this group are UN/CEFACT Core Components Technical Specification (CCTS) [19], World Customs Organization Data Model (WCO DM)[2], ISO 20022 [4], and NIEM [15]. They are key to Government-to-Government (G2G) and Government-to-Business (G2B) interactions.

In all aforementioned approaches, an electronic document is viewed as a hierarchy of data elements. In the 2000s, with growing popularity of Semantic Web, a step change from data hierarchy exchange to fact exchange was made. These approaches are mostly used in rather complicated specialized environments, for example, in industry standards (ISO 15926 [5]).

However, in electronic commerce and G2G/G2B interactions, "traditional" (hierarchical) document exchange is still dominating. It is strongly associated with specifics of these types of interactions, as their entire standard framework is geared towards exchange of (electronic) documents rather than facts. Also, a concept of legal value applies to documents, not facts. Documents have strictly determined hierarchical structure, and they do not need flexibility of data presentation provided by RDF or OWL. Instead, strict control of electronic document content is necessary.

All the abovementioned suggests that in the nearest future, standards oriented on "traditional" exchange of electronic documents rather than facts will be used in electronic commerce and G2G/G2B interactions. However, current standards in this field have at least two drawbacks.

First, there are too many standards and they are not compatible. In this article, we shall present a method unifying approaches described in CCTS, WCO DM, ISO 20022, NIEM specifications. It allows to design electronic documents structures in compliance with any of these standards.

Second, no readily available and convenient tools for designing structures of electronic documents exist. Either general-purpose editors not really convenient for documents design (UML editors, XML Schema editors), or specialized commercial editors (e.g. GEFEG.FX[3]) are usually proposed as tools. In this article, we shall describe our free editor based on open standards and frameworks.

The paper is organized as follows. In Sect. 2, we shall evaluate four approaches to electronic document structure design. In Sect. 3, we shall propose our ontology-driven approach. Finally, we shall conclude this paper in Sect. 4.

## 2    Overview of Analogues

While standards in question (CCTS, WCO DM, ISO 20022, NIEM) differ in details, their conceptual approach to design of electronic document structures can be summarized in the following chart (Fig. 1).

---

[2] http://www.wcoomd.org.
[3] http://www.gefeg.com/en/gefeg.fx/fx_descr.htm.

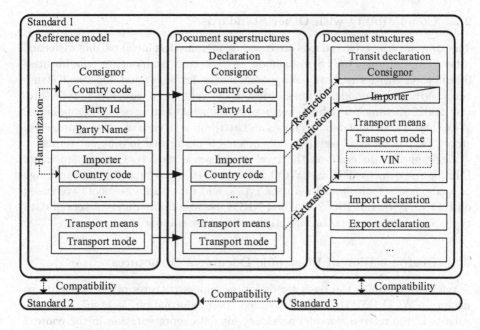

**Fig. 1.** General design chart of electronic document structures

## 2.1 Harmonization of Data Elements

All approaches assume creation of a library of data elements or core components. For clarity, let us call it a reference model. This model describes acceptable data elements that can be used in designing electronic document structures.

A reference model does not depend on an application context. In some information exchange, it may be sufficient to indicate a consignor's classification code in a customs declaration while their name may not be necessary. However, if the name of a consignor is necessary in some other document (outside the context of customs control), this element must be defined in a reference model.

The main purpose of a reference model is harmonization of data elements being used in different electronic documents. For example, country code element must have the same description and uncontroversial semantics regardless of the context where it is used.

As for the description of data elements, all standards in question are based on ISO 11179 [6]. It is the framework standard, which defines basic concepts (such as data element, value domain, etc.) and rules of metadata presentation.

As for the rest, specifics of these standards in harmonization are different. CCTS and ISO 20022 allow to harmonize data elements and structures conceptually. However, unlike NIEM, they do not allow to re-use associations, roles and properties of objects. WCO DM gives the least possibilities for harmonization. For example, it describes not a vehicle per se but a number of its characteristics.

## 2.2 Compatibility with Other Standards

The abovementioned standards are based on metamodels incompatible with each other. For example, CCTS metamodel defines core components and business information entities as basic component blocks for electronic document structures. NIEM metamodel defines object type, role type, association type, etc. ISO 20022 metamodel defines business components and message components. WCO DM metamodel defines classes and attributes. In each of these approaches, structural modelling of electronic documents is performed in different terms.

Differences also exist at the level of reference models. Core Components Library (used in CCTS) and WCO DM are partially compatible due to usage of a common set of unqualified data types, and many data elements are based on ISO 7372 [7]. But ISO 20022 and NIEM use absolutely different sets of data types and data elements, incompatible with other standards.

## 2.3 Customization of Electronic Document Structures

A reference model is the basis for designing superstructures of electronic documents. In WCO DM, they are also called "base information package". They are subsets of the reference model necessary for data representation in the context of some process.

Structures of electronic documents are designed on the basis of superstructures. As a rule, it amounts to exclusion of excessive data elements from superstructures (e.g. "Importer" in Fig. 1), posing restrictions on mandatory data elements requirements ("Consignor"), and posing restrictions on value domain.

CCTS specification describes two mechanisms of customization: qualification of core components and restriction of context of information entities use. WCO DM recommends to customize information packages at the level of XML schema. NIEM does not assume design of superstructures of electronic documents; new information exchange packages are created by copying and modifying the existing ones.

## 2.4 Extension of a Reference Model and Electronic Document Superstructures

In many cases, standard may not take into consideration national or other specificities of data exchange. For example, in some countries, a vehicle identification number should be noted in a customs declaration. But this data element is not defined in WCO DM. In this case, a standard must provide a mechanism for extending a reference model and superstructures.

WCO DM allows (but does not recommend) to extend information packages by inclusion of necessary elements into XML schema. ISO 20022 allows to include data with arbitrary XML schema into specific extension points of a message. CCTS and NIEM allow to describe extensions of electronic data structures at a higher (platform-independent) level of abstraction, not directly in XML schema. In CCTS, it is performed by qualification, which allows to describe several derived information entities on the basis of one core component in addition to restricting core components. In NIEM, possibilities of model extension are even wider. It allows to define new entities.

## 2.5   Design Tools

GEFEG.FX is a basic tool for design of electronic document structures on the basis of CCTS, WCO DM and ISO 20022. It is a closed-source software with proprietary model representation formats. For ISO 20022, Ecore-based metamodel is also accessible. On the basis of the metamodel, an Eclipse plugin for model viewing can be generated. ISO 20022 model is accessible in open XML Metadata Interchange (XMI) format [17]. For NIEM, a whole set of tools for designing information packages is available. However, they cannot be used to design electronic document structures based on other methodologies.

## 2.6   Comparison Results

We have summarized all the abovementioned and evaluated each of the standards using a five-grade scale in Table 1. All standards in question have limited possibilities of harmonization of data elements, they are not completely compatible with other standards, they have restrictions in customization and extension of electronic document superstructures, and they are not sufficiently equipped with design tools. Hereafter, we shall describe an approach that addresses some of these disadvantages.

**Table 1.** Evaluation of analogues

| Criteria | CCTS | WCO DM | ISO 20022 | NIEM |
| --- | --- | --- | --- | --- |
| Harmonization | 3 | 1 | 3 | 4 |
| Compatibility | 3 | 3 | 1 | 1 |
| Customization | 4 | 2 | 3 | 1 |
| Extension | 3 | 2 | 2 | 4 |
| Tools | 1 | 1 | 2 | 3 |
| Total | 14 | 9 | 11 | 13 |

## 3   Proposed Solution

In all of the examined approaches, electronic document structures are expected to be designed in a form of Platform-Independent Models (PIM). For example, that can be UML models [16] or other Ecore models [4]. After that, Platform-Specific Models (PSM) are generated from PIM, usually in a form of XML schema (Fig. 2).

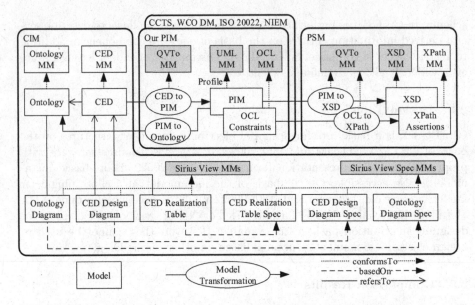

**Fig. 2.** The proposed approach to design of electronic document structures

A drawback of the existing approaches is a relatively limited mechanism of electronic document structures customization. For example, a transit declaration structure may be required to be built on the basis of some generalized customs declaration (Fig. 1). Herein, excessive data elements must be excluded from the structure, and conversely, other elements must be made mandatory. All studied approaches assume copying data sets available in the model and adding required changes to these copies (removing elements, changing multiplicity, etc.). More complicated restrictions (for example, on the summation value of some data elements or on a value of a data element depending on values of other data elements) are usually described in natural language and then programmers manually implement them in a code.

Our proposal is to replace creating copies with the usage of existing structures accompanied by describing all necessary restrictions in formal Object Constraint Language (OCL) [11]. During implementation of information exchange, these restrictions must be automatically transformed into expressions in some platform-specific language (XPath, SQL, Java, etc.). Examples of OCL constraints and appropriate XPath assertions will be presented in Sects. 3.3 and 3.4. This approach reduces amount of duplicated structures in a model, and minimizes human factor impact on implementation of information exchange [8].

We have been successfully using our approach to support cross-border information exchange between authorities of several countries [10]. However, integration of our information system with other systems based on different standards (CCTS, WCO DM, ISO 20022, and NIEM) is necessary in many cases. In those cases, it is usually proposed to create mappings between document structures.

Instead of mapping, we propose to unify all studied approaches to design of electronic document structures by adding one more level of abstraction, Computation Independent Model (CIM). This level is designated for describing Conceptual structures of Electronic Documents (CED), which are not dependent on any reference model (libraries of data elements, libraries of core components), and they are not dependent on the standard planned to be used during implementation of information exchange.

CED must be designed based on a uniform ontology. This ontology is a generalization of different reference models, but it does not describe data elements or core components used for design of electronic document structures. It describes all real-world objects, which details could be transmitted in electronic documents. It also describes all possible properties and relations of these objects. Our approach is conceptually different from the studied approaches as it clearly separates the ontological level (Sect. 3.1) and the level of data elements and data sets (Sect. 3.2).

To work with ontology and conceptual structures of electronic documents, we propose a tool [9], based on open standards (Meta Object Facility (MOF) [14], XMI [17]) and free frameworks (Eclipse Modeling Framework (EMF) [18], Sirius [20]). Using this tool, a developer can build a conceptual structure of an electronic document, choose a required standard, and automatically generate an electronic document structure compliant to the selected standard using MOF Query/View/Transformation (QVT) [13]. At the moment, only our standard is supported. First, it is a free alternative to commercial tools for designing structures of electronic documents (GEFEG.FX). Second, the need for mapping electronic documents structures is eliminated because they are based on a uniform ontology.

Designing a structure of an electronic documents based on a defined ontology is rather easy. The main challenge of this approach is creation and actualization of an ontology. We have developed a QVT transformation, which helps a developer to create an initial version of ontology based on an already existing reference model [9]. This ontology may contain many duplicated entities; some defined entities may result from conceptual errors. After automatic generation, the ontology must be harmonized manually.

## 3.1   Ontology

**Metamodel.** Like any model, an ontology must conform to some metamodel. We decided not to use RDF or OWL [12] and developed our own metamodel (Fig. 3). Being less universal than RDF, it allows to only describe well-defined types of facts but it is easier to work with. If necessary, the ontology can be automatically transformed into RDF representation. Our ontology is very similar to an Entity-Relationship (ER) model, but does not replicate it. The main difference is the ability to reuse properties and roles of objects.

Table 2 describes correlation of structures used for building our ontological model, and those described in other standards.

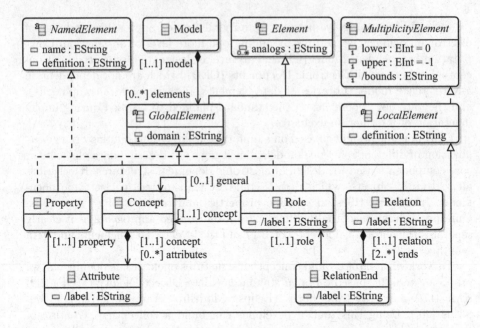

**Fig. 3.** The proposed ontology metamodel

**Table 2.** Correlation of structures defined in various metamodels

| Ontology | ISO 11179 | Our PIM [10] | CCTS | ISO 20022 | NIEM | WCO DM |
|---|---|---|---|---|---|---|
| Element | Administered item | — | Core component | Business concept | — | — |
| Concept | Concept | Complex type | Aggregate core component | Business component | Object type | Class |
| Property | Property | Simple element | Basic core component property | — | Property holder | Property term |
| Attribute | Data element concept | Component | Basic core component | Business attribute | Property | Attribute |
| Relation | Concept relationship | — | — | — | Association type | — |
| Role | — | Complex element | Association core component property | — | Role type | — |
| Relation end | — | Component | Association core component | Business association end | — | — |

**Sample Model.** Figure 4 presents a small fragment of our ontology. For simplicity, repeatedly used properties and roles are not depicted. For example, country code attributes (in a business entity and in a subject address) are both based on one global property. Consignor and consignee roles can also be reused. There are different notations for ontologies (RDF graphs, EXPRESS-G, VOWL), but we consider this simplified notation as the most convenient at the moment.

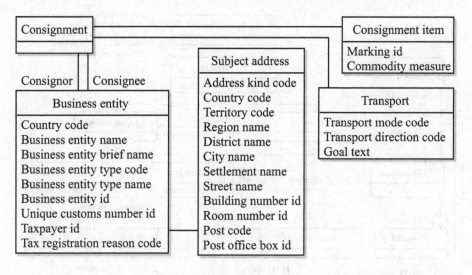

**Fig. 4.** An ontology fragment

## 3.2   Conceptual Structure of Electronic Document

**Metamodel.** An electronic document structure consists of data elements and data sets (Fig. 5). Data sets can include data elements and other data sets.

A data set must be based on some object role as defined in the ontology. Nesting other data sets based on roles connected with the basic role of the first data set is acceptable. It is also acceptable to use data elements based on properties of the object, which role defines this data set. In these cases, a structure of an electronic document complies with the ontology.

If objects, roles, relations or properties are not yet defined in the ontology, and their details must be transmitted in electronic document, then a developer can define new data elements or data sets. In this case, an electronic document structure will not comply with the ontology. The latter must be actualized afterwards.

**Sample Model.** Figure 6 presents an example of a conceptual structure of an electronic document. In the design view, a developer determines data sets and data elements necessary for this document based on a defined ontology (Fig. 4). Then, in the implementation view, he can (1) specify multiplicity of components; (2) specify their definitions; (3) indicate, which PIM objects they must be implemented with.

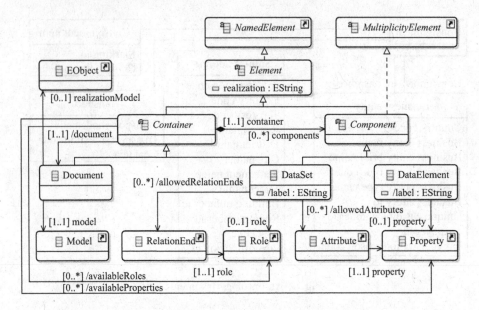

**Fig. 5.** The metamodel of conceptual structures of electronic documents

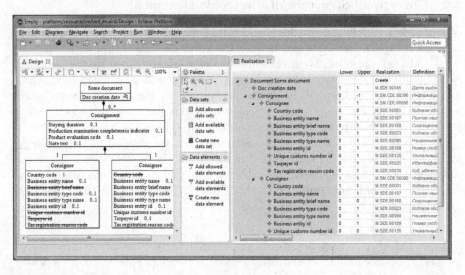

**Fig. 6.** The example of design and implementation views of an electronic document conceptual structure

## 3.3   Platform-Independent Conceptual Structure of Electronic Document

**Metamodel.** Once an analyst has designed a conceptual structure of an electronic document and has determined ways of implementation of its components, it only takes starting the QVT transformation and generating a structure of electronic document based on one of the standards. In the future, it will be possible to generate structures with different methodology (CCTS, ISO 20022, WCO DM, etc.), but for now, only our methodology is supported. To comply with it, an electronic document structure must be a UML model based on UML profile, which is described in [10].

**Sample Model.** The QVT transformation results in a UML model similar to the one below (Fig. 7).

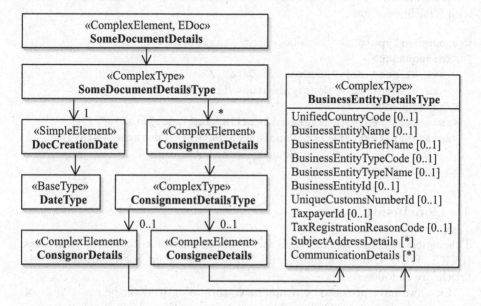

**Fig. 7.** The example of a platform-independent model of an electronic document

When designing a conceptual structure of the electronic document, an analyst defined required and prohibited details of consignor and consignee (Fig. 6). However, it is not reflected in PIM in Fig. 7. These requirements can be accommodated in one of two ways: (1) by creating separate composite data types for consignor and consignee, where appropriate multiplicity of data elements is specified; (2) by using common type for consignor and consignee while describing additional restrictions in some formal language. We use the latter approach and describe additional requirements in OCL. Examples of these restrictions in natural language and in OCL are presented below.

Element ConsignmentDetails/ConsignorDetails is required:

```
ConsignmentDetails.value.ConsignorDetails.value->notEmpty()
```

Element ConsignmentDetails/ConsigneeDetails/CountryCode is prohibited:

```
ConsignmentDetails.value.ConsigneeDetails.value
.CountryCode.value->isEmpty()
```

Constraints can be complicated; their detailed analysis is out of scope of this article.

## 3.4 Platform-Specific Model

The next stage is generation of PSM (XML schema, ER model, etc.) on the basis of PIM. With that, OCL expressions are translated in some platform-specific language (XPath, SQL, Java, etc.).

We transform UML models with OCL constraints into XML schemas 1.1 with XPath assertions:

```
<xs:complexType name="SomeDocumentDetailsType">
  <xs:sequence>
    <xs:element ref="DocCreationDate" />
    <xs:element ref="ConsignmentDetails" />
  </xs:sequence>
  <xs:assert test="ConsignmentDetails/ConsignorDetails" />
  <xs:assert test="fn:not(ConsignmentDetails/ConsigneeDetails/
                        CountryCode)" />
</xs:complexType>
```

## 4 Conclusion

This article studies four standards for design of electronic document structures (CCTS, WCO DM, ISO 20022, and NIEM). The main contributions of this article are as follows.

First, we have presented our approach, proposing to describe constraints of electronic documents in OCL and then to translate these constraints into some platform-specific language, for example, XPath [8].

Second, we have proposed to use ontology for description of real-world objects, which details can be transmitted in electronic documents. Based on this ontology, a developer must design conceptual structures of electronic documents, and then he must transform them into structures conforming to one of the studied standards.

Third, we have proposed a free tool [9] to work with ontologies and conceptual structures of electronic documents based on open standards (MOF [14], XMI [17]) and frameworks (EMF [18], Sirius [20]).

We plan to enhance our approach and tool in the following directions.

First, we plan to introduce additional relation types (for example, mereological) and concept types (subjects, objects, events, etc.) into the ontology metamodel.

Second, we plan to refine the tool so that it will allow to generate platform-independent models based on different standards (CCTS, ISO 20022, WCO DM, NIEM, etc.).

Third, we plan to develop several QVT transformations, which will add new entities to an ontology from different reference models (Core Components Library, WCO DM, etc.).

The presented approach can be applied to design of electronic document structures being used in electronic commerce, in G2G or G2B interactions.

# References

1. ANSI: Health Level Seven Standard Version 2.6 – An Application Protocol for Electronic Data Exchange in Healthcare Environments. ANSI/HL7 V 2.6, American National Standards Institute (2007)
2. Glushko, R., McGrath, T.: Document Engineering: Analyzing and Designing Documents for Business Informatics & Web Services. MIT Press, Cambridge (2005)
3. ISO: Electronic data interchange for administration, commerce and transport (EDIFACT) – Application level syntax rules. ISO 9735:2002, International Organization for Standardization (2002)
4. ISO: Financial services – Universal financial industry message scheme – Part 1: Metamodel. ISO 20022–1:2003, International Organization for Standardization (2003)
5. ISO: Industrial automation systems and integration – Integration of life-cycle data for process plants including oil and gas production facilities – Part 1: Overview and fundamental principles. ISO 15926–1:2004, International Organization for Standardization (2004)
6. ISO: Information technology – Metadata registries – Part 1: Framework. ISO/IEC 11179–1:2004, International Organization for Standardization (2004)
7. ISO: Trade data interchange – Trade data elements directory. ISO 7372:2005, International Organization for Standardization (2005)
8. Nikiforov, D.A.: UML to XML Schema 1.1 Transformation, November 2013. http://dx.doi.org/10.5281/zenodo.16151
9. Nikiforov, D.A.: Conceptual Electronic Document Editor, February 2016. http://dx.doi.org/10.5281/zenodo.46610
10. Nikiforov, D.A., Lisikh, I.G., Sivakov, R.L.: An approach to multi-domain data model development based on the model-driven architecture and ontologies. In: Khachay, M.Y., Konstantinova, N., Panchenko, A., Delhibabu, R., Spirin, N., Labunets, V.G. (eds.) Supplementary Proceedings of the 4th International Conference on Analysis of Images, Social Networks and Texts (AIST 202015), Yekaterinburg, Russia, 9–11 April 2015. CEUR Workshop Proceedings, vol. 1452, pp. 106–117. CEUR-WS.org (2015)
11. OMG: Object Constraint Language (OCL), version 2.4. Specification, Object Management Group (2014)
12. OMG: Ontology Definition Metamodel (ODM), version 1.1. Specification, Object Management Group (2014)
13. OMG: Meta Object Facility (MOF) 2.0 Query/View/Transformation, version 1.2. Specification, Object Management Group (2015)
14. OMG: Meta Object Facility (MOF), version 2.5. Specification, Object Management Group (2015)
15. OMG: UML Profile for National Information Exchange Model (NIEM), version 3.0. Specification, Object Management Group (2015)

16. OMG: Unified Modeling Language (UML), version 2.5. Specification, Object Management Group (2015)
17. OMG: XML Metadata Interchange (XMI), version 2.5.1. Specification, Object Management Group (2015)
18. Steinberg, D., Budinsky, F., Paternostro, M., Merks, E.: EMF: Eclipse Modeling Framework 2.0, 2nd edn. Addison-Wesley Professional, Amsterdam (2009)
19. UN/CEFACT: Core Components Technical Specification, version 3.0. Specification, United Nations Centre for Trade Facilitation and Electronic Business (2009)
20. Viyović, V., Maksimović, M., Perisić, B.: Sirius: a rapid development of DSM-graphical editor. In: 2014 18th International Conference on Intelligent Engineering Systems (INES), pp. 233–238, July 2014

# Data Augmentation for Training of Noise Robust Acoustic Models

Tatiana Prisyach[1,2], Valentin Mendelev[2(✉)], and Dmitry Ubskiy[3]

[1] STC-Innovations Ltd., St. Petersburg, Russia
[2] Speech Technology Center, St. Petersburg, Russia
{prisyach,mendelev}@speechpro.com
[3] ITMO-University, St. Petersburg, Russia
ubskiy@speechpro.com

**Abstract.** In this paper we analyse ways to improve the acoustic models based on deep neural networks with the help of data augmentation. These models are used for speech recognition in a priori unknown possibly noisy acoustic environment (with the presence of office or home noise, street noise, babble, etc.) and may deal with both the headset and distant microphone recordings. We compare acoustic models trained on speech corpora with artificially added noises of different origins and reverberation. At various test sets, word recognition accuracy improvement over the baseline model trained on clean headset recordings reaches 45%. In real-life environments like a meeting room or a noisy open space, the gain varies from 10 to 40%.

**Keywords:** Data augmentation · Robust speech recognition · Deep neural network

## 1 Introduction

Recently, multilayer neural networks (deep neural networks, DNNs) have found a widespread use for acoustic modeling in speech recognition [1]. In many cases the DNNs demonstrate better generalization capabilities as compared with the conventional Gaussian mixture models (GMMs). But in the case where the conditions for training and testing (usage) of the DNN mismatch the recognition quality may degrade significantly. In order to compensate this mismatch, various techniques are used to increase the quality of the speech and decrease the influence of noises.

This research is concerned with methods to improve the DNN based acoustic models using bottleneck features [2] and speech data augmentation [3].

The initial training dataset includes clean headset recordings, whereas the trained acoustic model is intended to be used for recognition in noisy open space or in a meeting room.

The general problem which arises in the case where the training and testing corpora mismatch is to construct a recognition system which is robust to acoustic environment variability.

D.I. Ignatov et al. (Eds.): AIST 2016, CCIS 661, pp. 17–25, 2017.
DOI: 10.1007/978-3-319-52920-2_2

To solve that problem, techniques are utilized which compensate the mismatch between the testing and training corpora with the help of:

1. special features (application of noise robust features such as PNCC [4] and RASTA [5], feature normalisation [6], feature compensation—correction of features in the frequency domain—spectral subtraction [7], Wiener filtering [8]) or acoustic model parameters transformation (standard statistical techniques such as the maximum a posteriori (MAP) estimators [9], SAT+CMLLR [10]);
2. a priori knowledge about the environment (utilization of stereo data [11] to train the mapping from the noisy to clean speech; here the advantage depends on how close the training corpora is to the testing environment; multi-condition training, construction of noise dictionaries (cluster adaptive training, CAT [12]); combination of pre-trained acoustic models with the use of non-negative matrix factorisation (NMF [13]));
3. application of explicit and implicit noise models (vector Taylor series [14]);
4. addition of various kinds of noise with different SNRs, which may occur in the testing corpus (data augmentation) [15–17].

Many of the above approaches use a priori information to estimate the parameters for specific conditions and fail when no environment-specific data are present. The data augmentation based approach provides a considerable advantage because it works well even when no target data is available.

There are several ways to augment the training data: semi-supervised training [15], multi-lingual training [18], transformation of acoustic data [19], speech synthesis [20, 21].

The semi-supervised training approach assumes the use of the text produced by an automatic speech recognition system to train acoustic models. The advantage of this approach is that we are able to use, say, radio or TV broadcasts featuring various kinds of speakers and noises; the obvious drawback is the presence of recognition errors in the texts.

The important advantage of synthesized datasets lies in the ability to approximate the required recognition conditions and get the necessary amount of training data. In addition, this method allows to obtain a precise alignment of noised data using known text transcriptions and the corresponding clean recordings.

The methods based on transformation of acoustic features include the variation of the vocal tract length on the stage of extracting the standard features [17] and stochastic feature mapping (SFM) [20].

The family of techniques based on recording transformations includes such methods as the audio signal speed alteration [19], applying noises, introduction of artificial reverberation into the records [22].

To transform the data we apply the artificial reverberation with the use of binaural room impulse response (BRIR) [21] and several kinds of noise (street noise, office or home noise, babble) with various signal-to-noise ratio (SNR). The initial training dataset includes headset recordings. The problem consists of training the acoustic model which can be applied both to headset and to distant microphone recordings under various noises and reverberation conditions. We demonstrate that the bottleneck feature extractor trained on the augmented train datasets is more robust to the noise and increases the recognition accuracy.

In the second section, we describe acoustic features and the DNN structure used in training. The third section includes the description of the train and test datasets, as well as the datasets resulting from data augmentation. The fourth section presents the results and discussion of the study, and the conclusion follows in the fifth section.

## 2    Bottleneck Features and DNN Structure

The bottleneck features extracted from a multilayer neural network have found a wide use in automatic speech recognition systems. Such features have been successfully used in [23,24] to solve the recognition problem under the testing and training corpora mismatch conditions. All acoustic models in our presentation are trained on this kind of features. The bottleneck features are generated from the DNN which has a hidden layer of smaller dimension as compared with the other layers.

In this paper we consider two bottleneck feature extractors:

1. the extractor trained on the initial training dataset including clean headset voice records only;
2. the extractor trained on the same corpora after applying data augmentation.

In Fig. 1, the general structure of deep neural networks used for training is shown. The first DNN is trained on plain MFCC features [25] (the left and right context length is equal to 15) to produce the bottleneck features. The network contains four fully connected hidden layers of dimension 2048 and a bottleneck layer of dimension 80.

The second DNN is trained on bottleneck features with context of length 5 and left/right spacing 3. The network contains four fully connected layers of dimension 2048 and a final classification layer with 2857 outputs.

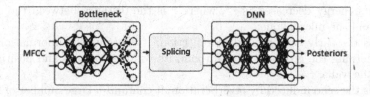

**Fig. 1.** The general DNN structure used to train the acoustic model

## 3   Speech Datasets for Training and Testing

In order to decrease the mismatch between the training and testing conditions, we make use of various transformations of the initial sound files preserving the state alignment unaltered. The difficulty consists in constructing a corpus which matches the reverberation and noise conditions which are unknown at the train- . ing phase. Since this objective is unattainable, we augment the training dataset with some variations to make our acoustic model more robust.

The training and test datasets are compiled from the recordings made by the Speech Technology Center. The sets contain phonetically rich sentences recorded with the use of a headset and distant microphones.

We consider the following ways to augment the training dataset:

1. application of noises corresponding to certain acoustic conditions (babble, office, home, car, street) with SNR from a fixed interval;
2. artificial reverberation of speech recordings.

For convenience we label the training datasets by abbreviations that reflect the properties of data containing in them. The training set C (clean data) contains only clean headset recordings of more than a thousand of different speakers. The set NB (noise, babble) includes a subset of recordings from C mixed with office, street, car noises and background speech (babble). The background recordings were scaled before mixing them with the clean data to produce the desired signal to noise ratio.

For artificial reverberation, we use BRIR, which contains the information about the size of the room where the recording is carried out, the distance to the sound source and its direction. BRIR includes three basic components:

$$h(t) = h_{\mathrm{dp}}(t) + h_{\mathrm{ee}}(t) + h_{\mathrm{rev}}(t),$$

where

$h_{\mathrm{dp}}(t)$   reproduces the sound passing directly from the source to the microphone; it depends on the azimuth and height of the source and the microphone; its energy decreases as $r^2$, where $r$ is the distance between the source and the microphone;

$h_{\mathrm{ee}}(t)$   is the early echo related to reflection; it contains the information concerning the geometry of the room, its volume, number and positions of the walls;

$h_{\mathrm{rev}}(t)$   is the echo induced by reverberation, it contains a large number of reflections and dispersions of higher order.

We use two kinds of BRIR:

1. the distance from the source to the microphone is equal to 3 m, the azimuth is 0, the room parameters are $24 \times 15 \times 4.5$, which makes the reverberation time equal to 0.5 s;

**Table 1.** The description of the training and test datasets

| Dataset | Duration, hours | SNR, dB | RT60, sec. |
|---|---|---|---|
| Train datasets | | | |
| C (clean data) | 353 | [15; 30] | [0.1; 0.3] |
| R (reverb) | 222 | [15; 30] | [0.5; 1.5] |
| NB (noise, babble) | 250 | [−5; 10] | [0.2; 1.5] |
| Test datasets | | | |
| T1 | 1.4 | [15; 30] | [0.1; 0.3] |
| T2 | 1.4 | [7; 10] | [0.1; 0.3] |
| T3 | 1.4 | [−5; 7] | [0.1; 0.3] |
| T4 | 1.5 | [15; 30] | [0.1; 0.5] |
| T5 | 0.4 | [20; 45] | [0.1; 0.3] |

2. the distance from the source to the microphone is equal to 5.5 m, the azimuth is 90, the room parameters are $24 \times 15 \times 4.5$, which makes the reverberation time equal to 0.8 s.

Detailed description of the training and test datasets is presented in Table 1.
The test datasets are divided into five groups based on the SNR and noise types. Each test dataset contains recordings of several dozens of speakers which were not included in the training sets. The first three groups contain the recordings made with the use of the close (T1), medium (T2) and long (T3) range microphone respectively. The environments are the office, domestic, and street. T4 contains background speech. T5 contains headset recording with a high SNR. The SNR is calculated as in [27] with decisions made by our voice activity detection (VAD) algorithm. $RT_{60}$ denotes the reverberation time which is the time required for reflections of a direct sound to decay 60 dB.

The concluding table in this paper contains the results of comparison of acoustic models trained with the use of data augmentation on real-life datasets, which consist of recordings of dialogues in a meeting room and in a noisy open space at a peak rush of people. The recordings are characterized by a low SNR (10 dB on average), presence of background speech and noise of various kinds (the sales register printer, electronic queue alerts, phone rings, etc.).

The information concerning the datasets compiled from real-life data is presented in Table 2.

**Table 2.** The description of the train and test datasets derived from real recordings

| Dataset | Type of microphone | SNR, dB | Rev-time, sec. |
|---|---|---|---|
| R1 | Headset | [20; 35] | [0; 0.3] |
| R2 | Distant (1 m) | [10; 15] | [0; 0.5] |
| R3 | Distant (1 m) | [−10; 15] | [0; 0.6] |

R1 and R2 are done at the same time and at the same place but with the use of different devices.

## 4   Experimental Results

In order to test the acoustic models which utilize the data augmentation techniques, we train several DNNs on bottleneck features. All networks contain 4 fully connected hidden layers of dimension 2048 and are trained with the use of discriminative pre-training [29]. In Table 3, we show how the word accuracy (recognition accuracy, WAcc) depends on the properties of the train datasets compiled with the use of clean, noisy and reverberated recordings. Only the most interesting results were included in Table 3.

Word accuracy is defined as follows:

$$WAcc = 1 - WER = \frac{N - S - D - I}{N},$$

where WER – word error rate, N is the number of words in the reference, S is the number of substitutions, D is the number of deletions, I is the number of insertions.

**Table 3.** The recognition accuracy dependence from the datasets properties

| N | Name | Training data | | | | Features | | | Test accuracy | | | | |
|---|------|---|---|----|-----|------|------|------|------|------|------|------|------|
| | | C | R | NB | Hrs | MFCC | bn_C | bn_N | T1 | T2 | T3 | T4 | T5 |
| 1 | Baseline | + | | | 280 | + | | | 75.2 | 27.9 | 1.4 | 67.7 | 83.3 |
| 2 | C_bn_C | + | | | 353 | | + | | 78.9 | 51.2 | 8.4 | 78.6 | 87.2 |
| 3 | CR_bn_C | + | + | | 575 | | + | | 83.7 | 63.5 | 25.5 | 78.7 | 86.3 |
| 4 | CNBR_bn_C | + | + | + | 825 | | + | | **84.6** | 73.6 | 41.1 | 76.9 | 86.9 |
| 5 | CNBR_mfcc | + | + | + | 825 | + | | | 82.8 | 74.1 | 46.3 | 78.9 | **87.8** |
| 6 | CNBR_bn_N | + | + | + | 825 | | | + | 84.3 | **74.3** | **47.2** | **79.6** | 87.2 |

As a baseline we used the model trained with plain MFCC features on a subset of the C dataset (280 of 353 h).

From the Table 3 it is obvious that adding augmented data improves recognition accuracy a lot and that bottleneck features are more robust to the speaker and environment variability.

In Table 4, comparison results on real-life test sets are given.

One can see that at different test cases the increase of the recognition accuracy as compared with the baseline model is substantial and varies from 12 to 40%.

The test set Real_3 is a more challenging one, so the recognition accuracy gain obtained with the proposed methods is less than on Real_2. Recordings

**Table 4.** The recognition accuracy on real-life test cases

| N | Name | R1 | R2 | R3 |
|---|------|------|------|------|
| 1 | Baseline | 51.5 | 2.4 | 8.6 |
| 2 | C_bn_C | 60.5 | 14.4 | 25.7 |
| 3 | CR_bn_C | 62.2 | 29.7 | 37.9 |
| 4 | CNBR_bn_C | 62.3 | 34.4 | **38.6** |
| 5 | CNBR_mfcc | 59.5 | 38 | 37.9 |
| 6 | CNBR_bn_N | **63.9** | **42.7** | 37.9 |

in the Real_3 contain specific kinds of noise which we didn't use during the augmentation process and background speech. The latter is loud enough to be passed by the voice activity detection algorithm so the acoustic models recognize it as they become more robust to noisy environment and since the reference texts contain only words belonging to a target speaker a larger number of insertions occurs. Some reduction in WER may be achieved with a VAD algorithm tuned to work in adverse noisy environments.

The presented recognition accuracy values are low but they allow to successfully perform keyword search and solve certain speech analytics tasks.

We publish a Kaldi recipe[1] for building a speech recognition system for the Russian language. It is based on publicly available speech corpus (Voxforge) and may well serve as a starting point to study data augmentation and other techniques aimed at producing effective ASR solutions.

## 5 Conclusions

In this research, it has been shown experimentally that the application of data augmentation methods increases substantially the robustness of the DNN-based acoustic models. The bottleneck features themselves are more robust to perturbations of acoustic conditions, but when the extractor is trained on the augmented datasets the recognition accuracy increases even more. The increase of the recognition accuracy has been found to be as high as 45% at some test cases. Experiments with real-life recordings in a quiet meeting room and in a noisy open space with low SNR demonstrate that even in the case where we have only clean recordings from a low-range microphone for training purposes, certain data transformations allow us to significantly increase the recognition accuracy.

**Acknowledgements.** This work was financially supported by the Ministry of Education and Science of the Russian Federation, Contract 14.579.21.0057 (ID RFMEFI57914X0057).

---

[1] https://github.com/freerussianasr/recipes.

# References

1. Hinton, G., Deng, L., Yu, D., Dahl, G.E.: Deep neural networks for acoustic modeling in speech recognition: the shared views of four research groups. IEEE Sig. Process. Mag. **29**, 82–97 (2012)
2. Yaman, S., Pelecanos, J.W., Sarikaya, R.: Bottleneck features for speaker recognition. Odyssey **12**, 105–108 (2012)
3. Ragni, A., Knill, K.M., Rath, S.P., Gales, M.J.F.: Data augmentation for low resource languages. In: Proceedings of Interspeech 2014, pp. 810–814 (2014)
4. Kim, C., Stern, R.M.: Feature extraction for robust speech recognition based on maximizing the sharpness of the power distribution and on power flooring. In: Proceedings of ICASSP 2010, pp. 4574–4577 (2010)
5. Hermansky, H., Morgan, N., Bayya, A., Kohn, P.: Compensation for the effect of communication channel in auditory-like analysis of speech (RASTA-PLP). In: Proceedings of European Conference on Speech Technology 1991, pp. 1367–1370 (1991)
6. Viikki, O., Bye, D., Laurila, K.: A recursive feature vector normalization approach for robust speech recognition in noise. In: Proceedings of ICASSP 1998, pp. 733–736 (1998)
7. Boll, F.: Suppression of acoustic noise in speech using spectral subtraction. IEEE T-ASSP **27**(2), 113–120 (1979)
8. Mauuary, L.: Blind equalization in the cepstral domain for robust telephone based speech recognition. In: Proceedings of EUSPICO 1998, vol. 1, pp. 359–363 (1998)
9. Gauvain, J.-L., Lee, C.-H.: Maximum a posteriori estimation of multivariate Gaussian mixture observations of Markov chains. IEEE T-SAP **2**(2), 291–298 (1994)
10. Gales, M.J.F.: Maximum likelihood linear transformations for HMM-based speech recognition. Comput. Speech Lang. **12**, 75–98 (1998)
11. Deng, L., Acero, A., Jiang, L., Droppo, J., Huang, X.D.: High-performance robust speech recognition using stereo training data. In: Proceedings of ICASSP 2001, pp. 301–304 (2001)
12. Gales, M.J.F.: Cluster adaptive training of hidden Markov models. IEEE T-SAP **8**(4), 417–428 (2000)
13. Lee, D.D., Seung, H.S.: Algorithms for non-negative matrix factorization. In: Proceedings of NIPS 2000, pp. 556–562 (2000)
14. Deng, J., Li, L., Yu, D., Gong, Y., Acero, A.: High-performance HMM adaptation with joint compensation of additive and convolutive distortions via vector Taylor series. In: Proceedings of ASRU 2007, pp. 65–70 (2007)
15. Lamel, L., Gauvain, J.-L.: Lightly supervised and unsupervised acoustic model training. Comput. Speech Lang. **16**, 115–129 (2002)
16. Gales, M.J.F., Ragni, A., AlDamarki, H., Gautier, C.: Support vector machines for noise robust ASR. In: Proceedings of ASRU 2009, pp. 205–210 (2009)
17. Jaitly, N., Hinton, G.E.: Vocal tract length perturbation (VTLP) improves speech recognition. In: Proceedings of ICML 2013 (2013)
18. Burget, L., Schwarz, P., Agarwal, M., Akyazi, P.: Multilingual acoustic modeling for speech recognition based on subspace Gaussian mixture models. In: Proceedings of ICASSP 2010, pp. 4334–4337 (2010)
19. Ko, T., Peddinti, V., Povey, D., Khudanpur, S.: Audio augmentation for speech recognition. In: Proceedings of Interspeech 2015 (2015)

20. Cui, X., Goel, V., Kingsbury, B.: Data augmentation for deep neural network acoustic modeling. In: Proceedings of ICASSP 2014 (2014)
21. Jeub, M., Schaefer, M., Vary, P.: A binaural room impulse response database for the evaluation of dereverberation algorithms. In: Proceedings of 16th International Conference on Digital Signal Processing (DSP), Santorini, Greece (2009)
22. Peddinti, V., Chen, G., Povey, D., Khudanpur, S.L.: Reverberation robust acoustic modeling using i-vectors with time delay neural networks. In: Proceedings of Interspeech 2015, pp. 2440–2444 (2015)
23. Yu, D., Seltzer, M.L.: Improved bottleneck features using pretrained deep neural networks. In: Proceedings of Interspeech 2011, pp. 237–240 (2011)
24. Karafiát, M., Grézl, F., Burget, L., Szőke, I., Černoský, J.: Three ways to adapt a CTS recognizer to unseen reverberated speech in BUT system for the ASpIRE challenge. In: Proceedings of Interspeech 2015, pp. 2454–2458 (2015)
25. Picone, J.W.: Signal modeling techniques in speech recognition. Proc. IEEE **81**(9), 1215–1247 (1993)
26. Dean, D.B., Kanagasundaram, A., Ghaemmaghami, H., Rahman, M., Sridharan, S.: The QUT-NOISE-SRE protocol for the evaluation of noisy speaker recognition. In: Proceedings of the 16th Annual Conference of the International Speech Communication Association, Interspeech 2015, pp. 3456–3460 (2015)
27. Pollák, P.: Efficient and reliable measurement and simulation of noisy speech background. In: 2002 11th European Signal Processing Conference, pp. 1–4 (2002)
28. Löllmann, H.W., Yilmaz, E., Jeub, M., Vary, P.: An improved algorithm for blind reverberation time estimation. In: Proceedings of International Workshop on Acoustic Echo and Noise Control (IWAENC) (2010)
29. McDermott, E., Hazen, T., Roux, J.L., Nakamura, A., Katagiri, S.: Discriminative training for large vocabulary speech recognition using minimum classification error. IEEE Trans. Speech Audio Process **15**(1), 203–223 (2007)

# Performance of Machine Learning Algorithms in Predicting Game Outcome from Drafts in Dota 2

Aleksandr Semenov[1]($\boxtimes$), Peter Romov[2,3], Sergey Korolev[4,5], Daniil Yashkov[2,3], and Kirill Neklyudov[2,3]

[1] International Laboratory for Applied Network Research,
National Research University Higher School of Economics, Moscow, Russia
avsemenov@hse.ru
[2] Yandex Data Factory, Moscow, Russia
peter@romov.ru, k.necludov@gmail.com
[3] Moscow Institute of Physics and Technology, Moscow, Russia
daniil.yashkov@phystech.edu
[4] National Research University Higher School of Economics, Moscow, Russia
sokorolev@edu.hse.ru
[5] Institute for Information Transmission Problems, Moscow, Russia

**Abstract.** In this paper we suggest the first systematic review and compare performance of most frequently used machine learning algorithms for prediction of the match winner from the teams' drafts in Dota 2 computer game. Although previous research attempted this task with simple models, weve made several improvements in our approach aiming to take into account interactions among heroes in the draft. For that purpose we've tested the following machine learning algorithms: Naive Bayes classifier, Logistic Regression and Gradient Boosted Decision Trees. We also introduced Factorization Machines for that task and got our best results from them. Besides that, we found that model's prediction accuracy depends on skill level of the players. We've prepared publicly available dataset which takes into account shortcomings of data used in previous research and can be used further for algorithms development, testing and benchmarking.

**Keywords:** Online games · Predictive models · Dota 2 · Factorization machines · MOBA

## 1 Introduction

Cybersport or eSports has became really popular in recent years, turning into a spectacular entertainment and sport discipline where professional teams regularly participate in tournaments with huge money prizes [1]. Due to such popularity it generates huge amounts of behavioral data on players and matches which is often accessible via API (Application Programming Interface). This data allows researchers to apply different machine learning algorithms to enhance gaming

© Springer International Publishing AG 2017
D.I. Ignatov et al. (Eds.): AIST 2016, CCIS 661, pp. 26–37, 2017.
DOI: 10.1007/978-3-319-52920-2_3

AI [2], develop optimal strategies [3], detect game balance issues [4] and identify players with multiple aliases [5].

Dota 2 is an online multiplayer video game and its first part, DotA (after "Defense of the Ancients") created a new genre, called Multiplayer Online Battle Arena (MOBA). It is played by two teams, called Radiant and Dire which consist of five players each. The main goal of the game is to destroy other team's "Ancient", located at the opposite corners of the map after destroying all the towers on three different lanes that lead to that Ancient.

Each of the players choose one hero to play with from a pool of 113 heroes. A hero has a set of particular features that define his role in the team and playstyle. Among these features there are his basic attribute (Strength, Agility or Intelligence), unique set of 4 (or for some heroes even more) skills which serve for a wide variety of purposes from healing and increasing stats of friendly units to different types of damage, stun and slow down of enemy heroes. Besides that, there is a lot of items a player can buy for his hero, which increase some of his stats, skills or add new effects or spells. Skills and items allow each hero to fill several roles in the team, such as "damage dealer" (hero, whose role is to attack the enemies in the fight), "healer" (hero, who mostly heals and otherwise helps his teammates), "caster" (hero, who mostly relies on his spells) etc. However, besides roles there is another way of classification the heroes in the teams which is based on their functions and is believed to be the most efficient and balanced. According to that classification, the optimal composition of the team is the following: "Mid" (player, who starts from the middle lane and responsible for ganking attempts on the other lanes), "Carry" (damage dealer who is supposed to kill enemy heroes), "Offlaner" or "Hardlaner" (hero who either starts on the "hard lane" which is bottom lane for Dire and top lane for the Radiant, or roams between lanes and in the jungle) and two "Supports" (heroes who are responsible for buying items for the team, like observer wards, smokes, couriers etc.).

The game has several modes: All Pick, Captains Draft, Random Pick etc. which define the way and order the players choose heroes. In All Pick, for example, players can choose from all the pool of heroes without well defined order of pick which means that the one, who was quicker gets the hero he wants to play. Captains Draft on the other hand make only one person in the team responsible for picks and bans which are made in consecutive order. This two opposite regimes of play create different strategies for players in both choosing a hero and playing it. For example in All Pick player tries to take the hero he likes the most which sometimes isn't good for the balance in the team. Moreover, some heroes are over represented in All Pick while this stats is smoothed in other modes (like Random Draft, where each player gets his hero randomly).

There is a Matchmaking Rating system in Dota 2 called MMR, which allows the algorithm to put players with similar skill level into the same match for more balanced gameplay. It is also possible to make your rating visible to others via participating in such modes as Ranked All Pick, Ranked Random Draft and Ranked Captain's Mode which give all the members of the winning team some MMR points and removes the same amount of points from the lost team.

Currently only a dozen of players have MMR $\geq$ 8,000, while everything below 4,000 is considered as low-level, and 6,000–7,000 is a level of the professional player who participates in esport championships.

Different gaming modes affect on the hero pick popularity, win rate and other parameters. Moreover, the same hero's stats for kills, deaths, win rate etc. may differ dramatically between gamers from different skill brackets, measured by MMR system.

From this brief introduction to the game's mechanics it must already be clear that this system leads to a huge number of possible combinations of heroes and their spells, items and roles. For example the amount of different teams for 5 heroes each is equal to 140,364,532. In the next section we'll review how researches dealt with this level of complexity in prediction of winning team.

## 2   Previous Research

Dota 2 got attention from researchers only recently. First articles were mostly descriptive, general and theoretical, investigating, for example, rules and fair play maintenance of the games [6] or correlation of leadership styles of players with roles in the game (carry, support, jungler etc.) they choose to play [7]. In the first quantitative research of Dota 2, authors analyzed cooperation withing teams, national compositions of players, role distribution of heroes and some other stats based on information from its web forums [8].

Later researchers discovered the potential of the data provided by the game itself soon after that and started using to test hypothesis, detect patterns and make predictions. For example Rioult et al. [9] analyzed topological patterns of DotA teams based on area, inertia, diameter, distance and other features derived from their positions and movements of the players around the map to identify which of them are related with winning or loosing the game. Drachen et al. used Neural Networks and Genetic Algorithms to analyze and optimize patterns of heroes movements on the map in DotA [10]. Eggert et al. [11] applied classification algorithms to heroes game statistics to identify their roles. However most researched topic was win prediction from heroe drafts.

Conley and Perry were the first to demonstrate the importance of information from draft stage of the game with Logistic Regression and k-Nearest Neighbors (kNN) [6]. They got 69.8% test accuracy on 18,000 training dataset for Logistic Regression, but it could not capture the synergistic and antagonistic relationships between heroes inside and between teams. To address that issue authors used kNN with custom weights for neighbors and distance metrics with 2-fold cross-validation on 20,000 matches to choose $d$ dimension parameter for kNN. For optimal d-dimension $= 4$ they got 67.43% accuracy on cross-validation and 70% accuracy on 50,000 test datasets. Based on that results authors built a recommendation engine with web interface. However one of its drawbacks was it's slow speed: for $k = 5$ kNN took 4 h and 12 h for cross-validation.

Although their work was the first to show the importance of draft alone, the interaction among heroes within and between teams were hard to capture with such a simplistic approach. Agarwala and Pearce tried to take that into account including the interactions among heroes into the logistic regression model [12]. To define a role of each hero and model their interactions they used PCA analysis of the heroes' statistics (kills, deaths, gold per minute etc.). However, their results showed inefficiency of such approach, because it got them only 57% accuracy while the model without interactions got 62% accuracy. But its worth noticing that although the PCA-based models couldn't match predictive accuracy of logistic regression, the composition of teams they suggested looked more balanced and reasonable from the game's point of view. Another caveat of their approach was that they took data from different sources: the data on match statistics was taken from public games while stats on heroes were based on professional games. This might bias the results because public games are completely different from professional ones and match stats from the former should not be mixed with heroes stats from the letter. In short, they didn't use heroes roles directly and replaced them with PCA components to model the balance of teams. Besides that, they tried to find some meaningful strategies with K-Means clustering on end-game statistics but couldn't find clusters which means that no patterns of gameplay could be detected on their data.

Another approach to that problem of modeling heroes' interactions was proposed by Song et al. [13]. They took 6,000 matches and manually added 50 combinations of 2 heroes to the features set and used forward stepwise regression for feature selection. They chose data from the "All Pick", "Ranked All Pick" and "Random Draft" without leavers and zero kills. 10-fold CV logreg: 3,000 matches total: 2,700 vs. 300. Training error 28%, test error – 46% . They concluded that only addition of particular heroes improves the model while the others might cause the prediction go wrong.

Kalyanaraman was the first one to implicitly introduced the roles of the heroes as a feature in the model of win prediction [14]. Author took 30,426 matches from the "All Pick", "Random Draft", "Single Draft", "All Random", "Least Played" and "Captain Draft" game types because in theory it should represent all the heroes in the best way since appearance of any particular hero depends on game type. They filtered the matches by MMR to select only skilled players and took the games which were at least 900 s and used ensemble of Genetic Algorithms and logistic regression on 220 matches. Logistic regression alone return 69.42% and ensemble with Genetic Algorithm and logistic regression approached 74.1% accuracy on the test set. Although it's the highest result among all the articles in the review, lack of ROC AUC information and small sample of matches, chosen for the Genetic Algorithm, hampers its reliability.

Another attempt to include interaction among heroes was done by Kinkade and Lim, who took 62,000 matches with "very high" skill level without leavers and game duration at least 10 min [15]. 52,000 training, 5,000 testing and 5,000 validation. Tried Logistic Regression and Random Forest with such feature of a pairwise winrate for Radiant and Dire. The feature could capture such relationships as matchup, synergy and countering and each of them increased the quality of the model up to 72.9%. Logistic Regression and Random Forest on picks data only. Got 72.9% test accuracy for Logistic Regression and overfitted Random Forest which gave them after tuning only 67% test accuracy. It is worth mentioning that their baseline, which included highest combined individual win rate for the heroes, had 63% accuracy.

Some authors expanded the scope of win prediction from draft information to other data from the game. Johansson and Wikstrom wrote a thesis where they trained Random Forest on the information from the game (such as amount of gold for each hero, his kills, deaths assists for each minute etc.) which had 82.23% accuracy at the five minute point [16]. Although such accuracy seem to be very high, that fact that it's based on data from the game events makes its use very limited, because it demand real-time data to be practically useful.

From the previous research we've found the following shortcomings:

- vague data acquisition strategies (for example, its not clear why authors filter players by their skill level and only use games with high MMR);
- not enough details on the the quality of results (for example, most papers reported only precision, without ROC AUC, recall, etc.);
- small or incorrect samples of data (sometimes datasets were gathered during periods when some changes in the game mechanics were introduced or the samples were mere thousands of matches).

Hence our contribution is:

- mining and preparing of large and consistent dataset of Dota 2 matches for prediction modeling tasks;
- test the methods suggested previously on this dataset with standard performance metrics;
- introduce Factorization Machines algorithm for match outcome prediction based on interactions among heroes;
- make this dataset publicly available[1].

The rest of the article is organized as follows. In the next part we describe our dataset and our approach to the representations of hero drafts. Then we introduce machine learning algorithms we chose to test on this data. And finally we demonstrate the results of our comparison and their implication for the future research.

## 3   Game Outcome Prediction

We've set out to estimate the quality of a range of machine learning algorithms for prediction of the match outcome given each team hero drafts. Given this subset of 5 heroes per team we try to predict the result of the match, assuming that there is no ties, so $P(radiant\ wins) = 1 - P(dire\ wins)$.

---

[1] http://dotascience.com/papers/aist2016.

## 4   Dataset

We have collected dataset using Steam API. It contains 5,071,858 matches from Captains Mode, Random Draft and Ranked All Pick modes, played between $11^{th}$ February 2016 10:50:04 GMT and $2^{nd}$ March 2016 14:07:10 GMT, including skill levels of players. During this period there were no changes to the core mechanics of the game, such as major patches, which makes this dataset especially appropriate for algorithm development and testing. Another key feature of this data is augmenting it with players' MMR for ranking the games into several brackets depending on the players skills.

The distributions of number of matches for skill levels and game modes in the dataset are (Table 1):

**Table 1.** Data distribution

|  | Normal skill | High skill | Very high skill | Total |
|---|---|---|---|---|
| Captains Mode | 33,037 | 5,599 | 8,840 | 47,476 |
| Random Draft | 86,472 | 15,560 | 39,407 | 141,439 |
| Ranked All Pick | 2,937,087 | 917,001 | 1,028,855 | 4,882,943 |
| Total | 3,056,596 | 938,160 | 1,077,102 | 5,071,858 |

In the end we used three representations of hero drafts as input of the algorithms. First is just "bag of heroes" technique, where each draft is encoded as a binary vector of length $2 \times N$ where $N$ is the size of hero pool with

$$x_i = \begin{cases} 1, \text{ if } i \leq N \text{ and hero } i \text{ was in the radiant team} \\ \quad \text{or if } i > N \text{ and hero } i - N \text{ was in the dire team} \\ 0, \text{ otherwise} \end{cases}$$

We used it as input to the naive bayes, since we wanted to use this result as the baseline for the most straightforward type of draft encoding. We also used it with logistic regression to compare it's performance between this encoding and the second one. Also, factorization machines require binary vectors as input for the model, so we used it with this data as well.

Second is "bag of heroes" with team symmetry for equal weights of Logistic Regression where

$$x_i = \begin{cases} 1, & \text{if hero } i \text{ was in radiant team} \\ -1, & \text{if hero } i \text{ was in dire team} \\ 0, & \text{otherwise} \end{cases}$$

This way of data representation allows logistic regression to use same weights for the same hero picks on radiant or dire side so as to force the symmetry of the game mechanics.

Third is the same "bag of heroes" as in first one, but with added features for number of carries, pushers, supports and other roles in the radiant and dire team. Our hope was that these features would provide tree-based model with explicit information about the strong and weak sides of the given draft based on the composition of heroes' roles in the team.

## 5   Methods

For our tests we've chosen the following models: Naive Bayes classifier, Logistic Regression, Factorization Machines and Gradient Boosting of Decision Trees. The quality of prediction was measured by AUC and Log-Loss (Cross Entropy) on 10-fold cross-validation.

Naive Bayes and Logistic Regression were chosen to replicate results of previous works on our dataset and to set the baseline performance. Since we assume that combinations and interactions between different heroes and their roles should increase the quality of the model, we chose Factorization Machines and Decision Trees for their ability to model complex interactions on the sparse data.

### 5.1   Naive Bayes

This model assumes the independence of variables (picking particular heroes), compute univariate probability estimates from training set $P(x_j|y)$ and then use bayesian rule to infer win probability of the draft:

$$P(y|x) \propto P(x_1|y) \dots P(x_p|y)P(y)$$

The final decision rule for the model is

$$\hat{y} = \underset{k \in \{0,1\}}{\operatorname{argmax}} \, p(C_k) \prod_{i=1}^{n} p(x_i|C_k)$$

where $C_k$ is the possible outcome, $n$ is the number of features.

### 5.2   Logistic Regression

Logistic Regression is a linear model that tries to estimate the probabilities for given classes using a logistic function:

$$P(win) = \sigma(w_0 + \sum_{i=1}^{p} w_i x_i),$$

where $\sigma(a) = (1 + \exp(-a))^{-1}$ is an activation function.

Similar to the Naive Bayes approach this model can not distinguish interactions between heroes and estimate possible combinations and their significance for the match outcome. As such it can only estimate individual picks importance for the result of the match.

We tested Logistic Regression on both first and second type of draft encodings and compared the results.

## 5.3   Factorization Machines

Factorization Machines proposed in [17] models some real-valued target as:

$$\hat{y}(x) = w_0 + \sum_{j=1}^{P} w_j x_j + \sum_{j=1}^{P} \sum_{j'=p+1}^{P} P x_j x_{j'} \sum_{f=1}^{k} v_{j,f} v_{j',f}$$

where $\Theta = (w_0, w, V)$ — set of model parameters. For binary probability prediction, bayesian inference is used.

In other words, Factorization Machines compute predicted probability using pairwise interactions of the second order between chosen heroes.

We have used bayesian Factorization Machines [18] implemented in FastFM library [19].

## 5.4   Gradient Boosting of Decision Trees

We have used XGBoost library [20] for implementation of gradient boosting algorithm. It minimizes the following regularized objective:

$$\mathcal{L}(\phi) = \sum_{i} l(\hat{y}_i, y_i) + \sum_{k} \Omega(f_k)$$

$$\text{where } \Omega(f) = \gamma T + \frac{1}{2}\lambda||w||^2,$$

$\hat{y}_i$ is model prediction, $y$ is the true value, $l$ is the loss function, $T$ is the number of leaves in the tree, each $f_k$ corresponds to an independent tree structure and leaf weights $w$ and $\lambda$ is an $L_2$ regularization parameter.

The prediction $\hat{y}_i$ is the sum of predictions of trees $f_k(x_i)$:

$$\hat{y}_i = \sum_{k=1}^{k} f_k(x_i)$$

# 6   Results

We've ran all of the models described above on 10-fold cross-validation for ROC AUC and log-loss estimation on the respective datasets and achieved following results (Figs. 1, 2 and 3).

In the Table 2 libFM is Factorization Machines classifier, XGBoost is boosting classifier on simple encoding, XGBoost_roles is boosting classifier on enhanced dataset with roles of heroes, LogReg_BoW is the logistic regression classifier on the second type of "bag of heroes" encoding.

We've decided to omit the confidence intervals from this table, since for each of the models the standard deviation was less than 0.0008.

First of all, we decided to check if mixing of different game modes affect the modeling quality. For that purposes we ran Naive Bayes classification algorithm

**Table 2.** Results

| Skill method | Normal | | High | | Very high | |
|---|---|---|---|---|---|---|
| | auc | log_loss | auc | log_loss | auc | log_loss |
| libFM | 0.706 | 0.898 | 0.670 | 0.933 | 0.660 | 0.940 |
| XGBoost | 0.701 | 0.903 | 0.664 | 0.937 | 0.654 | 0.944 |
| XGBoost_roles | 0.702 | 0.902 | 0.663 | 0.938 | 0.653 | 0.945 |
| LogReg | 0.687 | 0.916 | 0.656 | 0.943 | 0.643 | 0.952 |
| LogReg_BoW | 0.688 | 0.915 | 0.656 | 0.943 | 0.643 | 0.952 |
| Naive Bayes | 0.685 | 0.917 | 0.653 | 0.945 | 0.641 | 0.954 |
| Naive Bayes AP | 0.684 | 0.918 | 0.654 | 0.944 | 0.641 | 0.954 |
| Dummy | 0.500 | 0.996 | 0.500 | 0.999 | 0.500 | 0.999 |

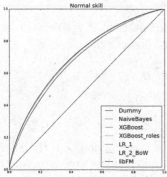

**Fig. 1.** Normal skill.

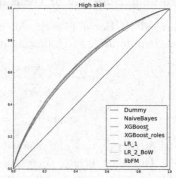

**Fig. 2.** High skill.

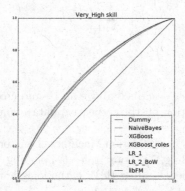

**Fig. 3.** Very high skill.

on 'All Pick' mode only (model Naive Bayes AP in the table) and compared the results with the same algorithm applied to the full dataset with all game modes. As we can see, the difference is at 0.001 level which means that mixing different game modes into one dataset almost doesn't affect prediction quality.

Besides that there is no difference between two logistic regression models performance, which means that there is already symmetry in the data so the model is able to infer the individual hero importance without the dataset augmentation. It suggests that side selection doesn't affect the performance of a player on the hero.

Final ranking of ROC AUC scores of the models are representative of the models ability to account for interactions among heroes from the data. Good performance of both Factorization Machines and XGBoost confirmed our assumptions that the ability to include interactions into the model results in significant increase of its performance. It's noteworthy that although gradient boosting decision trees can include interactions of more than second order while factorization machines work only with interactions of 2 heroes, the latter demonstrated better prediction quality than the former. The reasons behind that is unclear for as and might be the subject of research of its own.

It is also important to point out that the addition of role information to dataset held no improvement over "bag of heroes" data for XGBoost classifier. This finding seems counterintuitive since team's composition in terms of heroes' roles is considered to be important and common sense of the game suggests that having no carries in the game or having 5 of them is similarly bad for a team. It might be explained by the hypothesis that decision trees somehow found roles-like features by themselves, but we didn't investigate that hypothesis further.

Besides that we found out that the performance of the classifiers varies between different skill levels of players. More specifically, the higher the skill of the players is the harder it is to predict the outcome of a match. That might mean that low-skilled players depend on the pick more because they can only play on a limited amount of heroes. However for the more skilled player that's not the case since they know how to play and counter more heroes and interact with other teammates.

## 7   Discussion

In this article we introduced Factorization Machines and Gradient Boosted Decision Trees for the prediction of Dota 2 match winner. With them we managed to increase the accuracy of the winner prediction compared to the models, previously used to solve the problem of hero interactions in the drafts of Dota 2. We also mined the largest dataset of all studied before, for all levels of players' skills, demonstrated several data transformation techniques and used standard and consistent prediction quality measurements on it. Finally, we published the dataset and the code for our analysis online for reproducibility, algorithm development and benchmarking purposes. systems Although our results can be applied for prediction purposes it's still challenging to convert it into recommendation system which would suggest teams optimal pick in real time based on the currently picked/banned heroes because optimization of factorization machines is a very complex computational and analytic problem which demands further investigation.

Such work might be promising and useful for the game developers to access the balance of the game, and for the professional teams, because it will allow them to make data driven decisions for the drafts during the training and preparation for the tournaments. Besides that building applications, based on recommender systems for casual players also looks promising both from scientific and commercial point of view.

**Acknowledgements.** This paper was prepared within the framework of a subsidy granted to HSE by the Government of Russian Federation for implementation of the Global Competitiveness Program.

# References

1. Taylor, T.: Raising the Stakes: E-sports and the Professionalization of Computer Gaming. MIT Press, Cambridge (2012)
2. Ontanón, S., Synnaeve, G., Uriarte, A., Richoux, F., Churchill, D., Preuss, M.: A survey of real-time strategy game AI research and competition in starcraft. IEEE Trans. Comput. Intell. AI Games **5**(4), 293–311 (2013)
3. Synnaeve, G., Bessiere, P.: A Bayesian model for opening prediction in RTS games with application to StarCraft. In: IEEE Conference on Computational Intelligence and Games (CIG), pp. 281–288. IEEE (2011)
4. Bosc, G., Tan, P., Boulicaut, J.-F., Raissi, C., Kaytoue, M.: A pattern mining approach to study strategy balance in RTS games. IEEE Trans. Comput. Intell. AI Games **PP**(99), 1–1 (2015)
5. Cavadenti, O., Codocedo, V., Boulicaut, J.-F., Kaytoue, M.: When cyberathletes conceal their game: clustering confusion matrices to identify avatar aliases. In: 2015 IEEE International Conference on Data Science and Advanced Analytics (DSAA), pp. 1–10. IEEE, October 2015
6. Conley, K., Perry, D.: How does he saw me? A recommendation engine for picking heroes in Dota 2. Technical report (2013)
7. Nuangjumnonga, T., Mitomo, H.: Leadership development through online gaming. In: 19th ITS Biennial Conference: Moving Forward with Future Technologies: Opening a Platform for All, Bangkok, pp. 1–24 (2012)
8. Pobiedina, N., Neidhardt, J.: On successful team formation. Technical report (2013)
9. Rioult, F., Métivier, J.-P., Helleu, B., Scelles, N., Durand, C.: Mining tracks of competitive video games. AASRI Procedia **8**, 82–87 (2014). SECS
10. Drachen, A., Yancey, M., Maguire, J., Chu, D., Wang, I. Y., Mahlmann, T., Schubert, M., Klabajan, D.: Skill-based differences in spatio-temporal team behaviour in defence of the Ancients 2 (Dota 2). In: Games Media Entertainment (GEM), Dota 2, vol. 2, pp. 1–8. IEEE (2014)
11. Eggert, C., Herrlich, M., Smeddinck, J., Malaka, R.: classification of player roles in the team-based multi-player game Dota 2. In: Chorianopoulos, K., Divitini, M., Hauge, J.B., Jaccheri, L., Malaka, R. (eds.) ICEC 2015. LNCS, vol. 9353, pp. 112–125. Springer, Heidelberg (2015). doi:10.1007/978-3-319-24589-8_9

12. Agarwala, A., Pearce, M.: Learning Dota 2 team compositions. Technical report, Stanford University (2014)
13. Song, K., Zhang, T., Ma, C.: Predicting the winning side of DotA2. Technical report, Stanford University (2015)
14. Kalyanaraman, K.: To win or not to win? A prediction model to determine the outcome of a DotA2 match. Technical report, University of California San Diego (2014)
15. Kinkade, N., Jolla, L., Lim, K.: Dota 2 win prediction. Technical report, University of California, San Diego (2015)
16. Johansson, F., Wikström, J., Johansson, F.: Result prediction by mining replays in Dota 2. Ph.D. thesis, Blekinge Institute of Technology (2015)
17. Rendle, S.: Factorization machines. In: Proceedings of the IEEE International Conference on Data Mining, ICDM 2010, Washington, DC, USA, pp. 995–1000. IEEE Computer Society (2010)
18. Freudenthaler, C., Schmidt-Thieme, L., Rendle, S.: Bayesian factorization machines. In: Workshop on Sparse Representation and Low-rank Approximation, Neural Information Processing Systems (NIPS-WS) (2011)
19. Bayer, I.: fastFM: a library for factorization machines. CoRR, abs/1505.00641 (2015)
20. Chen, T., Guestrin, C.: XGBoost: a Scalable Tree Boosting System. arXiv e-prints, March 2016

# Machine Learning and Data Analysis

# Vote Aggregation Techniques in the Geo-Wiki Crowdsourcing Game: A Case Study

Artem Baklanov[1,2,3(✉)], Steffen Fritz[1], Michael Khachay[2,3],
Oleg Nurmukhametov[2], Carl Salk[1], Linda See[1], and Dmitry Shchepashchenko[1]

[1] International Institute for Applied Systems Analysis (IIASA), Laxenburg, Austria
{baklanov,fritz,salk,see,schepd}@iiasa.ac.at
[2] Krasovsky Institute of Mathematics and Mechanics, Ekaterinburg, Russia
mkhachay@imm.uran.ru, oleg.nurmuhametov@gmail.com
[3] Ural Federal University, Ekaterinburg, Russia

**Abstract.** The Cropland Capture game (CCG) aims to map cultivated lands using around 170000 satellite images. The contribution of the paper is threefold: (a) we improve the quality of the CCG's dataset, (b) we benchmark state-of-the-art algorithms designed for an aggregation of votes in a crowdsourcing-like setting and compare the results with machine learning algorithms, (c) we propose an explanation for surprisingly similar accuracy of all examined algorithms. To accomplish (a), we detect image duplicates using the perceptual hash function pHash. In addition, using a blur detection algorithm, we filter out unidentifiable images. In part (c), we suggest that if all workers are accurate, the task assignment in the dataset is highly irregular, then state-of-the-art algorithms perform on a par with Majority Voting. We increase the estimated consistency with expert opinions from 77% to 91% and up to 96% if we restrict our attention to images with more than 9 votes.

**Keywords:** Crowdsourcing · Image processing · Votes aggregation

## 1 Introduction

Crowdsourcing is a new approach for solving data processing problems for which conventional methods appear to be inaccurate, expensive, or time-consuming. Nowadays, the development of new crowdsourcing techniques is mostly motivated by so called Big Data problems, including problems of assessment and clustering of large datasets obtained in aerospace imaging, remote sensing, and even in social network analysis. For example, by involving volunteers from all over the world, the Geo-Wiki project tackles the problems of environmental monitoring with applications to flood resilience, biomass data analysis and forecasting, etc. The Cropland Capture game, which is a recently developed Geo-Wiki game, aims to map cultivated lands using around 170000 satellite images from the Earth's surface. Despite recent progress in image analysis, the solution to these problems is hard to automate since human-experts still outperform the majority of learnable machines and other artificial systems in this field. Replacement

© Springer International Publishing AG 2017
D.I. Ignatov et al. (Eds.): AIST 2016, CCIS 661, pp. 41–50, 2017.
DOI: 10.1007/978-3-319-52920-2_4

of rare and expensive experts by a team of distributed volunteers seems to be promising, but this approach leads to challenging questions: how can we aggregate individual opinions optimally, obtain confidence bounds, and deal with the unreliability of volunteers?

The main goals of the Geo-Wiki project are collecting land cover data and creating hybrid maps [15]. For example, users answer 'Yes' or 'No' to the question: 'Is there any cropland in the red box?' in order to validate the presence or absence of cropland [14]. In the paper [2], which is related to use of Geo-Wiki data, researchers studied the problem of using crowdsourcing instead of experts. The research showed that it is possible to use crowdsourcing as a tool for collecting data, but it is necessary to investigate issues such as how to estimate reliability and confidence.

This paper presents a case study that aims to compare the performance of several state-of-the-art vote aggregation techniques specifically developed for the analysis of crowdsourcing campaigns using the image dataset obtained from the Cropland Capture game. As a baseline, some classic machine learning algorithms such as Random Forest, AdaBoost, etc., augmented with preliminary feature selection and a preprocessing stage, are used.

The rest of the paper is structured as follows. In Sect. 2, we give a brief overview of the vote aggregation algorithms involved in our case study. In Sect. 3, we describe the general structure of the dataset under consideration. In Sect. 4, we propose quality improvements for the initial image dataset and introduce our vote aggregation heuristic. Finally, in Sect. 5, we present our benchmarking results.

## 2    Related Work

In the theoretical justification of crowdsourcing image-assessment campaigns, there are two main problems of interest. The first one is the problem of ground truth estimation from crowd opinion. The second one, which is equally important, deals with the individual performance assessment of the volunteers who participated in the campaign. The solution to this problem is in the clustering of voters with respect to their behavioural strategies into groups of *honest workers, biased annotators, spammers, malicious users*, etc. Note that a different approach is proposed in paper [1] that uses the biclustering to group the annotators based on their attempted questions.

Reflection of this posterior knowledge by reweighing of individual opinions of the voters can substantially improve the overall performance of the aggregated decision rule.

There are two basic settings of the latter problem. In the first setup, a crowdsourcing campaign admits some quantity of images previously labeled by experts (these labels are called *golden standard*). In this case, the problem can be considered as a supervised learning problem, and for its solution, conventional algorithms of ensemble learning (for example, boosting [7,11,20]) can be used. On the other hand, in most cases, researchers deal with the full (or almost full) absence of labeled images; ground truth should be retrieved simultaneously with estimation of voters' reliability, and some kind of unsupervised learning techniques should be developed to solve the problem.

Prior works in this field can be broadly classified in two categories: EM-algorithm inspired and graph-theory based. The works of the first kind extend results of the seminal paper [3], applying a variant of the well known EM-algorithm [4] to a crowdsourcing-like setting of the computer-aided diagnosis problem. For instance, in [13], the EM-based framework is provided for several types of unsupervised crowdsourcing settings (for categorical, ordinal and even real answers) taking into account different competency level of voters and different levels of difficulty in the assessment tasks. In [12], by proposing a special type of prior, this approach is extended to the case when most voters are *spammers*. Papers [8,10,17] develop the fully unsupervised framework based on Independent Bayesian Combination of Classifiers (IBCC), Chinese Restaurant Process (CRP) prior, and Gibbs sampling. Although EM-based techniques perform well in many cases, usually, they are criticized for their heuristic nature since in general there are no guarantees that the algorithm finds a global optimum.

Another approach applied to reliability of the voters is based on recent results obtained for random regular bipartite graphs. Karger et al. [6] obtained both an asymptotically optimal graph construction and an asymptotically optimal iterative inference algorithm on this graph. These results are extended in [9] by applying approximate variational methods including belief propagation and mean field.

Furthermore, in [5], an efficient reputation algorithm for identifying adversarial workers in crowdsourcing campaigns is elaborated. For some conditions, the reputation scores proposed are proportional to the reliabilities of the voters given that their number tends to infinity. Unlike the majority of EM-based techniques, the listed results have solid theoretical support, but conditions for which their optimality is proven (especially the graph-regularity condition) are too restrictive to apply them straightforward in our setup.

The aforementioned arguments have motivated us to carry out a case study on the applicability of several state-of-the-art vote aggregation techniques to an actual dataset obtained from the Cropland Capture game. Precisely, we compare the classic EM algorithm, methods proposed in [5,6], and a heuristic based on the computed reliability of voters. As a baseline, we use the simple Majority Voting (MV) heuristic and several of the most popular universal machine learning techniques.

## 3   Dataset

We carry out a benchmark of state-of-the-art vote aggregation techniques using the actual dataset obtained from the Cropland Capture game. The results of the game were captured as shown in two tables. The first table contains details of the images: *imgID* is an image identifier; *link* is the URL of an image; *latitude* and *longitude* are geo-coordinates which refer to the centroid of the image; *zoom* is the resolution of an image (values: 300, 500, 1000 m). The following table shows some sample of image data.

| imgID | link | latitude | longitude | zoom |
|-------|------|----------|-----------|------|
| 3009 | http://cg.tuwien.ac.at/~sturn/crop/img_-112.313_42.8792_1000.jpg | 42.8792 | −112.313 | 1000 |
| 3010 | http://cg.tuwien.ac.at/~sturn/crop/img_-112.313_42.8792_500.jpg | 42.8792 | −112.313 | 500 |
| 3011 | http://cg.tuwien.ac.at/~sturn/crop/img_-112.313_42.8792_300.jpg | 42.8792 | −112.313 | 300 |

All votes, i.e. 'a single decision by a single volunteer about a single image' [14], were collected in the second table: *ratingID* is a rating identifier; *imgID* is an image identifier; *volunteerID* is a volunteer's identifier; *timestamp* is the time when a vote was given; *rating* is a volunteer's answer. The possible values for *rating* are as follows: 0 ('Maybe'), 1 ('Yes'), −1 ('No'). The following table shows some sample of vote data.

| ratingID | imgID | volunteerID | timestamp | rating |
|----------|-------|-------------|-----------|--------|
| 75811 | 3009 | 178 | 2013-11-18 12:50:31 | 1 |
| 566299 | 3009 | 689 | 2013-12-03 08:10:38 | 0 |
| 641369 | 3009 | 1398 | 2013-12-03 17:10:39 | −1 |
| 3980868 | 3009 | 1365 | 2014-04-10 16:52:07 | 1 |

## 4 Methodology

### 4.1 Detection of Duplicates and Blurry Images

Since the dataset collected via the game was formed by combining different sources, it is possible that almost the same images can be referenced by different records. In order to check this, we download all 170041 .jpeg images (512 ∗ 512 size). The total size of all images is around 9 GB. Then we employ perceptive hash functions to reveal such cases. Examples of such functions are aHash (Average Hash or Mean Hash), dHash, and pHash [19]. Perceptual hashing aims to detect images such that a human cannot see the difference. We find that pHash performs much better than computationally less expensive dHash and aHash methods. Note that for a fixed image, the set of all images that is similar according to pHash will contain all images with the corresponding MD5 or SHA1 hash. To summarize, we detect duplicates for 8300 original images; votes for duplicates were merged.

Accepting the idea of the wisdom of the crowd, in order to make a better decision for an image, we need to collect more votes for each image. The detection of all similar images increases statistically significant effects and decreases the dimensionality of the data. In addition, if the detection is performed before the start of the campaign, there is a reduction in the workload of the volunteers.

A visual inspection of images shows the presence of illegible and blurry (unfocused) images. As expected, these images bewildered the volunteers. Thus, we apply automatic methods for blur detection. Namely, by using the Blur Detection algorithm [18], we detect 2300 poor quality images such that it is not possible to give the right answers even for experts. Note that for those images, voting inconsistency is high; volunteers and experts change their opinions frequently. After consultation with the experts, we remove all images of poor quality. Note that the image processing steps turn out to be crucial for decreasing the noise level and uncertainty in the dataset. Unfortunately, since the testing dataset is obtained after image processing, it is impossible to estimate direct impact of these steps on the accuracy of aggregated votes.

## 4.2 Majority Voting Based on Reliability

In this subsection we present a conjunction of majority voting and the widely used notion of reliability (see, for example, [5]). It is a standard to define reliability $w_i$ of worker $i$ as

$$w_i = 2p_i - 1$$

where $p_i$ is the probability that worker $i$ gives a correct answer (it is assumed that it does not depend on the particular task); obviously, $w_i \in [-1, 1]$. We use traditional weighted MV with weights obtained by the above rule. The heuristic

---

**Algorithm 1.** Weighted MV

---

**Input:** $V$ is the set of all volunteers;
$I$ is the set of all images with at least 1 vote;
$R = \left(r_{v,i}\right)_{v=1,i=1}^{|V|,|I|}$ is the rating matrix (see (2));
$E$ is the set of images with ground truth labels;
$(e_i)_{i \in E} \in \{-1; 1\}^{|E|}$ are ground truth labels for images from $E$.
**Output:** the predicted labels $\{y_1, y_2, ..., y_{|I|}\}$
**Initialization:**
    for $v \in V$ : do
        if $\sum_{i \in I \cap E} \mathbf{I}(r_{v,i} \neq 0) \neq 0$ then
            $w_v \leftarrow 2 \times \frac{\sum_{i \in I \cap E} \mathbf{I}(r_{v,i}=e_i)}{\sum_{i \in I \cap E} \mathbf{I}(r_{v,i}\neq0)} - 1$
        else
            $w_v \leftarrow 0$
**Repeat**
Calculate penalties for volunteers according to Algorithm 2 [5]. The algorithm takes $I, V, R$ as inputs and gives a vector $(p_v)_{v \in V} \in [0, 1]^{|V|}$ as output. For volunteer $\hat{v}$ with the highest penalty, we set

$$w_{\hat{v}} \leftarrow 0,$$

$$r_{\hat{v},i} \leftarrow 0 \ \ \forall i \in I.$$

**Until** reaching a pre-specified number of iterations
**Output:** the predictions $(y_i)_{i \in I}$

$$y_i = argmax_{k \in \{-1; 1\}} \sum_{v \in V} w_v \mathbf{I}(r_{v,i} = k). \tag{1}$$

---

admits a refinement; one may iteratively remove a volunteer with the highest penalty, then recalculate penalties, and obtain new results for the weighted MV.

The proposed heuristic is presented in Algorithm 1. Note that mapping $\mathbf{I}$ : $\{False; True\} \rightarrow \{0; 1\}$ is defined by the rule: $\mathbf{I}(True) = 1, \mathbf{I}(False) = 0$.

## 5   Experiments

During the crowdsourcing campaign, around 4.6 million votes were collected. The voting protocol was converted to a rating matrix. The matrix consists of ratings given to images (matrix columns) by the volunteers (matrix rows)

$$R = \left(r_{v,i}\right)_{v=1,i=1}^{|V|,|I|}, \tag{2}$$

$V$ is the set of all volunteers ($|V|$=2783);
$I$ is the set of all images with at least 1 vote ($|I|$=161752);
$r_{v,i}$ is a vote given by a volunteer to an image.

Due to an unclear definition, the *'Maybe'* answer is hard to interpret. As a result, we treat *'Maybe'* as a situation when the user has not seen the image; both situations are coded as 0. If a volunteer has multiple votes for the same image, then *only the last vote is used.*

To evaluate the volunteers' performance, a part of the dataset (854 images) was annotated by an expert after the campaign took place. For these images 1813 volunteers gave 16,940 votes in total. Then we sampled two subsets for training and testing (70/30 ratio).

**The baseline.** We treat columns of the rating matrix as feature vectors of images. To use some conventional machine learning algorithms, *we first apply SVD to the whole dataset* to reduce dimensionality. A study of the explained variance helps us to make an appropriate choice for the number of features: 5, 14, 35. Then we transform the feature space of the testing and training subsets accordingly. On the basis of 10-fold cross-validation of the training subset, we fit parameters for the AdaBoost and Random Forest algorithms. For Linear Discriminant Analysis (LDA), we use default parameters. The accuracy of the algorithms with fitted parameters was estimated using the testing subset; see *Table* 1.

**Benchmarking of algorithms for an aggregation of crowd votes** is performed as follows. We feed the expert dataset to the algorithms and check their accuracy on the same test subset as above. Note that the transformation of a feature space is not required in this case. In this section, we experimentally test the heuristic based on reliability and compare it with the state-of-art algorithms designed for crowdsourcing. We use publicly available code[1] that was developed for experiments in [5]. The code implements the iterative algorithm in [6] referred

---

[1]  https://github.com/ashwin90/Penalty-based-clustering.

**Table 1.** Baseline algorithms

| Number of features | Random forest | LDA | AdaBoost |
|---|---|---|---|
| 5 | **89.92** | 87.60 | 89.15 |
| 14 | 89.14 | **90.70** | 89.92 |
| 35 | 88.37 | 89.53 | **91.08** |

**Table 2.** Accuracy for 'crowdsourcing' algorithms without image-vote thresholding

| Iteration | MV | EM | KOS | KOS+ | Weighted MV |
|---|---|---|---|---|---|
| Base | 89.81 | 89.81 | 88.99 | 89.81 | 90.63 |
| 1 | 90.05 | 90.16 | 88.88 | 90.16 | **91.45** |
| 2 | 90.05 | 90.05 | 88.64 | 90.16 | **91.45** |
| 3 | 89.67 | 89.58 | 88.17 | 89.70 | 91.22 |
| 4 | 89.34 | 89.46 | 88.17 | 89.22 | 90.98 |
| 5 | 89.93 | 89.81 | 88.41 | 89.58 | 91.10 |
| 6 | 89.81 | 89.93 | 88.52 | 89.58 | 90.98 |
| 7 | 90.16 | 90.05 | 88.64 | 89.46 | 90.98 |
| 8 | 90.16 | 89.93 | 88.88 | 89.58 | 90.87 |
| 9 | 90.16 | 89.81 | 89.11 | 89.70 | 90.75 |

to as the KOS and EM algorithms [3]; both are implemented in conjunction with reputation Algorithm 2 in [5] (also called Hard penalty). Note that KOS+ is a normalized version (see [5]) of KOS. This version may be more suitable for arbitrary graphs (KOS is developed for regular graphs). During each iteration, the reputation algorithm helps to exclude the volunteer with the highest penalty and recalculates the penalties for the remaining volunteers. The accuracy of the compared algorithms on the test sample is presented in Table 2. Note that the first row (*Base*) corresponds to results before the exclusion of volunteers. Surprisingly, all crowdsourcing algorithms perform on par with Majority voting.

**Table 3.** Accuracy for 'crowdsourcing' algorithms with image-vote thresholding. Only images with at least 4 votes are left in the expert dataset. In this case we have 729 images annotated by 1812 volunteers.

| Iteration | MV | EM | KOS | KOS+ | Weighted MV |
|---|---|---|---|---|---|
| Base | 90.95 | 91.08 | 90.12 | 91.08 | 91.63 |
| 1 | 91.08 | 91.36 | 90.26 | 91.36 | 92.18 |
| 2 | 91.08 | 91.36 | 90.12 | 91.36 | 92.18 |
| 3 | 91.63 | 91.36 | 90.26 | 91.36 | **92.32** |
| 4 | 91.22 | 91.08 | 89.71 | 91.08 | 91.77 |
| 5 | 91.22 | 91.22 | 89.71 | 91.22 | 92.04 |
| 6 | 91.08 | 91.36 | 90.26 | 91.36 | 91.91 |
| 7 | 91.08 | 91.36 | 90.40 | 91.36 | 91.91 |
| 8 | 91.08 | 91.08 | 90.53 | 90.81 | 91.91 |
| 9 | 90.81 | 91.08 | 90.40 | 90.81 | 91.91 |

**Table 4.** Accuracy for 'crowdsourcing' algorithms with image-vote thresholding. Only images with at least 10 votes are left in the expert dataset. In this case we have 404 images annotated by 1777 volunteers.

| iteration | MV | EM | KOS | KOS+ | Weighted MV |
|---|---|---|---|---|---|
| Base | 94.55 | 94.55 | 94.06 | 94.55 | 95.05 |
| 1 | 94.55 | 94.55 | 93.81 | 94.55 | 95.05 |
| 2 | 94.55 | 94.55 | 93.81 | 94.55 | 95.05 |
| 3 | 94.55 | 94.55 | 94.06 | 94.55 | 95.05 |
| 4 | 94.55 | 94.55 | 94.06 | 94.55 | 95.05 |
| 5 | 94.55 | 94.55 | 94.06 | 94.55 | 95.05 |
| 6 | 94.55 | 94.80 | 94.06 | 94.55 | 95.30 |
| 7 | 94.55 | 94.80 | 94.06 | 94.80 | 95.30 |
| 8 | 94.55 | 94.80 | 94.06 | 94.80 | 95.30 |
| 9 | 94.80 | 94.80 | 94.06 | 95.05 | **95.54** |

A possible explanation is the irregular task assignment leading, in particular, to a high percentage of images with only a few votes. To deal with this issue, we continue our analysis using *image thresholding by the number of votes received* (or simply image-vote thresholding). Namely, we perform the same benchmarking for two subsets of the expert dataset. The subsets were obtained by filtering images with the number of votes less than the threshold; see Tables 3 and 4. Note that the training and the testing sets are different in the experiments reflected in Tables 2, 3 and 4.

Another possible explanation is that we mostly deal with reliable volunteers, and thus, crowdsourcing algorithms cannot profit from the detection of spammers or from flipping votes of malicious voters. To analyze this hypothesis, we classify volunteers according to their performance. In this regard, we use notation introduced in [12]. Namely, as it was suggested, in Fig. 1, we depict the Receiver Operating Characteristic (ROC) plot containing details of individual performance. Each plot in Fig. 1 depicts two values for each volunteer: *the sensitivity* and *the specificity*. If the true label is 1, then the sensitivity is defined as the probability that the volunteer votes 1 (this probability corresponds to the

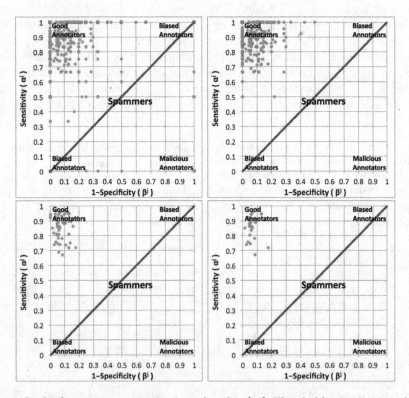

**Fig. 1.** In the figure we use notation introduced in [12]. Threshold = 0, 12, 44, and 100 votes. These thresholds leave 1813, 262, 52, and 24 volunteers, respectively. ROCs of spammers lie on the red line. (Color figure online)

true positive rate). If the true label is $-1$, then the specificity is defined as the probability that the volunteer votes $-1$. Since the task assignment was highly irregular, it is important to study how voting activity of volunteers influences the ROC. Namely, Fig. 1 contains not one, but four ROCs, where each of them is obtained according to a different level of volunteer thresholding. This thresholding helps to remove volunteers that had a total number of votes less than that defined by the threshold. Note that the definition of spammer introduced in [12] may differ from an intuitive one. Namely, spammer is a volunteer voting randomly and independently of true classes of images. Figure 1 provides plausible observations: there are no spammers among voters with more than 12 votes; good annotators prevail over all other types of annotators; there are frequently voting volunteers (more than 100 votes) showing better accuracy than any examined algorithm. These are the reasons why algorithms detecting spammers do not outperform the baseline noticeably.

# 6     Conclusions

Comparing the results in Tables 1 and 2, we conclude that 'general purpose' learning algorithms only slightly outperform 'special purpose' crowdsourcing algorithms. Surprisingly, the proposed simple heuristic (see Algorithm 1) based on reliability shows the best result. Also, numerical experiments show that Majority Voting performs on par with all other algorithms. The analysis of the ROCs of the volunteers suggests that surprisingly high accuracy of frequently voting volunteers coupled with the absence of spammers is a possible explanation for this result. The highly irregular task assignment in the dataset with a high percentage of images with a low number of votes may also contribute to this fact. Note that image-vote thresholding helps to improve the results of the 'crowdsourcing' algorithms (see Tables 2, 3 and 4) although the results are still on a par with Majority Voting. This parity differs from an observation obtained in comprehensive benchmark [16] where *'MV was often outperformed by some other method.'*

In the future we plan to benchmark the remaining state-of-the-art methods for the aggregation of votes and include 'Maybe' votes into consideration.

**Acknowledgments.** This research was supported by Russian Science Foundation, grant no. 14-11-00109, and the EU-FP7 funded ERC CrowdLand project, grant no. 617754.

# References

1. Chatterjee, S., Bhattacharyya, M.: A biclustering approach for crowd judgment analysis. In: Proceedings of the Second ACM IKDD Conference on Data Sciences. pp. 118–119. ACM (2015)

2. Comber, A., Brunsdon, C., See, L., Fritz, S., McCallum, I.: Comparing expert and non-expert conceptualisations of the land: an analysis of crowdsourced land cover data. In: Tenbrink, T., Stell, J., Galton, A., Wood, Z. (eds.) COSIT 2013. LNCS, vol. 8116, pp. 243–260. Springer, Heidelberg (2013). doi:10.1007/978-3-319-01790-7_14

3. Dawid, A.P., Skene, A.M.: Maximum likelihood estimation of observer error-rates using the EM algorithm. Appl. Stat. **28**, 20–28 (1979)

4. Dempster, A.P., et al.: Maximum likelihood from incomplete data via the EM algorithm. JRSS Ser. B **39**, 1–38 (1977)

5. Jagabathula, S., et al.: Reputation-based worker filtering in crowdsourcing. In: Advances in Neural Information Processing Systems, pp. 2492–2500 (2014)

6. Karger, D.R., Oh, S., Shah, D.: Iterative learning for reliable crowdsourcing systems. In: Advances in Neural Information Processing Systems, pp. 1953–1961 (2011)

7. Khattak, F.K., Salleb-Aouissi, A.: Improving crowd labeling through expert evaluation. In: 2012 AAAI Spring Symposium Series (2012)

8. Kim, H.C., Ghahramani, Z.: Bayesian classifier combination. In: International conference on Artificial Intelligence and Statistics, pp. 619–627 (2012)

9. Liu, Q., Peng, J., Ihler, A.T.: Variational inference for crowdsourcing. In: Advances in Neural Information Processing Systems, pp. 692–700 (2012)

10. Moreno, P.G., Teh, Y.W., Perez-Cruz, F., Artés-Rodríguez, A.: Bayesian nonparametric crowdsourcing. arXiv preprint arXiv:1407.5017 (2014)

11. Pareek, H., Ravikumar, P.: Human boosting. In: Proceedings of the 30th International Conference on Machine Learning (ICML2013), pp. 338–346 (2013)

12. Raykar, V.C.: Eliminating spammers and ranking annotators for crowdsourced labeling tasks. JMLR **13**, 491–518 (2012)

13. Raykar, V.C., et al.: Learning from crowds. J. Mach. Learn. Res. **11**, 1297–1322 (2010)

14. Salk, C.F., Sturn, T., See, L., Fritz, S., Perger, C.: Assessing quality of volunteer crowdsourcing contributions: lessons from the cropland capture game. Int. J. Digit. Earth **9**, 410–426 (2015)

15. See, L., et al.: Building a hybrid land cover map with crowdsourcing and geographically weighted regression. ISPRS J. Photogramm. Remote Sens. **103**, 48–56 (2015)

16. Sheshadri, A., Lease, M.: Square: a benchmark for research on computing crowd consensus. In: First AAAI Conference on Human Computation and Crowdsourcing (2013)

17. Simpson, E., Roberts, S., Psorakis, I., Smith, A.: Dynamic Bayesian combination of multiple imperfect classifiers. In: Guy, T.V., Karny, M., Wolpert, D. (eds.) Decision Making and Imperfection, pp. 1–35. Springer, Heidelberg (2013)

18. Tong, H., Li, M., Zhang, H., Zhang, C.: Blur detection for digital images using wavelet transform. In: 2004 IEEE International Conference on Multimedia and Expo, ICME 2004, vol. 1, pp. 17–20. IEEE (2004)

19. Zauner, C.: Implementation and benchmarking of perceptual image hash functions. Ph.D. thesis (2010)

20. Zhu, X., et al.: Co-training as a human collaboration policy. In: AAAI (2011)

# On Complexity of Searching a Subset of Vectors with Shortest Average Under a Cardinality Restriction

Anton V. Eremeev[1,2]([✉]), Alexander V. Kel'manov[3,4], and Artem V. Pyatkin[3,4]

[1] Omsk Branch of Sobolev Institute of Mathematics,
Siberian Branch of Russian Academy of Sciences, Omsk, Russia
eremeev@ofim.oscsbras.ru
[2] Omsk State University n.a. F.M. Dostoevsky, Omsk, Russia
[3] Sobolev Institute of Mathematics,
Siberian Branch of Russian Academy of Sciences, Novosibirsk, Russia
{kelm,artem}@math.nsc.ru
[4] Novosibirsk State University, Novosibirsk, Russia

**Abstract.** In this paper, we study the computational complexity of the following subset search problem in a set of vectors. Given a set of $N$ Euclidean $q$-dimensional vectors and an integer $M$, choose a subset of at least $M$ vectors minimizing the Euclidean norm of the arithmetic mean of chosen vectors. This problem is induced, in particular, by a problem of clustering a set of points into two clusters where one of the clusters consists of points with a mean close to a given point. Without loss of generality the given point may be assumed to be the origin.

We show that the considered problem is NP-hard in the strong sense and it does not admit any approximation algorithm with guaranteed performance, unless P = NP. An exact algorithm with pseudo-polynomial time complexity is proposed for the special case of the problem, where the dimension $q$ of the space is bounded from above by a constant and the input data are integer.

**Keywords:** Vectors sum · Subset selection · Euclidean norm · NP-hardness · Pseudo-polymonial time

## 1 Introduction

In this paper, we study a discrete extremal problem of searching a subset of vectors with shortest average under a cardinality restriction. The goal of the study is finding out the computational complexity of this problem and its approximability. The research is motivated by significance of the problem in many applications (see below). The Subset with the Shortest Average under Cardinality Restriction (SSA) problem is formulated as follows.

*Given*: a set $\mathcal{Y} = \{y_1, \ldots, y_N\}$ of points (vectors) from $\mathbb{R}^q$ and a positive integer $M$.

© Springer International Publishing AG 2017
D.I. Ignatov et al. (Eds.): AIST 2016, CCIS 661, pp. 51–57, 2017.
DOI: 10.1007/978-3-319-52920-2_5

*Find*: a subset $C \subseteq \mathcal{Y}$ such that $|C| \geq M$ and

$$\frac{1}{|C|} \left\| \sum_{y \in C} y \right\| \to \min, \tag{1}$$

where $\| \cdot \|$ denotes the Euclidean norm.

Note that the above formulation involves a norm of a sum of elements of the desired subset $C$. Therefore this problem may be viewed as an optimal summation problem and has an obvious geometrical interpretation. At the same time, this problem may be considered as a problem of clustering a set of points into two clusters ($C$ and $\mathcal{Y} \backslash C$) when one of the clusters consists of points with a mean close to the origin. Obviously, given any other vector instead of the origin, the problem can be easily reduced to the mentioned above. This type of 2-clustering problems can be used for censoring the input data, if the expectation of an observed variable is known in advance.

Also SSA problem has applications in the diverse and multidisciplinary area of Data Mining (see e.g. [1,2,13]). One of the central problems in this area consists in approximation of data by some mathematical model which allows to interpret the data adequately and explain their emergence. In particular, such a model may be expressed as a statistical hypothesis that the input data $\mathcal{Y}$ are sampled from a mixture of several distributions and at least $M$ observations correspond to a distribution with zero mean. First one can solve SSA problem with the given data, after that the classical methods of statistical hypothesis testing may be applied to the obtained SSA solution and finally the data interpretation may be done on the basis of hypothesis testing results.

Another area where the SSA problem emerges is the trading hubs construction for electricity markets under locational marginal pricing [4].

It can be seen from the form of the optimization criterion (1) that the problem under consideration may be easily interpreted as a version of important classical problems in physics that ask for a balanced subset of forces (vectors). Besides that, if the given points of the Euclidean space correspond to people so that the coordinates of points are equal to some characteristics of these people (w.r.t. some matters), then the formulated problem may be treated as a problem of finding a balanced group (a subset) of people.

The formulation of SSA problem is resembling the formulation of optimal summation problems with a *maximization* criterion which first arose in studying the problem of noise-proof off-line search for an unknown repeating fragment in a discrete signal [17]. The maximization criterion in [17] has a different scaling compared to (1):

$$\frac{1}{|C|} \left\| \sum_{y \in C} y \right\|^2 \to \max. \tag{2}$$

The strong NP-hardness of the maximization problems with criterion (1) was proved in [3,9,10,22,23] under different restrictions on the cardinality of the desired set. These problems, their generalizations and special cases were also studied in [5,6,8,10–12,14–16,18–21,24]. In particular, it was proved in

[11,20] that in the case of the fixed dimension $q$ of the space, the problems with criterion (2) are polynomially solvable in time $\mathcal{O}(N^{2q})$.

The complexity and approximability status of SSA problem was not completely known up to now. An equivalent single hub selection problem was studied in [4] where it was shown to be NP-hard in the 2-dimensional Euclidean space. A modification of the single hub selection problem, where the size of the sought subset $\mathcal{C}$ is given in the input, was shown to be strongly NP-hard in [25]. SSA problem may be transformed to $\mathcal{O}(N)$ instances of the problem from [25] but this does not help to identify the complexity status of SSA in the general case. In the next section, we provide a detailed study of computational complexity of the SSA problem and its approximability.

## 2    Analysis of Computational Complexity and Approximability

Note that in the general formulation of the SSA problem given above, the dimension $q$ of the space is a part of the input data. The following theorem states the complexity status of this problem.

**Theorem 1.** *SSA problem is NP-hard in the strong sense.*

*Proof.* Let us prove the strong NP-completeness of the equivalent decision problem, which implies the strong NP-hardness (see e.g. [7]). Let us formulate SSA problem in the form of decision problem.

*Instance*: A set $\mathcal{Y} = \{y_1, \ldots, y_N\}$ of points from $\mathbb{R}^q$, a positive integer $K$ and a positive integer $M$.

*Question*: Is there a nonempty subset $\mathcal{C} \subseteq \mathcal{Y}$ of size at least $M$, such that the value of objective function (1) is at most $K$?

SSA decision problem obviously belongs to class NP. In what follows we will consider a special case of this problem, where $K = 0$, denoting it by SSA0. Let us reduce a classical NP-complete problem [7] EXACT COVER BY 3-SETS to SSA0.

EXACT COVER BY 3-SETS.
*Instance*: A finite set $\mathcal{Z}$ such that $|\mathcal{Z}| = 3n$ and a collection $\mathcal{X} = \{\mathcal{X}_1, \mathcal{X}_2, \ldots, \mathcal{X}_k\}$ of 3-element subsets of the set $\mathcal{Z}$.

*Question*: Does $\mathcal{X}$ contain an exact cover for set $\mathcal{Z}$, i.e. a collection $\{\mathcal{X}_{i_1}, \mathcal{X}_{i_2}, \ldots, \mathcal{X}_{i_n}\} \subseteq \mathcal{X}$ such that $\cup_{j=1}^{n} \mathcal{X}_{i_j} = \mathcal{Z}$?

Given an instance of EXACT COVER BY 3-SETS, let us construct an equivalent instance of SSA0 problem. Put $q = 3n$ and $M = n + 1$. For each subset $\mathcal{X}_i$, $i = 1, \ldots, k$, a $3n$-dimensional point $y_i$ is assigned, whose $j$-th coordinate $(j = 1, 2, \ldots, 3n)$ is defined as $y_i^{(j)} = 1$, if $j \in \mathcal{X}_i$, and $y_i^{(j)} = 0$ otherwise. Let $y_{k+1} = (-1, \ldots, -1)$, $N = k + 1$ and $\mathcal{Y} = \{y_1, y_2, \ldots, y_k, y_{k+1}\}$.

Note that the objective function of SSA problem equals zero iff $z := \sum_{y \in \mathcal{C}} y = 0$.

If the instance of EXACT COVER BY 3-SETS has the answer "Yes", then obviously the subset $\mathcal{C} = \{y_{i_1}, \dots, y_{i_n}, y_N\}$ turns the objective function (1) into 0.

Now let the optimal value of the objective function in SSA problem be equal to 0. Then subset $\mathcal{C}$ must contain the point $y_N = (-1, \dots, -1)$, because otherwise all coordinates of point $z$ are non-negative and at least one of them is positive. In this case, the rest of the points in the subset $\mathcal{C}$ altogether should contain exactly one 1 in each coordinate, so there should be exactly $n$ such points and the subsets corresponding to them form an exact cover. Note that the equality $|\mathcal{C}| = n + 1 = M$ holds.

Finally, the strong NP-hardness of SSA problem follows from the fact that an NP-complete in the strong sense problem EXACT COVER BY 3-SETS is reduced to a special case of SSA problem with binary input and the objective function values are bounded by a polynomial in $n$.     □

Now let us consider complexity and approximability of SSA problem when the dimension $q$ is fixed.

Let $\rho > 1$. A polynomial-time algorithm that finds a feasible solution to a minimization problem, such that the value of objective function in this solution is at most $\rho$ times the optimal value (if the problem is solvable) is called a $\rho$-approximation algorithm. The corresponding feasible solution is called a $\rho$-approximate solution.

Below we denote by $\mathbb{N}$ the set of positive integers.

**Theorem 2.** *For any function $r : \mathbb{N} \to (1, \infty)$, the problem of searching an $r(N)$-approximate solution to SSA problem is NP-hard even in the special case of $q = 2$.*

*Proof.* Let us reduce the following modification of the NP-complete PARTITION problem (see e.g. [7]), which we call BOUNDED PARTITION, to the decision problem SSA0.

BOUNDED PARTITION

*Instance*: An even number $n$ of positive integers $\alpha_j$, $j = 1, 2, \dots, n$.

*Question*: Is there a subset $\mathcal{I} \subset \{1, 2, \dots, n\}$ such that $|\mathcal{I}| = n/2$ and $\sum_{i \in \mathcal{I}} \alpha_i = \frac{1}{2} \sum_{i=1}^{n} \alpha_i$?

Given a set of integers $\alpha_1, \alpha_2, \dots, \alpha_n$ we construct an instance of SSA0 with $q = 2$, $N = n + 1$ and $M = n/2 + 1$. Let $L = \sum_{i=1}^{n} \alpha_i$. Put $y_i = (L, \alpha_i)$ for $i = 1, 2, \dots, n$, $y_{n+1} = (-Ln/2, -L/2)$ and for each subset $\mathcal{I} \subseteq \{1, 2, \dots, N\}$ denote $S(\mathcal{I}) = \sum_{i \in \mathcal{I}} y_i$.

If the set $\mathcal{I}$ required in BOUNDED PARTITION problem exists, then it is easy to see that $S(\mathcal{I} \cup \{n + 1\}) = 0$, and therefore the objective function (1) turns into zero.

Suppose there exists a set $\mathcal{C}^*$ of cardinality at least $M$ such that the value of the objective function on this set is zero. Let $z$ denote the sum of elements of $\mathcal{C}^*$. Now since the first coordinate of $z$ equals 0, we have $|\mathcal{C}^*| = n/2 + 1$ and $y_{n+1} \in \mathcal{C}^*$. Then, due to zero value in the second coordinate of $z$ we have $\sum_{i \in \mathcal{I}} \alpha_i = L/2 = \frac{1}{2} \sum_{i=1}^{n} \alpha_i$, where $\mathcal{I} = \{i \mid y_i \in \mathcal{C}^*\} \setminus \{n + 1\}$.

The observed properties of the reduction imply the NP-completeness of SSA0 problem for $q = 2$. Under this reduction, the objective function value of an optimal solution to the SSA problem instance equals zero iff the BOUNDED PARTITION problem instance has the answer "Yes", and the same applies to any $r(N)$-approximate solution to SSA problem. Finally, since the objective function of SSA problem is efficiently computable, the problem of searching an $r(N)$-approximate solution is NP-hard.                              □

Theorem 2 implies that unless $P = NP$, SSA problem does not admit approximation algorithms with any non-trivial guaranteed approximation ratio, and, in particular it does not admit a fully polynomial time approximation scheme (FPTAS).

SSA problem with a fixed $q \geq 2$ can not be solved by a polynomial-time algorithm, unless $P = NP$. Nevertheless, as shown below, it is solvable in a pseudo-polynomial time, provided that all points of set $\mathcal{Y}$ have integer coordinates and the dimension $q$ of the space is fixed.

For any two sets $\mathcal{P}, \mathcal{Q} \subset \mathbb{R}^q$ we introduce the following rule of summation:

$$\mathcal{P} + \mathcal{Q} = \{x \in \mathbb{R}^q \mid x = y + y', \ y \in \mathcal{P}, \ y' \in \mathcal{Q}\}. \tag{3}$$

For any positive integer $r$ we denote by $\mathcal{B}(r)$ the set of integer points in $\mathbb{R}^q$ with absolute values of all coordinates at most $r$. Then $|\mathcal{B}(r)| \leq (2r + 1)^q$.

Let us denote the maximal absolute value of coordinates of the input points $y_1, y_2, \ldots, y_N$ by $b$. The proposed algorithm for solving SSA problem consists in consequent computing of subsets $\mathcal{S}_k \subseteq \mathcal{B}(bk)$, $k = 0, 1, \ldots, M$, that can be obtained by summing at most $k$ different elements of the set of points $y_1, y_2, \ldots, y_k$. First we assume $\mathcal{S}_0 = \{0\}$. After that we compute $\mathcal{S}_k = \mathcal{S}_{k-1} + \{0, y_k\}$ for all $k = 1, 2, \ldots, N$ using formula (3). For each element $z \in \mathcal{S}_k$ we store an integer parameter $n_z$, equal to the maximum number of addends that can be used to produce $z$ and the $n_z$-element set of these addends $\mathcal{C}_z \subseteq \mathcal{Y}$.

Finally, find in the subset $\mathcal{S}_N$ an element $z \in \mathcal{S}_N$ with $n_z \geq M$ and the minimum value of $\|z\|/n_z$ and output the subset $\mathcal{C}_z$ corresponding to such $z$.

Computation of $\mathcal{S}_k$ takes $\mathcal{O}(q \cdot |\mathcal{S}_{k-1}|)$ operations. Therefore the following theorem holds

**Theorem 3.** *If the coordinates of the points of input set $\mathcal{Y}$ are integer and $b$ is the maximum absolute value of these coordinates then SSA problem is solvable in $\mathcal{O}(qN(2bN + 1)^q)$ time.*

In the case of fixed dimension $q$, i.e. $q = \mathcal{O}(1)$, the complexity of the algorithm presented above is $\mathcal{O}(N(bN)^q)$ and SSA problem is solvable in a pseudo-polynomial time in this special case.

## 3   Conclusion

The obtained results imply that there exist no exact polynomial or pseudo-polynomial algorithms for SSA problem, unless $P = NP$.

In the case when the dimension of the space is not a part of the input (i.e. the dimension is fixed), SSA problem is NP-hard even on the plane and no approximation algorithms with non-trivial guaranteed approximation ratio exist for this problem, unless $P = NP$. SSA problem is solvable, however, within a pseudo-polynomial time if the coordinates of the input points are all integer and the dimension is fixed.

The obtained results indicate that in spite of simplicity of formulation of the considered problem, efficient algorithms finding an exact or even an approximate solution to it are unlikely to exist. An exception is the special case where the space dimension is bounded by a constant and coordinates of the input points are bounded by a polynomial in $N$. We expect that obtaining "positive" results for SSA would require analysis of the special cases, which reflect the specifics of applications area.

**Acknowledgements.** This research is supported by RFBR, projects 15-01-00462, 16-01-00740 and 15-01-00976.

# References

1. Aggarwal, C.C.: Data Mining: The Textbook. Springer International Publishing, Switzerland (2015)
2. Bishop, M.C.: Pattern Recognition and Machine Learning. Springer Science+Business Media, LLC, New York (2006)
3. Baburin, A.E., Gimadi, E.K., Glebov, N.I., Pyatkin, A.V.: The problem of finding a subset of vectors with the maximum total weight. J. Appl. Ind. Math. **2**(1), 32–38 (2008)
4. Borisovsky, P.A., Eremeev, A.V., Grinkevich, E.B., Klokov, S.A., Vinnikov, A.V.: Trading hubs construction for electricity markets. In: Kallrath, J., Pardalos, P.M., Rebennack, S., Scheidt, M. (eds.) Optimization in the Energy Industry. Energy Systems, pp. 29–58. Springer, Heidelberg (2009)
5. Dolgushev, A.V., Kel'manov, A.V.: An approximation algorithm for solving a problem of cluster analysis. J. Appl. Ind. Math. **5**(4), 551–558 (2011)
6. Dolgushev, A.V., Kel'manov, A.V., Shenmaier, V.V.: Polynomial-time approximation scheme for a problem of partitioning a finite set into two clusters. Trudy Instituta Matematiki i Mekhaniki UrO RAN **21**(3), 100–109 (2015). (in Russian)
7. Garey, M.R., Johnson, D.S.: Computers and Intractability. A Guide to the Theory of $NP$-Completeness. W.H. Freeman and Company, San Francisco (1979)
8. Gimadi, E.K., Glazkov, Y.V., Rykov, I.A.: On two problems of choosing some subset of vectors with integer coordinates that has maximum norm of the sum of elements in euclidean space. J. Appl. Ind. Math. **3**(3), 343–352 (2009)
9. Gimadi, E.K., Kel'manov, A.V., Kel'manova, M.A., Khamidullin, S.A.: Aposteriori finding a quasiperiodic fragment with given number of repetitions in a number sequence (in Russian). Sibirskii Zhurnal Industrial'noi Matematiki **9**(25), 55–74 (2006)
10. Gimadi, E.K., Kel'manov, A.V., Kel'manova, M.A., Khamidullin, S.A.: A posteriori detecting a quasiperiodic fragment in a numerical sequence. Pattern Recogn. Image Anal. **18**(1), 30–42 (2008)

11. Gimadi, E.K., Pyatkin, A.V., Rykov, I.A.: On polynomial solvability of some prob-lems of a vector subset choice in a Euclidean space of fixed dimension. J. Appl. Ind. Math. **4**(4), 48–53 (2010)

12. Gimadi, E.K., Rykov, I.A.: A randomized algorithm for finding a subset of vectors. J. Appl. Ind. Math. **9**(3), 351–357 (2015)

13. Hastie, T., Tibshirani, R., Friedman, J.: The Elements of Statistical Learning: Data Mining, Inference, and Prediction. Springer, New York (2001)

14. Kel'manov, A.V.: Off-line detection of a quasi-periodically recurring fragment in a numerical sequence. Proc. Steklov Inst. Math. **263**(S2), 84–92 (2008)

15. Kel'manov, A.V.: On the complexity of some data analysis problems. Comput. Math. Math. Phys. **50**(11), 1941–1947 (2010)

16. Kel'manov, A.V.: On the complexity of some cluster analysis problems. Comput. Math. Math. Phys. **51**(11), 1983–1988 (2011)

17. Kel'manov, A.V., Khamidullin, S.A., Kel'manova, M.A.: Joint finding and eval-uation of a repeating fragment in noised number sequence with given number of quasiperiodic repetitions (in Russian). In: Book of Abstracts of the Russian Con-ference "Discret Analysis and Operations Research" (DAOR-2004), p. 185. Sobolev Institute of Mathematics SB RAN, Novosibirsk (2004)

18. Kel'manov, A.V., Khandeev, V.I.: A 2-approximation polynomial algorithm for a clustering problem. J. Appl. Ind. Math. **7**(4), 515–521 (2013)

19. Kel'manov, A.V., Khandeev, V.I.: A randomized algorithm for two-cluster parti-tion of a set of vectors. Comput. Math. Math. Phys. **55**(2), 330–339 (2015)

20. Kel'manov, A.V., Khandeev, V.I.: An exact pseudopolynomial algorithm for a problem of the two-cluster partitioning of a set of vectors. J. Appl. Ind. Math. **9**(4), 497–502 (2015)

21. Kel'manov, A.V., Khandeev, V.I.: Fully polynomial-time approximation scheme for a special case of a quadratic Euclidean 2-clustering problem. Comput. Math. Math. Phys. **56**(2), 334–341 (2016)

22. Kel'manov, A.V., Pyatkin, A.V.: On the complexity of a search for a subset of "similar" vectors. Doklady Math. **78**(1), 574–575 (2008)

23. Kel'manov, A.V., Pyatkin, A.V.: On a version of the problem of choosing a vector subset. J. Appl. Ind. Math. **3**(4), 447–455 (2009)

24. Kel'manov, A.V., Pyatkin, A.V.: Complexity of certain problems of searching for subsets of vectors and cluster analysis. Comput. Math. Math. Phys. **49**(11), 1966–1971 (2009)

25. Tarasenko, E.: On complexity of single-hub selection problem. In: Proceedings of 24-th Regional Conference of Students "Molodezh tretjego tysacheletija", pp. 45–48. Omsk State University, Omsk (2010). (in Russian)

# The Problem of the Optimal Packing of the Equal Circles for Special Non-Euclidean Metric

Alexander L. Kazakov[1], Anna A. Lempert[1(✉)], and Huy L. Nguyen[2]

[1] Matrosov Institute for System Dynamics and Control Theory SB RAS, Irkutsk, Russia
{kazakov,lempert}@icc.ru
[2] Irkutsk National Research Technical University, Irkutsk, Russia
nguyenhuyliem225@gmail.com

**Abstract.** The optimal packing problem of equal circles (2-D spheres) in a bounded set $P$ in a two-dimensional metric space is considered. The sphere packing problem is to find an arrangement in which the spheres fill as large proportion of the space as possible. In the case where the space is Euclidean this problem is well known, but the case of non-Euclidean metrics is studied much worse. However there are some applied problems, which lead us to use other special non-Euclidean metrics. For instance such statements appear in the logistics when we need to locate a given number of commercial facilities and to maximize the overall service area. Notice, that we consider the optimal packing problem in the case, where $P$ is a multiply-connected domain. The special algorithm based on optical-geometric approach is suggested and implemented. The results of numerical experiment are presented and discussed.

**Keywords:** Optimal packing problem · Equal circles · Non-Euclidean space · Multiply-connected domain · Numerical algorithm · Computational experiment

## Introduction

The optimal circle packing problem [1] is one of the classical problems of combinatorial geometry. It is of interest both from a theoretical point of view and in connection with a wide variety of applications.

The circle packing problem has a long history, for example, one of the famous statements relating to the packing of spheres in three-dimensional Euclidean space is called "Kepler conjecture" and was formulated more than 400 years ago. It says that no arrangement of equally sized spheres filling space has a greater average density than that of the cubic close packing (face-centered cubic) and hexagonal close packing arrangements. An introduction to its history can be found in [2].

© Springer International Publishing AG 2017
D.I. Ignatov et al. (Eds.): AIST 2016, CCIS 661, pp. 58–68, 2017.
DOI: 10.1007/978-3-319-52920-2_6

In the literature there are a lot of formulations of the problem: from the different variants of the knapsack problem [3] to considering it in the abstract $n$-dimensional space [4].

Note that nearly always (except for some special cases) the packing problem is NP-complete. So the problem of constructing efficient numerical algorithms is extremely urgent [5].

Apparently, the most popular problems is the problem of the 2-D optimal circle packing of equal radius to a closed set with smooth or piecewise smooth boundary (circles, squares, rectangles, etc.). For example, in the papers [2,6,7] authors deal with the problem to maximize the radius associated with the $n$ circles when the container is the unit square. The number of circles is from 1 to 200. In the case where the number of packing elements is small (up to 36, inclusive), the problem is solved analytically. In other words, it is proved that the constructed packing is the best.

In papers [8–10] the problem of packing identical circles of unit radius in the circle is considered. The results for number of packing elements up to 81 are obtained. Birgin and Gentil [11] consider the problem of packing identical circles of unit radius in a variety of containers (circles, squares, rectangles, equilateral triangles and strips of fixed height) to minimize the size of the latter.

The linear model for the approximate solution of the problem of packing of the maximum number of identical circles to the closed bounded set is suggested in [12]. The problem of packing of various circles of a given radius in order to maximize the number (or weight), or to minimize the waste is considered in [13,14].

Lopez and Beasley present a heuristic algorithm based on the formulation space search method to solve the packing problem for equal [15] and unequal [16] circles. Finally, a substantial and original class of packing problems where circles may be placed either inside or outside other circles, the whole set being packed in a rectangle is considered in [17].

Completing the survey part of the article, we note that the authors, who deal with the packing problem, were not limited by the case when elements are circles. Thus, in [18] authors consider the packing of rectangles (with the possibility of rotation) into triangles.

The survey of publications could be continued because there are hundreds of notable publications. The vast majority of books and articles devoted to the study of the problem of packing in Euclidean space. This is not accidental because such is the most natural formulation. However, sometimes the problems arise in applications, where in order to define the distance between two points it is necessary to use another metrics. For instance such statements appear in the logistics when we need to locate a given number of commercial facilities and to maximize the overall service area [19].

Coxeter [20] and Boroczky [21] deal with congruent circles packing problem for multidimensional spaces of constant curvature (elliptic and hyperbolic) and assess the maximum packing density. Besides above, this problem was studied in a series of papers by Szirmai. In [22,23] he presents a method that determines

the data and the density of the optimal ball and horoball packings Coxeter
tiling (Coxeter honeycomb) in the hyperbolic 3-, 4- and 5-spaces and based on
the projective interpretation of the hyperbolic geometry. The goal of [24] is to
extend the problem of finding the densest geodesic ball (or sphere) packing for
the other 3-dimensional homogeneous geometries (Thurston geometries).

In this paper we consider the circle packing problem in a bounded set with
piecewise smooth boundary in a special metric, which, generally speaking, is
not an Euclidean except one particular case. The container is not required to be
convex, and even simply connected. We present a numerical algorithm for solving
this problem and perform computational experiment, the results of which show
the effectiveness of the suggested approach.

## 1   Formulation

Let $X$ is a metric space, $C_i$, $i = 1, ..., n$ are congruent circles with centers in
$s_i = (x_i, y_i)$, $P$ is closed multiply-connected set.

$$P = \text{cl} \left( D \setminus \bigcup_{k=1}^{m} B_k \right) \subset X \subseteq \mathbb{R}^2.$$

Here $D \subset X$ is the bounded set, $B_k \subset D, k = 1, ..., m$ are compact sets with
non-empty interior.

It is necessary to find vector $s = (s_1, ..., s_n) \in \mathbb{R}^{2n}$, which provides the
packing of the given number of circles with maximum radius $R$ in $P$.

The distance between the points of the space $X$ is determined as follows:

$$\rho(a, b) = \min_{G \in G(a,b)} \int_G \frac{dG}{f(x, y)}, \tag{1}$$

where $G(a, b)$ is the set of all continuous curves, which belong $X$ and connect
the points $a$ and $b$, $0 < \alpha \leq f(x, y) \leq \beta$ is continuous function defined instantaneous speed of movement at every point of $P$. In other words, the shortest route
between two points is a curve, that requires to spend the least time.

It is easy to make sure that all the metric axioms are satisfied. In the particular case when $f(x, y) \equiv 1$ we have a Euclidean metric in the two-dimensional
space and the shortest route is a straight line.

Thus, we formulate the following problem:

$$R \rightarrow \max \tag{2}$$

$$\rho(s_i, s_j) \geq 2R, \, \forall i = \overline{1, n-1}, \forall j = \overline{i+1, n} \tag{3}$$

$$\rho(s_i, \partial P) \geq R, \, \forall i = \overline{1, n} \tag{4}$$

$$s_i \in P, \, \forall i = \overline{1, n} \tag{5}$$

Here $\partial P$ is the boundary of the set $P$, $\rho(s_i, \partial P)$ is the distance from a point
to a closed set.

The objective, Eq. (2), maximizes the radius associated with the circles. Equation (3) ensure that no circles overlap each other. Equations (4)–(5) are the constraints which ensure that every circle is fully inside the container.

For any vector $s$, satisfying the conditions (3)–(5) define the sets

$$P_i = \{p \in P : \rho(p, s_i) \le \rho(p, s_j), \forall j = 1, ..., n, i \ne j\}. \tag{6}$$

In the literature, such sets are called Dirichlet cells [25] for points $s_i$ on the set $P$. It's obvious that $P = \bigcup_{i=1}^{n} P_i$.

The solution of the problem above reduces to the solution of the following sequence of subproblems:

1. For every set $P_i$ find the point $\bar{s}_i \in P_i$ that $\rho(\bar{s}_i, \partial P_i) = \max_{p \in P_i} \rho(p, \partial P_i)$.
2. Find the guaranteed value of the radius satisfying the constraints (3)–(5):
   $R = \min_{i=1,...,n} \rho(\bar{s}_i, \partial P_i)$.
3. For the new vector $\bar{s} = (\bar{s}_1, ..., \bar{s}_n)$ redefine sets $P_i$ according to formula (6).

The steps 1–3 are carried out while the coordinates of $\bar{s}$ are changed.

## 2    Solution Method

To solve the described subproblems authors suggest methods based on the physical principles of Fermat and Huygens, which are used in geometric optics. The first principle says that the light in its movement chooses the route that requires to spend a minimum of time. The second one states that each point reached by the light wave, becomes a secondary light source.

Thus, in order to solve of the first subproblems we should carried out the construction of the light wave front, started from the border $\partial P_i$ for each $P_i$ to the time when the front degenerate into a point. It's coordinates are the required solution $\bar{s}_i, i = 1, ..., n$. To solve the third subproblems it's required to simultaneously initiate the light waves from the points $\bar{s}_i, i = 1, ..., n$, and to find such points of $P$, which are simultaneously reached by two or more waves. We presented two algorithms in [26, 27].

Let's go back to the first subproblem and suggest an algorithm for it.

*Algorithm "BorderWaveInside-BWI"*

1. Boundary of the considered set is approximated by the closed polygonal line with nodes at the points $A_i, i = \overline{0, m}$. $A_i$ are called the initial points.
2. For each pair of points $A_i, A_{i+1}$ we construct line segment $A_i A_{i+1}$. Then we construct line segments $A_i B_i'$ and $A_{i+1} B_i''$ which are perpendicular to $A_i A_{i+1}$. The length of the line segments are $f(A_i) \Delta t$ and $f(A_{i+1}) \Delta t$ respectively. Let **B** is a set of all these line segments. It is easy to see that the amount of new points is twice more than initial one.
3. If there is a pair of line segments $VW \in \mathbf{B}$ and $YZ \in \mathbf{B}$ that $W = Z$, then all initial points between $V$ and $Y$ are eliminated.

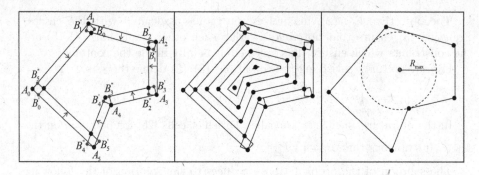

**Fig. 1.** Several iterations of $BWI$-algorithm

4. We construct straight lines $B_i'B_i''$, $i = \overline{0, m-1}$, by the points $B_i'$ and $B_i''$.
5. The points of intersection of $B_i'B_i''$ and $B_{i+1}'B_{i+1}''$, $i = \overline{0, m-2}$, form a set of the new front points.
6. If there is a pair of crossing line segments $VW \in \mathbf{B}$ and $YZ \in \mathbf{B}$, then all initial points between $V$ and $Y$ are eliminated. The point of intersection becomes a point of the new front.
7. If the constructed front is nonclosed line, then the solution is the "middle" of the line, namely the point the distance from which to the ends of line is the same. If the constructed front consists of one point, then this point is the solution. Otherwise, built front is taken as the initial and Go to Step No 1.

Note that after finding of the initial wave front the outer part of considered set becomes to be impassable. So, the perpendicular is directed to inside part.

The steps 3 and 6 provide the correct construction of the front when the "dovetail" problem arises.

Figure 1 illustrates $BWI$-algorithm. Left part shows the process of the first front constructing, in the middle there is a moment of the front splitting, on the right there is a packed circle of maximum radius.

In the case when the set $P_i$ is not simply connected, in other words, it contains impenetrable for the light wave barriers, in order to solve the first subproblem we need an additional algorithm. The algorithm allow to construct light wave fronts, propagating from the boundary of barrier in the outer area ($BorderWaveOutside$-$BWO$). This algorithm differs from $BWI$ only by the perpendicular direction.

Thus the algorithm for multiply connected set is follow.

*Algorithm (BorderWaveInside Multiply-connected set-BWI-MCS)*

1. By the algorithm $BWI$ we construct the fronts of the light wave, which is started from the border of the considered set. By the algorithm $BWO$ we construct the fronts of the light waves, which are started from barriers' borders. The algorithms work until the first contact $BWI$-wave with one of $BWO$-waves. Constructed fronts are saved.
2. The set of points, which are not reached neither of the waves, is divided into the maximal simply connected subsets $S_j$ (segments), which are saved in the list of segments $S$. Segments obtained in the previous iteration are removed.

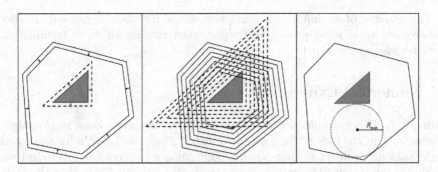

**Fig. 2.** Several iterations of $BWI$-$MCS$-algorithm

3. One iteration of algorithms $BWI$ and $BWO$, since the stored fronts, is performed. The process 2–3 continues while the set of points, which are not reached by any of the waves, is not empty.
4. All elements of the list of segments are analyzed: If the segment contains a single point, it is a potential solution. If the segment is open curve, its "middle" is a potential solution. In other cases, we find a potential solution by using an algorithm $BWI$, because the segment now is simply connected set.

   All received potential solutions $q_j$ are added to the list $Q$.
5. The list of potential solutions $Q$ is analyzed: If it contains one point, then this point is the desired solution. Otherwise, for each point $q \in Q$ by using the algorithm proposed in [28], the value of $r_j = \rho(q_j, \partial X_i)$ is calculated, where $\partial X_i$ are sets the boundaries of the $P_i$. The desired solution is a point $q^* \in Q$, for which $r^* = \max\limits_{j} r_j$.

Figure 2 illustrates $BWI$-$MCS$-algorithm. Left part shows the process of the first front constructing, in the middle there is the last iteration, on the right there is a packed circle of maximum radius. The solid line shows the wave front started from the boundary of set. The dashed line shows the wave front initiated from the border of barrier.

Now we are able to present the general algorithm for the problem (2)–(5).

*Algorithm of Equal Circles Packing – AECP-MCS*

1. Randomly generate an initial solution $s = (s_1, ..., s_n)$, which satisfies the constraint (5). The radius $R$ is assumed to be zero.
2. The set $P$ is divided into subsets $P_i$, $i = 1, ..., n$, according to the definition (6) by the authors' algorithm proposed in [28].
3. For each $P_i$, $i = 1, ..., n$, we solve the subproblem 1 by $BWI$-$MCS$ algorithm. As a result, for each $P_i$, $i = 1, ..., n$, we find the coordinates of the packed circle center $\bar{s}_i$ and its maximum radius $r_i$.
4. Calculate $R = \min\limits_{i=1,...,n} r_i$.

   Steps 2–4 are repeated until the $R$ increases, then the current vector $\bar{s}$ is saved as an approximation to a global maximum of the problem.

5. The counter of an initial solution generations *Iter* is incremented. If *Iter* becomes equal some preassigned value, then the algorithm is terminated. Otherwise, go to step 1.

# 3   Numerical Experiment

*Example 1.* This example illustrates how the given in the previous section algorithms work in the case of the Euclidean metric $f(x, y) \equiv 1$. We solve the equal circle packing problem in unit square. The number of circles is given and we maximize the radius. The results are presented in Table 1. Note, that the Best of known results were obtained from [29].

Considering Table 1 it is clear that our AECP-MCS algorithm produces low percentage deviations (less then 0.1%). In the case when $n \geq 50$ the deviations are retained, but the calculation time is significantly increased. So we can say that AECP-MCS algorithm allows to solve the equal circle packing problem for Euclidian metric, but it is not highly effective. It's advantages will be shown in next examples.

**Table 1.** AECP-MCS results for Euclidean metric

| $n$ | Best of known | | AECP-MCS | | Deviation | |
|---|---|---|---|---|---|---|
| | Radius ($R$) | Density ($d$) | Radius ($R$) | Density ($d$) | $\Delta R$ | $\Delta d$ |
| 1 | 0,50000000 | 0,78539816 | 0,50000000 | 0,78539816 | 0,00000000 | 0,00000000 |
| 2 | 0,29289322 | 0,53901208 | 0,29289140 | 0,53900539 | 0,00000182 | 0,00000669 |
| 3 | 0,25433310 | 0,60964481 | 0,25433090 | 0,60963429 | 0,00000220 | 0,00001052 |
| 4 | 0,25000000 | 0,78539816 | 0,25000000 | 0,78539816 | 0,00000000 | 0,00000000 |
| 5 | 0,20710678 | 0,67376511 | 0,20710390 | 0,67374635 | 0,00000288 | 0,00001875 |
| 6 | 0,18768060 | 0,66395691 | 0,18767851 | 0,66394208 | 0,00000210 | 0,00001483 |
| 7 | 0,17445763 | 0,66931083 | 0,17445600 | 0,66929832 | 0,00000163 | 0,00001251 |
| 8 | 0,17054069 | 0,73096383 | 0,17053746 | 0,73093617 | 0,00000323 | 0,00002766 |
| 9 | 0,16666667 | 0,78539816 | 0,16662136 | 0,78497118 | 0,00004531 | 0,00042698 |
| 10 | 0,14820432 | 0,69003579 | 0,14819925 | 0,68998856 | 0,00000507 | 0,00004723 |
| 11 | 0,14239924 | 0,70074158 | 0,14239800 | 0,70072940 | 0,00000124 | 0,00001218 |
| 12 | 0,13995884 | 0,73846822 | 0,13992800 | 0,73814277 | 0,00003084 | 0,00032545 |
| 15 | 0,12716655 | 0,76205601 | 0,12694119 | 0,75935742 | 0,00022536 | 0,00269859 |
| 16 | 0,12500000 | 0,78539816 | 0,12500000 | 0,78539816 | 0,00000000 | 0,00000000 |
| 2000 | 0,01172594 | 0,86392312 | 0,01151634 | 0,83331402 | 0,00020960 | 0,03060910 |
| 3000 | 0,00967451 | 0,88212297 | 0,009243172 | 0,80521747 | 0,00043134 | 0,07690550 |

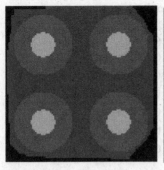 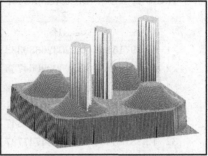

**Fig. 3.** Level curves (left) and 3-D view with barriers (right) of $f(x, y)$

*Example 2.* The metric is define by Eq. (1) where $f(x, y)$ has following form:

$$a_1(x, y) = \frac{(x-2.5)^2 + (y-2.5)^2}{1 + (x-2.5)^2 + (y-2.5)^2}, f_1(x, y) = \begin{cases} 0, a_1(x, y) \geq 0.8 \\ a_1(x, y) \end{cases}$$

$$a_2(x, y) = \frac{(x-2.5)^2 + (y-7.5)^2}{1 + (x-2.5)^2 + (y-7.5)^2}, f_2(x, y) = \begin{cases} 0, a_2(x, y) \geq 0.8 \\ a_2(x, y) \end{cases}$$

$$a_2(x, y) = \frac{(x-7.5)^2 + (y-2.5)^2}{1 + (x-7.5)^2 + (y-2.5)^2}, f_3(x, y) = \begin{cases} 0, a_3(x, y) \geq 0.8 \\ a_3(x, y) \end{cases}$$

$$a_3(x, y) = \frac{(x-7.5)^2 + (y-7.5)^2}{1 + (x-7.5)^2 + (y-7.5)^2}, f_4(x, y) = \begin{cases} 0, a_4(x, y) \geq 0.8 \\ a_4(x, y) \end{cases}$$

$$F(x, y) = f_1(x, y) + f_2(x, y) + f_3(x, y) + f_4(x, y)$$

$$f(x, y) = \begin{cases} 0.4, 0 < F(x, y) \leq 0.4 \\ F(x, y) \\ 0.8, F(x, y) = 0 \end{cases}$$

Figure 3 shows level curves of function $f(x, y)$ and location of barriers which is superimposed on the 3-D view of $f(x, y)$.

The metrics like described above arise in infrastructure logistics when we want to locate some objects in the highlands. Here speed of movement depends on the angle of ascent or descent. Therefore the wave fronts are strongly distorted.

The computational results are presented in Table 2. Here $R_{\max}$ is the radius, $d_{\max}(C)$ is the density of package, $d(B)$ is the density of barriers, $t$ is the computing time.

Figure 4 shows the solutions associated with Table 2 for $n = 1, 2, 4, 8, 16, 32$ when the container is multi-connected set in the case where the form of the wave fronts is unknown.

With respect to Fig. 4 it is clear that AECP-MCS algorithm gives acceptable results even for quite complicated metric. Note that, as in the previous example, in the given metric presented on Fig. 4 "circles" have the same radius.

**Table 2.** AECP-MCS results for multiply-connected set

| $n$ | $R_{\max}$ | $d_{\max}(C)$ | $d(B)$ | $t$ |
|---|---|---|---|---|
| 1 | 52,09715102 | 0,220326330383 | 0,0542817532 | 12,3 |
| 2 | 50,39798735 | 0,395009064733 | 0,0542817532 | 105,2 |
| 4 | 29,65098366 | 0,440226085102 | 0,0542817532 | 151,4 |
| 8 | 25,35214222 | 0,623120400981 | 0,0542817532 | 315,3 |
| 16 | 17,21536804 | 0,592833528847 | 0,0542817532 | 2088,7 |
| 32 | 12,88156661 | 0,665244747787 | 0,0542817532 | 3150,9 |

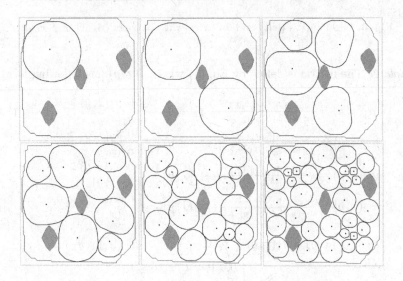

**Fig. 4.** AECP-MCS results for $n = 1, 2, 4, 8, 16, 32$

## 4   Conclusion

The presented algorithm is a modification of the Lloyd algorithm (widely known in machine learning and data mining community as k-means). The difference between the traditional k-means method and proposed approach is that the distance of each object to the centroids is not Euclidean.

We suppose that the scope of the clustering and classification problems may be significantly expanded by using special non-Euclidean metrics. So, authors will try to apply it to solve the problem of machine learning and data mining on the completely new aspect.

In conclusion we note that the further development of proposed approach involves consideration of the 3-D packing problem [30].

**Acknowledgements.** The reported study was particulary funded by RFBR according to the research projects No. 14-07-00222 and No. 16-06-00464.

# References

1. Conway, J., Sloane, N.: Sphere Packing, Lattices and Groups. Springer Science and Business Media, New York (1999)
2. Szabo, P., Specht, E.: Packing up to 200 equal circles in a square. In: Torn, A., Zilinskas, J. (eds.) Models and Algorithms for Global Optimization. Optimization and Its Applications, vol. 4, pp. 141–156. Springer, Heidelberg (2007)
3. Kellerer, H., Pferschy, U., Pisinger, D.: Knapsack Problems. Springer, Berlin (2004)
4. Levenshtein, V.: On bounds for packing in n-dimensional Euclidean space. Sov. Math. Dokl. **20**(2), 417–421 (1979)
5. Garey, M., Johnson, D.: Computers and Intractability. A Guide to the Theory of NP-Completeness. W.H. Freeman & Co., New York (1979)
6. Casado, L., Garcia, I., Szabo, P., Csendes, T.: Packing equal circles in a square ii. New results for up to 100 circles using the tamsass-pecs algorithm. In: Giannessi, F., Pardalos, P., Rapcsac, T. (eds.) Optimization Theory: Recent Developments from Matrahaza, vol. 59, pp. 207–224. Kluwer Academic Publishers, Dordrecht (2001)
7. Markot, M., Csendes, T.: A new verified optimization technique for the "packing circles in a unit square" problems. SIAM J. Optim. **16**(1), 193–219 (2005)
8. Goldberg, M.: Packing of 14, 16, 17 and 20 circles in a circle. Math. Mag. **44**(3), 134–139 (1971)
9. Graham, R., Lubachevsky, B., Nurmela, K., Ostergard, P.: Dense packings of congruent circles in a circle. Discrete Math. **181**(1–3), 139–154 (1998)
10. Lubachevsky, B., Graham, R.: Curved hexagonal packings of equal disks in a circle. Discrete Comput. Geom. **18**, 179–194 (1997)
11. Birgin, E., Gentil, J.: New and improved results for packing identical unitary radius circles within triangles, rectangles and strips. Comput. Oper. Res. **37**(7), 1318–1327 (2010)
12. Galiev, S., Lisafina, M.: Linear models for the approximate solution of the problem of packing equal circles into a given domain. Eur. J. Oper. Res. **230**(3), 505–514 (2013)
13. Litvinchev, I., Ozuna, E.: Packing circles in a rectangular container. In: Proceedings of the International Congress on Logistics and Supply Chain, pp. 24–25 (2013)
14. Litvinchev, I., Ozuna, E.: Integer programming formulations for approximate packing circles in a rectangular container. Math. Probl. Eng. (2014)
15. Lopez, C., Beasley, J.: A heuristic for the circle packing problem with a variety of containers. Eur. J. Oper. Res. **214**, 512–525 (2011)
16. Lopez, C., Beasley, J.: Packing unequal circles using formulation space search. Comput. Oper. Res. **40**, 1276–1288 (2013)
17. Pedroso, J., Cunha, S., Tavares, J.: Recursive circle packing problems. Int. Trans. Oper. Res. **23**(1), 355–368 (2014)
18. Andrade, R., Birgin, E.: Symmetry-breaking constraints for packing identical rectangles within polyhedra. Optim. Lett. **7**(2), 375–405 (2013)
19. Lempert, A., Kazakov, A.: On mathematical models for optimization problem of logistics infrastructure. Int. J. Artif. Intell. **13**(1), 200–210 (2015)
20. Coxeter, H.S.M.: Arrangements of equal spheres in non-Euclidean spaces. Acta Math. Acad. Scientiarum Hung. **5**(3), 263–274 (1954)
21. Boroczky, K.: Packing of spheres in spaces of constant curvature. Acta Math. Acad. Scientiarum Hung. **32**(3), 243–261 (1978)

22. Szirmai, J.: The optimal ball and horoball packings of the coxeter tilings in the hyperbolic 3-space. Beitr. Algebra Geom. **46**(2), 545–558 (2005)
23. Szirmai, J.: The optimal ball and horoball packings to the coxeter honeycombs in the hyperbolic d-space. Beitr. Algebra Geom. **48**(1), 35–47 (2007)
24. Szirmai, J.: A candidate for the densest packing with equal balls in thurston geometries. Beitr. Algebra Geom. **55**(2), 441–452 (2014)
25. Preparata, F., Shamos, M.: Computational Geometry. An Introduction. Springer, New York (1985)
26. Lempert, A., Kazakov, A., Bukharov, D.: Mathematical model and program system for solving a problem of logistic object placement. Autom. Remote Control **76**(8), 1463–1470 (2015)
27. Kazakov, A., Lempert, A., Bukharov, D.: On segmenting logistical zones for servicing continuously developed consumers. Autom. Remote Control **74**(6), 968–977 (2013)
28. Kazakov, A., Lempert, A.: An approach to optimization in transport logistics. Autom. Remote Control **72**(7), 1398–1404 (2011)
29. Specht, E.: Packomania. http://www.packomania.com/. Accessed 28 Oct 2015
30. Stoyan, Y., Yaskov, G.: Packing congruent spheres into a multi-connected polyhedral domain. Int. Trans. Oper. Res. **20**(1), 79–99 (2013)

# Random Forest Based Approach
# for Concept Drift Handling

Aleksei V. Zhukov[1,2(✉)], Denis N. Sidorov[1,2(✉)], and Aoife M. Foley[3(✉)]

[1] Energy Systems Institute SB RAS, Irkutsk, Russia
zhukovalex13@gmail.com
[2] Irkutsk State University, Irkutsk, Russia
dsidorov@isem.irk.ru
[3] Queens University Belfast, Belfast, UK
a.foley@qub.ac.uk

**Abstract.** Concept drift has potential in smart grid analysis because the socio-economic behaviour of consumers is not governed by the laws of physics. Likewise there are also applications in wind power forecasting. In this paper we present decision tree ensemble classification method based on the Random Forest algorithm for concept drift. The weighted majority voting ensemble aggregation rule is employed based on the ideas of Accuracy Weighted Ensemble (AWE) method. Base learner weight in our case is computed for each sample evaluation using base learners accuracy and intrinsic proximity measure of Random Forest. Our algorithm exploits ensemble pruning as a forgetting strategy. We present results of empirical comparison of our method and other state-of-the-art concept-drfit classifiers.

**Keywords:** Machine learning · Decision tree · Concept drift · Ensemble learning · Classification · Random forest

## 1 Introduction

Ensemble methods of classification (or, briefly, ensembles) employ various learning algorithms to obtain better predictive accuracy comparing with individual classifiers. Ensembles are much used in research and most of the papers is devoted to stationary environments where the complete datasets are available for learning classifiers and transfer functions of dynamical systems are not changing as time goes on. For real world applications (e.g. in power engineering [1], [2]) learning algorithms are supposed to work in dynamic environments with data continuously generated in the form of a stream on not necessarily equally spaced time intervals. Data stream processing commonly relies on single scans of the training data and implies restrictions on memory and time. Changes caused by dynamic environments (e.g. consumer behaviour in future smart grids) can be categorised into sudden or gradual concept drift subject to appearance of novel classes in a stream and the rate of changing definitions of classes.

D.I. Ignatov et al. (Eds.): AIST 2016, CCIS 661, pp. 69–77, 2017.
DOI: 10.1007/978-3-319-52920-2_7

One of the most generally effective ensemble classifier is a Random Forest. This algorithm employs bagging [3] and principles of the random subspace method [4] to build a highly decorrelated ensemble of decision trees [5]. There are some attempts [6] to adapt Random Forest to handling concept-drift, but this approach is not fully discovered.

The objective of this paper is to propose a novel approach to classification to adapt to concept drifts. This novel classification method can be applied to consumer behaviour in future smart grids to predict and classify the random behaviour of humans when they interact with smart appliances in the home and charging and discharging of electric vehicles. Likewise in these methods can be used to forecast wind power and indeed solar power.

We propose to compute the base learner weight for samples evaluation using base learners accuracy and intrinsic proximity measure of random forest.

It is to be noted that concept drifting is related to non-stationary dynamical systems modelling [7], where the completely supervised learning is employed using a special set of input test signals. For more details concerning integral dynamical models theory and applications refer to monograph [8] and its bibliography.

This paper is organised into five sections. In Sect. 2, a brief review of ensemble streaming classifiers based on random forest is given. Section 3 delivers detailed presentation of proposed algorithm. In Sect. 4 experiments on both machine learning and tracking tasks are provided. Finally, the paper concludes with Sect. 5 where the main results are discussed.

## 2    Related Work

The use of drifting concepts for huge datasets analysis is not unfamiliar to the machine learning and systems identification communities [9]. In this work we restrict ourselves to considering decision tree ensemble classification methods only.

Let us briefly discuss methods most related to our proposal and employed in the experiments. For a more detailed overview of the results in this area, including online incremental ensembles, readers may refer to monograph [10] and review [11]. Bayesian logistic regression was used in [12] to handle drifting concepts in terms of dynamical programming. Concept drifting handling is close to methodology of on-line random forest algorithm [6] where ideas from on-line bagging, extremely randomised forests and on-line decision tree growing procedure are employed. Accuracy Weighted Ensemble (AWE) approach was proposed in [9]. The main idea is to train a new classifier on each incoming dataset and use it to evaluate all the existing classifiers in the ensemble. We incorporate this idea in our approach.

Analysis of ensemble methods with decision tree base learners is interesting topic to be addressed in this paper. The recursive nature of decision trees makes on-line learning a difficult task due to the hard splitting rule, errors cannot be corrected further down the tree.

# 3    Proximity Driven Streaming Random Forest

In paper [13] the authors clearly demonstrated that a classifier ensemble can outperform a single classifier in the presence of concept drifts when the base classifiers of the ensemble are adherence weighted to the error similar to current testing samples. We propose another approach which exploits Random Forest properties.

To produce a novel classification algorithm capable of handling concept drift the following questions need to be answered:

- How to adapt original Random Forest for data streaming?
- How to define the base classifier weighting function?
- How to choose forgetting strategy?

These questions are considered in the following subsections.

## 3.1    Streaming Classifier Based on Random Forest

Methodologically ensemble approaches allow concept-drift to be handled in the following ways: base classifier adaptation, changing in training dataset (such as Bootstrap [3] or RSM [4]), ensemble aggregation rule changing or changing in structure of an ensemble (pruning or growing). In this paper we propose Proximity Driven Streaming Random Forest (PDSRF) which exploit combinations of these approaches. Besides some methods are already incorporated to the original Random Forest. Contrary to conventional algorithms we use weighted majority voting as an aggregation rule of ensemble. This allows us to adapt the entire classifier by changing the weights of the base learners. In order to obtain the classifiers weight estimation we should store samples. For this purpose we use a sliding windows approach which is used in the periodically updated Random Forest [14]. The length of this window is fixed and can be estimated by cross-validation. For the sake of time and memory optimization Extremely randomized trees [15] is used as a base learner instead of original randomized trees.

## 3.2    Base Classifier Weighting Function

We employ the assumption that the base classifiers make similar errors on similar samples even under concept-drift. Conventional Random Forest employ the so called proximity measure. It uses a tree structure to obtain similarity in the following way: if two different sample are in the same terminal node, their proximity will be increased by one. At the end, proximities are normalised by dividing by the number of trees [5].

Following the AWE approach proposed in [13] we use an error rate to produce weights (1) of classifiers, where $E$ is an new block testing error for $i$-th classifier, $\varepsilon$ is a small parameter.

$$w_i = 1/(E^2 + \varepsilon) \tag{1}$$

### 3.3   Forgetting Strategy

One of the main problems in concept-drifting learning is to select the proper forgetting strategy and forgetting rate [11]. The classifier should be adaptive enough to handle changes. In this case different strategies can be more appropriate to different types of drift (for example, sudden and gradual drifts). In this paper we focus on gradual changes only.

In this paper we propose ensemble pruning technique to handle the concept-drift. This technique uses the classic *replace-the-looser* approach [11] to discard trees with high error on new block samples.

### 3.4   Algorithm

To predict the sample we propose the following algorithm. First we use a stored window to find similar items using the specified similarity metric. Second we evaluate our current ensemble on similar examples. Then we compute weights adherence to errors on $k$ similar samples.

On every chunk the algorithm tests all the trees to choose the poorest base learner and replace it with new one trained on new block data. This process is iterative while the ensemble error on new block samples is higher than a specified threshold.

## 4   Experimental Evaluation

In this experiment, we evaluate our algorithm on publicly available dataset CoverType [16], and compared it to the most popular concept-drift classifiers. Our algorithm was implemented natively in C++ according to the same testing methodology. With adherence to this methodology classication accuracy was calculated using the *data block evaluation method*, which exploits the *test-then-train* approach. The data block evaluation method reads incoming samples without processing them, until they form a data block of size $d$. Each new data block is first used to test the existing classifier, then it updates the classifier [17].

Proposed proximity driven streaming random forest was tested with different sizes of block, window, nearest neighbours number and ensemble size (Table 2). In order to make the results more interpretable we also test original random forest with incorporated *"replace-the-looser"* forgetting.

### 4.1   Datasets

We follow the literature on adaptive ensemble classifiers and select the publicly available benchmark datasets with concept drift. In this paper the proposed method was tested on the *cover type dataset* from Jock A. Blackard, Colorado State University [16]. *Cover type dataset* contains the forest cover type for $30 \times 30$ m cells obtained from US Forest Service Region 2 Resource Information System (RIS) data. It contains 581 012 instances and 54 attributes, and it has been used as a benchmark in several papers on data stream classification.

## 4.2 Results

The proposed approach shows results similar to the above mentioned AUE2 [18]. Next comparison (ref. Fig. 1) of the original random forest with incorporated "replace-the-looser" forgetting and proposed proximity driven streaming random forest are presented. In Table 1 the mean accuracy is shown. It must be noted that in case of high window size optimal number of nearest neighbours tends to one (Fig. 2).

*Remark 1.* For more details of our approach evaluation readers may refer to http://mmwind.github.io/pdsrf.

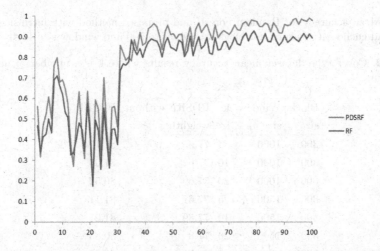

**Fig. 1.** Original random forest with incorporated "replace-the-looser" and PDSRF accuracy on first 100 chunks of the dataset.

**Table 1.** Cover type dataset mean accuracy results. Some of the results obtained using settings from [18].

| Method | Mean accuracy |
|---|---|
| PDSRF | 87.42 |
| HOT | 86.48 |
| AUE2 | 85.20 |
| PDSRF(without weighting) | 82.79 |
| AUE1 | 81.24 |
| Lev | 81.04 |
| DWM | 80.84 |
| Oza | 80.40 |
| AWE | 79.34 |
| Win | 77.19 |

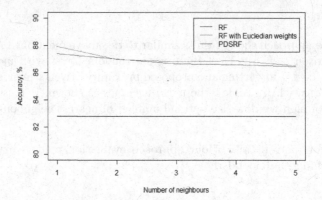

**Fig. 2.** Mean accuracy for Random Forest, and proposed method with intrinsic proximity and euclidean measures for block size equal to 500 and window size 1500.

**Table 2.** Cover type dataset mean accuracy results where k is a number of nearest neighbours.

| Block size | Window size | k | PDSRF without weighting | PDSRF |
|---|---|---|---|---|
| 300 | 1000 | 5 | 77.65 | 81.15 |
| 300 | 1000 | 10 | 77.69 | 81.05 |
| 300 | 1000 | 20 | 77.69 | 80.52 |
| 300 | 1500 | 5 | 77.67 | 81.11 |
| 300 | 1500 | 10 | 77.89 | 81.21 |
| 300 | 1500 | 20 | 77.8 | 80.83 |
| 500 | 500 | 5 | 82.76 | 86.38 |
| 500 | 500 | 10 | 82.73 | 86.16 |
| 500 | 500 | 20 | 82.68 | 86.02 |
| 500 | 1000 | 5 | 82.76 | 86.45 |
| 500 | 1000 | 10 | 82.87 | 86.27 |
| 500 | 1000 | 20 | 82.75 | 86.04 |
| 500 | 1500 | 5 | 82.75 | 86.49 |
| 500 | 1500 | 10 | 82.74 | 86.29 |
| 500 | 1500 | 20 | 82.7 | 85.96 |

## 5   Discussion and Conclusions

In comparison to other concept-drift approaches like Online Random Forest and the AWE, our approach needs more computational resources and thus more time for both the training and prediction stages. But PDSRF can outperform the most of modern methods as it is demonstrated in the experiment. The proposed

**Table 3.** Mean accuracy on first 100 blocks of cover type dataset.

| Block size | Window size | k | Number of trees | PDSRF (without weighting) | PDSRF |
|---|---|---|---|---|---|
| 500 | 1000 | 1 | 5 | 75.03 | 79.25 |
| 500 | 1000 | 1 | 7 | 74.37 | 79.42 |
| 500 | 1000 | 1 | 10 | 74.34 | 78.76 |
| 500 | 1000 | 1 | 13 | 74.22 | 79.56 |
| 500 | 1000 | 1 | 15 | 74.00 | 79.72 |
| 500 | 1000 | 1 | 17 | 74.18 | 80.09 |
| 500 | 1000 | 1 | 20 | 74.23 | 79.90 |
| 500 | 1000 | 2 | 5 | 74.96 | 78.77 |
| 500 | 1000 | 2 | 7 | 74.36 | 79.33 |
| 500 | 1000 | 2 | 10 | 74.14 | 79.51 |
| 500 | 1000 | 2 | 13 | 74.03 | 79.02 |
| 500 | 1000 | 2 | 15 | 74.02 | 79.51 |
| 500 | 1000 | 2 | 17 | 74.35 | 79.68 |
| 500 | 1000 | 2 | 20 | 74.09 | 79.85 |
| 500 | 1000 | 3 | 5 | 74.83 | 77.78 |
| 500 | 1000 | 3 | 7 | 74.66 | 79.59 |
| 500 | 1000 | 3 | 10 | 74.34 | 79.14 |
| 500 | 1000 | 3 | 13 | 74.35 | 79.17 |
| 500 | 1000 | 3 | 15 | 74.09 | 79.32 |
| 500 | 1000 | 3 | 17 | 74.09 | 79.68 |
| 500 | 1000 | 3 | 20 | 74.13 | 80.22 |
| 500 | 1000 | 5 | 5 | 74.59 | 78.70 |
| 500 | 1000 | 5 | 7 | 74.29 | 78.76 |
| 500 | 1000 | 5 | 10 | 74.26 | 78.66 |
| 500 | 1000 | 5 | 13 | 73.94 | 79.52 |
| 500 | 1000 | 5 | 15 | 74.37 | 80.16 |
| 500 | 1000 | 5 | 17 | 74.09 | 79.41 |
| 500 | 1000 | 5 | 20 | 74.01 | 80.16 |

approach is highly parallelizable and can be implemented using the GPGPU. As it was shown in Table 3 high accuracy can be achieved with relatively small number of base classifiers. It must be noted that the proposed approach can be efficiently applied only to gradual concept drifts. PDSRF is sensible to all the parameters changes and all of these parameters must be accurately tuned. As it shown the proposed approach significantly exceeds the original random forest with incorporated "replace-the-looser" forgetting. Although the presented results show that the accuracy is higher than AUE2, the approach has some promising directions is that the Random Forest can be used in unsupervised and allows to work with missing data, which is an issue with smart grid datasets and wind power forecasting where telecommunications signals and data recording is not 100% robust.

**Acknowledgment.** This work is funded by the RSF grant No. 14-19-00054 and by the International science and technology cooperation program of China, project 2015DFR70850, NSFC Grant No. 61673398.

# References

1. Zhukov, A., Kurbatsky, V., Tomin, N., Sidorov, D., Panasetsky, D., Foley, A.: Ensemble methods of classification for power systems security assessment. arXiv, Artificial Intelligence (cs.AI), pp. 1–6. arXiv:1601.01675 (2016)
2. Tomin, N., Zhukov, A., Sidorov, D., Kurbatsky, V., Panasetsky, D., Spiryaev, V.: Random forest based model for preventing large-scale emergencies in power systems. Int. J. Artif. Intell. **13**, 211–228 (2015)
3. Breiman, L.: Bagging predictors. Mach. Learn. **24**(2), 123–140 (1996)
4. Ho, T.K.: The random subspace method for constructing decision forests. IEEE Trans. Pattern Anal. Mach. Intell. **20**(8), 832–844 (1998)
5. Breiman, L.: Random forests. Mach. Learn. **45**(1), 5–32 (2001)
6. Saffari, A., Leistner, C., Santner, J., Godec, M., Bischof, H.: On-line random forests. In: 2009 IEEE 12th International Conference on Computer Vision Workshops (ICCV Workshops), 1393–1400. IEEE (2009)
7. Sidorov, D.: Modelling of non-linear dynamic systems by Volterra series. In: Attractors, Signals, and Synergetics Workshopp, vol. 2000, pp. 276–282. Pabst Science Publication, USA-Germany (2002)
8. Sidorov, D.: Integral Dynamical Models: Singularities, Signals and Control. World Scientific Publishing, Singapore (2015)
9. Wang, H., Fan, W., Yu, P.S., Han, J.: Mining concept-drifting data streams using ensemble classifiers. In: Proceedings of SIGKDD, 24–27 August 2003, Washington, DC, USA, pp. 226–235 (2003)
10. Gama, J.: Knowledge discovery from data streams. CRC Press Publishing, Singapore (2010)
11. Kuncheva, L.: Classier ensembles for changing environment. In: Roli, F., Kittler, J., Windeatt, T. (eds.) 2004 5th International Workshop on Multiple Classier Systems, pp. 1–15. Springer, Heidelberg (2004)
12. Turkov, P., Krasotkina, O., Mottl, V.: Dynamic programming for Bayesian logistic regression learning under concept drift. In: Maji, P., Ghosh, A., Murty, M.N., Ghosh, K., Pal, S.K. (eds.) PReMI 2013. LNCS, vol. 8251, pp. 190–195. Springer, Heidelberg (2013)
13. Wang, H., Fan, W., Yu, P.S., Han, J.: Mining concept-drifting data streams using ensemble classifiers. In: Proceedings of the Ninth ACM SIGKDD International Conference on Knowledge Discovery and Data Mining, pp. 226–235. ACM (2003)
14. Zhukov, A., Kurbatsky, V., Tomin, N., Sidorov, D., Panasetsky, D., Spiryaev, V.: Random forest based model for emergency state monitoring in power systems. In: Mathematical Method for Pattern Recognition: Book of abstract of the 17th All-Russian Conference with Interneational Participation, p. 274. TORUS PRESS, Svetlogorsk (2015)

15. Geurts, P., Ernst, D., Wehenkel, L.: Extremely randomized trees. Mach. Learn. **63**(1), 3–42 (2006)
16. Blake, C.L., Merz, C.J.: UCI repository of machine learning databases (1998)
17. Brzezinski, D.: Mining data streams with concept drift. Dissertion MS thesis. Department of Computing Science and Management, Poznan University of Technology (2010)
18. Brzezinski, D., Stefanowski, J.: Reacting to different types of concept drift: the accuracy updated ensemble algorithm. IEEE Trans. Neural Netw. Learn. Syst. **25**(1), 81–94 (2014)

# Social Networks

# The Detailed Structure of Local Entrepreneurial Networks: Experimental Economic Study

Dmitry B. Berg, Rustam H. Davletbaev, Yulia Y. Nazarova,
and Olga M. Zvereva$^{(\boxtimes)}$

Ural Federal University, Ekaterinburg, Russia
{bergd, r.davletbaev, nazarova_yukiru}@mail.ru,
OM-Zvereva2008@yandex.ru

**Abstract.** Economic agents' behavior during the last 40 years had tremendously changed from perfect competition to cooperation between them, and coopetition phenomenon was revealed. This phenomenon is always based on the certain entrepreneurial network. The paper is focused on entrepreneurial networks which are geographically localized. Such networks are formed as a result of two different types of cooperation: production cluster cooperation and cooperation in a community. The main goal of the present study is to find differences between internal structures of these two types entrepreneurial networks. Data was collected using experimental economic techniques, it was represented in the form of transactions between network agents and was aggregated over the certain time period. Social Network Analysis (SNA) methods and instruments were used in this research. Detailed structure analysis was based on the set of quantitative parameters such as density, diameter, clustering coefficient, different kinds of centrality, and etc. The entrepreneurial networks of two production clusters and three cooperative communities were under investigation. These networks were compared with each other and also with random Bernoulli graphs of the corresponding size and density. It was found that cooperative community networks are more random and dense than the production cluster ones and their other parameters also differ. Discovered variations of network structures are explained by the peculiarities of agents functioning in these two type networks.

**Keywords:** Economic network · Entrepreneurial network · Social network analysis · Experimental economics · Communications · Coopetition · Localization · Local payment system

## 1 Introduction

Economic agents' behavior during the last 40 years had tremendously changed from the perfect competition to cooperation between them. One can find different forms of cooperation relationships. If they are localized in the certain territory in order to supply a special complex product with a high level of added value, one might identify this structure as a "cluster" [1]. If agents are mostly interested in meeting each other

© Springer International Publishing AG 2017
D.I. Ignatov et al. (Eds.): AIST 2016, CCIS 661, pp. 81–90, 2017.
DOI: 10.1007/978-3-319-52920-2_8

demands, it is a cooperative community [2]. If agents are distributed geographically, they become parts of a vertically integrated production structure with crossed share-holdings, than we call this structure an "international corporation" [3]. One more form of cooperation between agents is a barter network [4]. Even if employees are distributed geographically, more and more often they become to work in the same company. Such great variety of cooperative relationships now is considered in the ranks of the coopetition concept [5].

"Coopetition or Co-opetition (sometimes spelled "coopertition" or "co-opertition") is a neologism coined to describe a kind of cooperative competition. Coopetition is a portmanteau of cooperation and competition, emphasizing the "petition"-like nature of the joint work" [6].

Embedded ties [7] at formally competitive markets became another aspect of this phenomenon. Now it is absolutely clear that the composition of competition and cooperation determines agents' behavior in the modern markets. Competition in different forms (perfect, imperfect, monopolistic, etc.) nowadays is the well studied phenomenon. There is the opposite situation if we consider cooperation. Cooperation occurs to be on the sidelines of the mainstream and most of economists pay it less attention because of its poor mathematical base. Nowadays, cooperation requires close attention and detailed study because of its significance both for economic theory and practice.

Cooperative relationships are institutionalized in recurring communications between agents – entrepreneurs. In every communication act they can exchange information, senses, money, services, industrial goods, energy, and etc. The set of communications forms a stable cooperative network with corresponding commodity and financial flows. Communications' dual nature makes it possible to design the network communication matrix using the information about agent transactions.

This paper is focused on the entrepreneurial networks localized geographically. They are based on production cluster cooperation and cooperation in a community. The main goal of the present study is to compare internal structures of these two type entrepreneurial networks.

## 2 Data Description

Exact data of entrepreneurial production network structure and its functioning can be obtained from the set of agents' (actors') bank statements. A bank statement contains all information required to design a communication matrix: the sum, the recipient/payer name and the date of transaction. One can investigate communications day-by-day or aggregate them for the certain period of time. A communication matrix with information about transaction dates makes it possible to study a network structure evolution. But an agent's bank statement is a trade secret, so it seems impossible to get real statements of agents of a network.

There are two basic sources to obtain data about a cooperative community entrepreneurial network. One of them is the same as for a production network. The main difficulties, which arise, were described above. If a community is using so-called "complementary" currency, or "local" currency [8] for payments and it is an electronic currency, all transactions are recorded in a local payment system. The required communication matrix can be easily exported (under permission of the operator) and utilized for the further analysis. Unfortunately, the operator's permission is also a great problem. Besides, there are almost no local payment systems in cooperative communities in Russia.

In order to solve this problem it was decided to use the experimental economics [9] approach. Two sets of experiments were performed and five different entrepreneurial networks were obtained as the result: three cooperative community networks and two production ones.

For relevant results obtaining, the experiments for cooperative community network generation were organized in three different regions of the country: Ufa city (Bashkortostan Republic of Russian Federation), Nabereznye Chelny city (Tatarstan Republic of Russian Federation) and Moscow city. Every group of entrepreneurs was comprised randomly using Facebook but from the local entrepreneurial community. Experimental network was formed on the basis of their face-to-face communications. Each experiment lasted for two hours: during the first hour participants read the rules and regulations, the second hour was spend in communications and network building activity. The main requirement of experiments was to exchange the real goods and services produced by the participants themselves. Exchange intensification and network formation was received through the negative cash fund interest rate [10] usage.

Production cooperative networks were designed in two different ways. The first production network was obtained from the municipal economy's model discussed in [11, 12]. It is based on the set of the 12 real business entities which have the partner relations in manufacturing and consuming their goods and services for 10000 residents. The set of municipal business entities includes farms (agricultural, poultry, meat and dairy), plants (dairy, meat processing), mills (flour, feed), bakery, factory, workshop, and etc.

The second network was generated by the group of students in economics (School of Economics and Management of Ural Federal University, Ekaterinburg) while developing the project task: to design a network of small business companies which will be able to supply the whole life cycle (for two weeks) of a summer tourist camp with educational and entertainment programs, with building, rent, logistics, catering, security, garbage collection, and other required services. The total number of virtual camp inhabitants was set to 120 including children and staffs. First two and two last days were spent for assembling/disassembling building constructions. The balanced matrix of payments between agents aggregated for 14 days is the base for the second production network under investigation.

Both production networks include population as the special agent which consumes the other agents' products and provides them with the necessary labor resource.

Networks are determined by the following parameters:

- $Nn$ - number of nodes (agents, actors);
- $Ne$ - number of edges (relations, communications, ties, links);
- $D$ - density (the proportion of all possible ties which are actually present), $D = Ne/Nn(Nn - 1)$;
- $Sum$ – total amount of transactions/payments (in rubles);
- $AvrCost$ - average transaction cost (in rubles), $AvrCost = Sum/Ne$

- *Ng* - number of different types of goods and services produced and consumed in the network;
- *Var* - variety of products in the network, *Var = Ng/Nn*.

Values of network parameters are listed in Table 1. The network graphs are shown in Fig. 1.

**Table 1.** Network parameters

| Parameter | Community networks | | | Production networks | |
|---|---|---|---|---|---|
| | *1* | *2* | *3* | *1* | *2* |
| *Nn* | 10 | 11 | 17 | 12 | 8 |
| *Ne* | 56 | 29 | 50 | 28 | 19 |
| *D* | 0.622 | 0.264 | 0.184 | 0.212 | 0.339 |
| *Sum* | 438000 | 80000 | 576000 | 306200000 | 2484599 |
| *AvrCost* | 7821.43 | 2758.62 | 11520 | 10935714 | 130768 |
| *Ng* | 42 | 29 | 44 | 14 | 8 |
| *Var* | 4.2 | 2.64 | 2.59 | 1.17 | 1 |

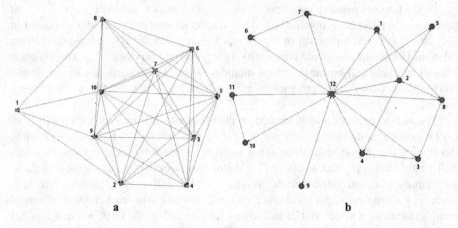

**Fig. 1.** Network Graphs of: (a) community 1 network; (b) production 1 network. Designed in UCINET 6 for Windows

All investigated networks are of the comparable size (*Nn* and *Ne*) and density (may be, except community 1 network, which density reflects the extremely high intensity of agent communications). The main differences of the community and production network sets are:

- in the total amounts of transactions (production networks demonstrates the higher values);
- number of goods and services consumed in the community networks exceeds two times or more the corresponding number in the production networks;

- variety of products in the community networks is also two of more times higher than in the production ones;
- average transaction cost in the community networks is at least 10 times lower than one was obtained in the production networks.

Such significant differences are the results of the network nature: the most number of production network transactions are localized in B2B segment, while a community network is focused on B2C communications. So, every production network agent tries to supply large volumes from the limited range of products in order to reduce product costs. Conversely, every community network agent is interested in production of a wide variety of products, even whether only one sample is demanded. Moreover, a community network agent can be both a product manufacturer and its final consumer.

## 3   The Calculation Technique and Quantitative Parameters

Network quantitative characteristic study is focused on four types of parameters: for a whole network, for an ego network (neighborhood), dyadic, and single actor parameters. Calculations were performed according to the formulas from Table 2 [13, 14].

**Table 2.** Network parameters and their calculations

| № | Parameter | Formula | Explanation |
|---|-----------|---------|-------------|
| 1 | $D$ | $D = \max\limits_{i,j=1,\dots,n} \left(d\left(n_i, n_j\right)\right)$ | $d(n_i, n_j)$ – the shortest (geodesic) path $n_i$ – actor $i$; $n_j$ – actor $j$ |
| 2 | $CC$ | $CC = \frac{1}{N}\sum\limits_{i=1}^{N} C_i,$ | $C_i$ – density of the $i$-th actor's neighborhood |
| 3 | $Tr$ | $Tr = \frac{N_t}{N_d},$ | $N_t$ – number of non-vacuous transitive ordered triples $N_d$ – number of triples in which ties go from actor $n_i$ to actor $n_j$ and from actor $n_j$ to actor $n_k$ |
| 4 | $Re$ | $Re = \frac{\sum L_p}{\sum L},$ | $L$ – dyadic tie $Lp$ - reciprocated dyadic tie |
| 5 | $IDCenz$ | $\dfrac{\sum_{i=1}^{n}(IDC^*-IDC_i)}{\max_{i=1,\dots,n}\sum_{i=1}^{n}(IDC^*-IDC_i)}$ | $IDC^*$ – in-degree centrality of the most central actor $IDC_i$ – in-degree centrality of the $i$-th actor |
| 6 | $ODCenz$ | $\dfrac{\sum_{i=1}^{n}(ODC^*-ODC_i)}{\max_{i=1,\dots,n}\sum_{i=1}^{n}(ODC^*-ODC_i)}$ | $ODC^*$ – outdegree centrality of the most central actor $ODC_i$ – outdegree centrality of the $i$-th actor |
| 7 | $BCenz$ | $\dfrac{\sum_{i=1}^{n}(BC^*-BC_i)}{\max_{i=1,\dots,n}\sum_{i=1}^{n}(BC^*-BC_i)}$ | $BC^*$ – berweenness centrality of the most central actor $BC_i$ – betweenness centrality of the $i$-th actor |

General network parameters are: nodes (actors) and edges numbers, density, diameter and clustering coefficient. Three of them have been already discussed and listed above in Table 1.

*Diameter (D)* is the largest geodesic path from one actor (node $n_i$) to another (node $n_j$). A geodesic path (or the shortest path) between nodes i and j is the path connecting these vertices with minimum length [43]. A diameter tells us how "compact" the network is (that is, how many steps at least are necessary to get from one node to another).

*Clustering coefficient* (CC) is the mean of neighborhood densities for all network actors [15]. Clustering coefficient *CC* indicates how "cohesive" a network is [16] proposes formulas for weighted networks.

For a directed network the *transitivity coefficient (Tr)* is often measured as the ratio of really transitive triad number in a network to the number of cases where a single link could complete the transitive triad. Transitive relationship in a triad means that if A directs a tie to B, and B directs a tie to C, then A also directs a tie to C. Transitive triads are argued by some scholars [17] to be the balanced, or natural, network patterns. So, transitivity indicates the potential ability of a network to become a "stable", or "natural", one.

One of the dyadic main characteristics is *reciprocity (Re)*. It indicates the proportion of connected actor pairs (dyads) having a reciprocated tie between them. Disconnected pairs (null relationship between actors) are usually ignored.

Different kinds of centrality are used to characterize a single actor. The chosen characteristics of centrality take into account only direct links between nodes because of the network specificity. *Centralization* indicates how unequal the distribution of actor connections (degree centrality) is in a network. There are several centrality measures, only some of them are under discussion.

If we consider a directed network, the total number of ties sent by an actor is called *out-degree centrality of the actor,* and the total number of ties received is called *in-degree actor centrality*. Actor in-degree centrality is considered to be the measure of its prestige, and its out-degree centrality characterizes the level of its expansivity.

*Betweenness* centrality is the extent to which an actor falls on the paths between other pairs of actors in the network. It stresses the control level or the capacity to interrupt relations [18].

Three different types of network centralization are calculated in the present study: *Indegree Centralization (IDCenz), Outdegree Centralization (ODCenz)* and *Betweenness Centralization (BCenz)*. To estimate the network centralization one must find the most central actor $C^*$, take its centrality score and subtract the centrality score of each other actor from it, add up the differences: $\Sigma(C^* - C_i)$, then divide this by what this sum would be under the largest possible centralization ($Max\ \Sigma(C^* - C_i)$).

The main idea of the study is to compare two types of networks with the corresponding random Bernoulli graphs and with each other in order to find differences between them for deep understanding of entrepreneurial network formation.

Each experimental entrepreneurial network has its own unique structure. For the aim of valid comparison random Bernoulli graphs of the same sizes and densities as the experimental network graphs were generated. Investigation of a network structure was performed according the following technique:

– calculations of the experimental network parameters;
– generation of the random Bernoulli graphs of the same size and density;
– calculations of the random Bernoulli graph parameters;
– comparison of the experimental network and the corresponding random Bernoulli graph parameters, estimation of their relative deviations;
– comparison of the relative deviations from Bernoulli graph among the community networks and the production networks separately and calculation of the relative deviation average values for each set;

– comparison of these relative deviation averaged values.

*Relative deviation (RD)* from a Bernoulli graph is calculated as follows:

$$RD = \frac{\Delta}{BV} \qquad (1)$$

where $\Delta = |V - BV|$, $V$– the experimental network parameter value, $BV$ – the corresponding Bernoulli graph parameter value.

All calculations of the network parameters listed in Table 2 were carried out utilizing framework UCINET 6 for Windows [19] which supports SNA (Social Network Analysis) methodology.

## 4   Results and Discussion

Calculated community network parameters (their real values ($V$) and relative deviations from Bernoulli graphs ($RD$)) are shown in Table 3. One can find that differences between "community 1" network and the two others are greater than between the "community 2" and the "community 3" networks. The main reasons of this fact were discussed above (see comments to Table 1). The most similar to the random Bernoulli graph is the "community 3" network – relative deviation values do not exceed 17% for all calculated parameters and for most of them they are in the range of 2%–13%. Centralization in general seems to be the most distinctive parameter of the community networks 1 and 2 and the corresponding random graphs: in the substantial number of cases the relative deviation reaches the values of 44% and greater. Summarizing results of comparison, it should be noted that community network graphs and the random ones differ significantly.

**Table 3.** Values of community networks parameters

|        | Community 1 |      | Community 2 |      | Community 3 |      |
|--------|------|------|------|------|------|------|
|        | V    | RD   | V    | RD   | V    | RD   |
| D      | 2    | 0.33 | 4    | 0.2  | 5    | 0.17 |
| Re     | 0.47 | 0.23 | 0.07 | 0.46 | 0.16 | 0.02 |
| Tr     | 0.61 | 0.06 | 0.23 | 0    | 0.14 | 0.12 |
| CC     | 0.64 | 0.07 | 0.28 | 0.52 | 0.23 | 0.07 |
| IDCenz | 0.71 | 0.46 | 0.26 | 0.13 | 0.20 | 0.13 |
| ODCenz | 0.17 | 0.46 | 0.15 | 0.21 | 0.20 | 0.13 |
| BCenz  | 0.03 | 0.70 | 0.15 | 0.44 | 0.17 | 0.15 |

The same comparison for production networks is shown in Table 4. Networks of this set seem more similar to each other than in the previous case. At the same time, they are tremendously differ from random Bernoulli graphs – the relative deviation exceeds 50% overwhelmingly and for the most centralization and reciprocity values is 100% and greater.

The real values of the calculated networks parameters and their relative deviations from the random Bernoulli graphs (Tables 3 and 4) were averaged for the qualitative comparison of the two investigated sets (Table 5). The networks and random graph parameters differ by more than 22% (with rare exceptions). It is the first significant result of the present study.

**Table 4.** Values of production networks parameters

|  | Production 1 | | Production 2 | |
|---|---|---|---|---|
|  | V | RD | V | RD |
| D | 4 | 0.33 | 3 | 0 |
| Re | 0.33 | 0.99 | 0.46 | 2.92 |
| Tr | 0.15 | 0.52 | 0.30 | 0.38 |
| CC | 0.52 | 0.53 | 0.60 | 0.40 |
| IDCenz | 0.46 | 0.27 | 0.59 | 0.38 |
| ODCenz | 0.86 | 4.2 | 0.76 | 1.85 |
| BCenz | 0.77 | 2.2 | 0.83 | 6.4 |

**Table 5.** Averaged values of community and production networks parameters

|  | Average values of parameters | | | |
|---|---|---|---|---|
|  | Community | | Production | |
|  | <V> | <RD> | <V> | <RD> |
| D | 3.7 | 0.23 | 3.5 | 0.17 |
| Re | 0.24 | 0.24 | 0.40 | 1.9 |
| Tr | 0.33 | 0.06 | 0.22 | 0.45 |
| CC | 0.38 | 0.22 | 0.56 | 0.47 |
| IDCenz | 0.21 | 0.24 | 0.53 | 0.33 |
| ODCenz | 0.18 | 0.27 | 0.81 | 3.1 |
| BCenz | 0.12 | 0.43 | 0.79 | 4.3 |

The second important result is the detection of the fundamental difference between two investigated networks: a community network graph is more similar to a random Bernoulli graph than a production one. It becomes absolutely evident after comparison of averaged RD-values. The reason is based on the different nature of two network sets: a production network is more determined by resource supply chains, and its ability to change suppliers and buyers is very restricted, while a community network is more flexible.

It means that coopetition phenomenon is accompanied by formation of agent networks. It is important that such networks are far from random ones. The network ("production" or "community" in the present study) specificity is reflected in its structure. One might suppose that different types of entrepreneurial networks will be discovered in the nearest future. It will influence investigation of modern markets and make it more complicated.

These important findings in the structure of local entrepreneurial networks determine the direction of the further detailed study including analysis of the larger volumes of experimental data and also usage of real economic data. In combination with research of networks functioning [20], it will impact our understanding of the coopetition phenomenon.

**Acknowledgements.**  Present study was carried out under financial support of the Russian Fund of Fundamental Research grant №. 15-06-04863 "Mathematical models of local payment system lifecycles".

# References

1. Porter, M.E.: The Competitive Advantage of Nations. Harvard Business Review, pp. 72–91 (1990)
2. Mc Millan, D.W., Chavis, D.M.: Sense of community: a definition and theory. Am. J. Commun. Psychol. **14**(1), 6–23 (1986)
3. Pitelis, C., Sugden, R.: The Nature of the Transnational Firm. Routledge, London (2000)
4. Birch, D., Liesch, P.W.: Moneyless business exchange: practitioners' attitudes to business-to-business barter in Australia. Ind. Mark. Manag. **27**, 329–340 (1998)
5. Gnyawali, D.R., Park, B.J.R.: Coopetition and technological innovation in small and medium-sized enterprises: a multilevel conceptual model. J. Small Bus. Manag. **47**(3), 308–330 (2009). doi:10.1111/j.1540-627X.2009.00273.x
6. Wikipedia. The Free Encyclopedia. https://en.wikipedia.org/wiki/Coopetition
7. Uzzi, B.: The sources and consequences of embeddedness for the economic performance of organizations: the network effect. Am. Sociol. Rev. **61**(4), 674–698 (1996)
8. Lietar, B.A.: The Future of Money: Creating New Wealth, Work and Wiser World. M. KRPA Olymp: AST: Astrel, Moscow (2007)
9. Davis, D.D., Holt, C.A.: Experimental Economics. Princeton University Press, Princeton (1993)
10. Gesell, S.: The Natural Economic Order (Translation by Philip Pye). Peter Owen Ltd., London (1958)
11. Popkov, V.V., Berg, D.B., Ulyanova, E.A., Selezneva, N.A.: Commodity and financial networks in regional economics r-economy, vol. 1, issue 2, pp. 305–314 (2015). doi:10. 17059/e-2015-2-13. http://r-economy.ru/?page_id=257
12. Berg, D.B., Zvereva, O.M.: Identification of autopoietic communication patterns in social and economic networks. In: Khachay, M.Y., Konstantinova, N., Panchenko, A., Ignatov, D.I., Labunets, V.G. (eds.) AIST 2015. CCIS, vol. 542, pp. 286–294. Springer, Heidelberg (2015). doi:10.1007/978-3-319-26123-2_28
13. Newman, M.E.J.: The structure and function of complex networks. SIAM Rev. **45**(2), 168–256 (2003)
14. Costa, L.D.F., Rodrigues, F.A., Travieso, G., Villas Boas, P.R.: Characterization of complex networks: a survey of measurements. Adv. Phys. **56**(1), 167–242 (2007)
15. Hanneman, R.A., Riddle, M.: Introduction to Social Network Methods. http://faculty.ucr. edu/ ~ hanneman/nettext/
16. Phan, B., Engo-Monsen, K., Fjeldstad, O.D.: Considering clustering measures: third ties, means, and triplets. Soc. Netw. **35**(3), 300–308 (2013)

17. Faust, K.: Comparing social networks: size, density and local structure. Metodološki Zvezki 3(2), 185–216 (2006)
18. Marsden, P.V.: Network data and measurement. Ann. Rev. Sociol. **16**, 435–446 (1990)
19. Borgatti, S.P., Everett, M.G., Freeman, L.C.: Ucinet for Windows: Software for Social Network Analysis. Analytic Technologies, Harvard (2002)
20. Berg, D., Zvereva, O., Shelomentsev, A., Taubayev, A.: Autopoietic structures in local economic systems. In: 15th International Conference Proceedings on Multidisciplinary Scientific Geoconference SGEM 2015, Ecology, Economics, Education and Legislation, Ecology and Environmental Protection, vol. I, pp. 109–117, 18–24 June, 2015, Albena, Bulgaria. Published by STEF92 Technology Ltd. (2015)

# Homophily Evolution in Online Networks: Who Is a Good Friend and When?

Sofia Dokuka[✉], Diliara Valeeva, and Maria Yudkevich

Center for Institutional Studies, Higher School of Economics, Moscow, Russia
sdokuka@hse.ru

**Abstract.** Homophily is considered by network scientists as one of the major mechanisms of social network formation. However, the role of dynamic homophily in the network growth process has not been investigated in detail yet. In this paper, we estimate the role of homophily by various attributes at different stages of online network formation process. We consider the process of online friendship formation in the Vkontakte social networking site among first-year students at a Russian university. We reveal that at the beginning of the network formation a similarity in gender and score in entrance exams plays the key role, while by the end of network establishment period the role of the same group affiliation becomes more important. We explain the results with the tendency of students to follow different strategies to control the information flow in their social environment.

**Keywords:** Network growth · Network evolution · Homophily · Online networks · Student networks

## 1    Introduction and Related Works

Homophily is considered by network scientists as one of the major covariate-based mechanisms in network formation and evolution processes [1]. Lazarsfel and Merton define homophily as a tendency of people to form ties between those who are alike in some designated respect [2]. Homophily was revealed in various network types such as those of friendship, advice, support, information transfer, co-membership etc. The basic characteristics that affect homophily are race, ethnicity, gender, age, religion, education, occupation, social class, network position, and behavior [3]. The existence of homophily phenomena was also revealed in online social networks. [4] is one of the most influential first studies of online homophily. Using data about online communication of political blogs, they show that people with similar political views tend to communicate and cite each other. [5] investigate the role of similarity in online ties formation in the **LiveJournal.com** blogging platform. The authors show that having even a few common interests makes friendship significantly more likely to happen. They also outline that geographical homophily (so-called proximity) in an online social network is less important in contrast to offline communities.

D.I. Ignatov et al. (Eds.): AIST 2016, CCIS 661, pp. 91–99, 2017.
DOI: 10.1007/978-3-319-52920-2_9

Most studies on student homophily are based on offline friendship networks [6,7]. However, there are a few studies where scholars are interested in online student communications and friendship based on their similarity. For example, [8,9] analyze a student friendship network in Facebook and show that students tend to form connections with similar others. On the one hand, they form connections based on their observable characteristics such as gender, race, home city, studying the same field or living in the same dorm. On the other hand, they tend to befriend others with similar nonobservable characteristics such as preferences in music, movies, and books. [10] also study a Facebook friendship network of students and reveal a similar phenomena found in various offline networks: students are strongly divided in smaller network communities based on their race and ethnicity. [11] investigate the process of student communications in a university's online platform and reveal the homophily based on academic performance. They show that over time high-achieving students tend to form a dense network core and communicate mostly with each other excluding their less able peers.

While homophily studies are some of the oldest in network science, questions still remain. First, there are not enough studies on how homophily changes over time [3]. For instance, when a person enters a new social environment, she or he starts to form ties based on individual or institutional characteristics (e.g., gender, group membership) or based on network tendencies (reciprocity, transitivity). Over time, she or he starts to be more selective and form or dissolve ties based on hidden characteristics such as psychological traits or level of abilities. While there are longitudinal network studies investigating homophily, its transformation is not analyzed in smaller time-steps. Homophily is perceived as a continuous process that remains relatively stable during the whole period of network evolution and does not change dramatically. Also, homophily is traditionally investigated in established network structures. However, its analysis from the very first network formation steps might give us more insights about the network evolution process.

Second, there are not enough studies on the multidimensionality of homophily [3]. A coincidence in only one attribute might be an important predictor for a tie formation but having several common attributes might explain a connection with a stronger power. Such multidimensionality was investigated for the first time in detail by Block and Grund in [1]. Studying friendship networks in eleven European and American schools, they show that students with more than one common attribute do not tend to continue and maintain their friendship connections while similarity in only one attribute influences friendship formation and evolution positively. It means that people who are very similar to each other do not tend to create or continue social interactions with each other. In general, homophily is a multidimensional and dynamic process evolving both with changing networks and attributes [3].

In this study, we ask the question: how does homophily change during the process of online friendship network formation? Using dynamic data about a student friendship network in the Russian online networking website Vkontakte, we analyze homophily evolution. We show that during the very first steps of network formation, when students do not know each other offline, they tend to form ties based on similarity in gender and school performance level. Over time, when they meet offline and begin to study, they connect based on the same study group affiliation and this similarity remains the strongest until the network stabilizes. We explain the results through the strategic behavior of students that allow them to form useful connections, gather important study-related information and be successful in the university's social and educational environment.

## 2  Data

### 2.1  Case Description

In this paper we investigate the dynamics of the online social network of students who were enrolled in the Economics department in one Russian university in the 2014–2015 academic year. Enrollment in Russian universities can be completed through three different trajectories. Winners and awardees of all-Russian competitions for talented school students (so-called Olympiads) can be automatically enrolled in any university without exams on tuition-free places. Other applicants have to submit the results of the Unified State Examination (USE) to the university. USE is a standardized test in different subjects and has a unified grading scale (0 to 100). Students with the highest scores can be accepted to the university on tuition-free places. Students with low USE scores might study at full tuition places. The study year lasts from September, 1 to June, 30. Lectures are usually delivered to the whole cohort simultaneously, while seminars are delivered to each group separately.

In the studied university, 97 students were accepted to the Economic department either on tuition-free or full tuition places in 2014. They were randomly divided into five study groups of up to 30 students.

### 2.2  Data Collection

Data for the analysis were gathered from two sources: the Vkontakte online social network and the University administrative database. Vkontakte is the most popular Russian online social networking site. The options and interface of Vkontakte are very similar to Facebook's. There are more than 340 million unique users. About 80 million unique users attend the site every day. Vkontakte is very popular among adolescents and youth.

The list of students matriculated to the university is publicly available on the University website and we gathered the student profiles in Vkontakte based on their first and second names. In cases when there are several profiles of users with the same first and second names we took into the consideration the age and the university affiliation of the account holder.

We gathered data in 13 waves with an average difference between waves in one week. Data collection was at important events that could change or influence the network structure. The dates of data collection are the following: July, 31; August, 5, 11, 27, 30; September, 1, 5, 10, 21, 29; October, 6, 14. Data collection stopped on October, 14, 2014.

We gathered data about student ego-networks using the Vkminer application[1]. The Vkontakte API system allows us to download open data about friends, pictures, audios, videos, and other types of open information from Vkontakte user pages. Using open API, the Vkminer application downloads relevant data based on the unique ID of each user. In this paper, we gather data on student friendship ties between each other. Information about student's gender, USE scores, home city and study group was collected from the University's administrative database.

**Table 1.** Description of time periods, number of nodes and edges

| Date | Event | Number of enrolled students | Number of students that participated in the analysis | Fraction of students that participated in the analysis |
|------|-------|-----------------------------|------------------------------------------------------|--------------------------------------------------------|
| July 31 | Public list of students that were enrolled according to Olympiad results | 7 | 9 | 78% |
| August 5 | Public list of students that were enrolled based on their USE results | 59 | 73 | 81% |
| August 11 | Public list of students that were enrolled on full tuition places | 64 | 97 | 66% |
| August 27 | | 63 | 97 | 65% |
| August 30 | | 64 | 97 | 66% |
| September 1 | Official beginning of the school year | 58 | 97 | 60% |
| September 5 | | 67 | 97 | 69% |
| September 10 | | 61 | 97 | 63% |
| September 21 | | 71 | 97 | 73% |
| September 29 | | 71 | 97 | 73% |
| October 6 | | 71 | 97 | 73% |
| October 14 | | 71 | 97 | 73% |

---

[1] http://linis.hse.ru/en/soft-linis/.

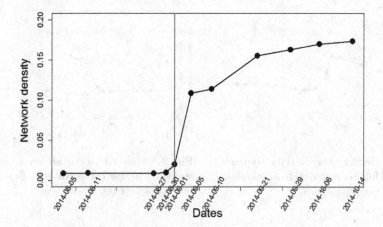

**Fig. 1.** The dynamics of network density. The vertical line corresponds to September 1, the start of the school year.

Data were analyzed in R statistical environment [12]. Additional information on the research design and dataset can be found in [13]. A database can be found on https://github.com/Sdokuka/aist2016.

## 2.3   Descriptive Statistics

There are 85% female students in our sample. Mean USE in Math is 74 ($SD = 5.6$). 70% of students are living in the same region where the university is. In Table 1 we present information on the main events connected with the enrollment and study.

In Fig. 1 we present the dynamics of network density. The growth of the social network (in other words, an increase in the number of edges between the nodes) can be divided into three different periods.

During the first period (from the beginning of enrollment until the official start of the school year), the density of the social network is very low and does not increase dramatically over time. It shows that students are not very active in the formation of online connections with their university peers. In the second period (from the official start of the school year until mid-September), the number of connections within the online social network multiplied dramatically. Such network behavior demonstrates that students prefer precede online connections by offline acquaintance. Within the third period (from mid-September until the end of data collection) students decrease their activity within the online social network and form few online connections.

## 3   Results

Using information about the social network and students characteristics, we calculated the homophily dynamics over time. The level of homophily was estimated

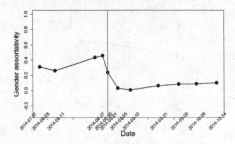 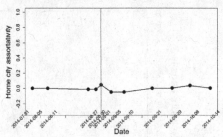

**Fig. 2.** Gender assortativity dynamics. Vertical line corresponds to September 1, start of the school year.

**Fig. 3.** Home city assortativity dynamics. Vertical line corresponds to September 1, start of the school year.

by assortativity index [14] (1).

$$Q = \frac{1}{2m}(2m - \sum_{ij} \frac{k_i * k_j}{2m} \delta(c_i, c_j)) \tag{1}$$

$Q$ is the network assortativity, $m$ is the total number of edges, $k_i$ and $k_j$ are the degree centralities of the nodes $i$ and $j$. $\delta(c_i, c_j)$ is the Kronecker's delta that equals 1 when there is a link between nodes $i$ and $j$ and equals 0 otherwise. Positive values of assortativity show that actors with similar characteristics tend to form ties with one another. Negative values indicate that actors prefer to form relationships with others who are dissimilar. We calculated the homophily level by gender, region, USE in Math and by being assigned to the same study group in the University (Figs. 2, 3, 4 and 5).

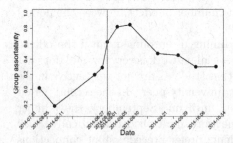 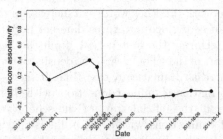

**Fig. 4.** Group assortativity dynamics. Vertical line corresponds to September 1, start of the school year.

**Fig. 5.** Math score assortativity dynamics. Vertical line corresponds to September 1, start of the school year.

Data shows that during the first period of online social network formation (from the beginning of enrollment until the official start of the school year) students form very few ties and these ties are mostly between people with the same gender and similar USE scores in Math. In the second period (from the official start of the school year until mid-September) students establish many connections within their groupmates. The homophily by group affiliation increases dramatically, which means that students prefer to form online relationships with their groupmates. Other types of homophily decrease, which shows the high importance of group affiliation for the tie formation.

Within the third period (from mid-September until the end of data collection) students decrease their activity within the online social network. Homophily by group affiliation also moderately decreases, while gender homophily slightly increases.

In order to investigate the dynamics of the observed social network and the role of actor's characteristics in network formation we use a Conditional Uniform Graph test (CUG-test) [15]. CUG allows us to analyze observed network graph-level indices with the same measures of the same-size random networks. We compare the assortativity index of the observed network with 10000 random networks of the same size.

It turns out that at each timepoint the observed homophily was significantly ($p < 0.05$) different than the homophily for random networks. In all the cases the simulated assortativity by different characteristics is very close to 0, while the observed results are significantly higher. These results are similar to [16] who states that each social network tends to have a relatively high homophily index, which indicates the segregation based on some actor characteristic.

## 4   Discussion

In this study, we explore the evolution of the online social network of students, paying attention to exogenous characteristics of students. Using the CUG-test we reveal that the level of homophily – the tendency of individuals with similar characteristics to be linked with each other – is higher in our sample than can be expected by chance. These results support the evidence found in other studies showing that actors in social networks tend to segregate based on different attributes (for example, gender or race).

Most homophily-based studies investigate homophily as a static index. In contrast to them, we measure homophily by different attributes during the whole period of online network formation. It gives us the option to reveal the changes in homophily level over time. We find out that at the beginning of online network formation (before the beginning of the official school year) students do not tend to arrange a lot of online connections with their peers. They prefer to form very few links with actors who are at some point similar to them. At this observation moment we traced homophily by gender and Unified State Examination in Math which can be at some point a proxy for the academic achievement level. We can call this network behavior *formation of comfortable environment*. It means that students link with those people whom they find similar.

During the next period of online network formation (two or three weeks after the beginning of the school year) we reveal the decreasing gender and Math exam homophily and a prevalence of same study group homophily. Forming a connection with students from the same group is of crucial importance for the freshmen. At the beginning of the school year students have very little information about the process in the university, so they have to arrange a lot of ties with peers and receive some information from them. Students from the same group also spend a lot of time together during both lectures and seminars. Formation of online friendship ties with them is very important for information exchange. We call this network behavior *formation of information environment*. The process of information environment formation, according to our observations, proceeds during the first two or three weeks. When students form their informational channels and receive all the important information about the learning process and administrative procedures, they switch their network behavior back to the formation of the comfortable environment.

In the case of formation of comfortable environment, actors take into consideration the similarity in such characteristics as gender and abilities. It means that the probability of tie creation between nodes $i$ and $j$ $P(x_{ij})$ is high when $i$ and $j$ are similar to each other and goes to 0 otherwise. In the period of formation of information environment the similarity in study group affiliation plays the major role. The probability of tie formation between nodes $i$ and $j$ $P(x_{ij})$ is high when $i$ and $j$ are from the same study group.

We see that after first couple of weeks the role of same study group homophily slowly decreases. It does not mean that students do not pay attention to study group-related issues. From our point of view, it shows that when students have a well-organized social network which can serve for information exchange, then they try to connect with people who are similar to them and who they find interesting.

Our results show that homophily should not be considered as a static index of the social network. We empirically show that homophily changes over time and propose exploratory mechanisms for homophily evolution. We argue that actors change their network behavior in different social contexts. As we can see from the freshmen data example, first year students switch from the strategy of tie formation with those who are similar to them by gender and abilities, to the strategy of forming connections with those who can be important in the information exchange process. These friendship ties allow actors to receive important study-related information [17] and feel part of the social group. Such online behavior can be an example of rational network behavior.

In this paper we show that social similarity through different attributes plays a very important role even in a homogenous social environment. Thus the investigation of the dynamics of homophily based on various attributes can reveal the inner mechanisms of social system formation.

**Acknowledgements.** The authors thank Olessia Koltsova, Sergey Koltsov, and Vladimir Filippov for the opportunity to use Vkminer application. We would like to thank Benjamin Lind for the discussion and feedback on this work. The financial support of the 5–100 Government Program and Basic Research Program at the National Research University Higher School of Economics (HSE) is greatly appreciated.

# References

1. Block, P., Grund, T.: Multidimensional homophily in friendship networks. Netw. Sci. **2**, 189–212 (2014)
2. Lazarsfeld, P.F., et al.: Friendship as a social process: a substantive and methodological analysis. Freedom Control Mod. Soc. **18**(1), 18–66 (1954)
3. McPherson, M., Smith-Lovin, L., Cook, J.M.: Birds of a feather: Homophily in social networks. Ann. Rev. Sociol. **27**, 415–444 (2001)
4. Adamic, L.A., Glance, N.: The political blogosphere and the 2004 US election: divided they blog. In: Proceedings of the 3rd International Workshop on Link Discovery, pp. 36–43. ACM (2005)
5. Lauw, H.W., Shafer, J.C., Agrawal, R., Ntoulas, A.: Homophily in the digital world: a livejournal case study. IEEE Internet Comput. **14**, 15–23 (2010)
6. Moody, J.: Race, school integration, and friendship segregation in America1. Am. J. Sociol. **107**, 679–716 (2001)
7. Goodreau, S.M., Kitts, J.A., Morris, M.: Birds of a feather, or friend of a friend? using exponential random graph models to investigate adolescent social networks*. Demography **46**, 103–125 (2009)
8. Lewis, K., Kaufman, J., Gonzalez, M., Wimmer, A., Christakis, N.: Tastes, ties, and time: a new social network dataset using facebook. com. Soc. Netw. **30**, 330–342 (2008)
9. Lewis, K., Gonzalez, M., Kaufman, J.: Social selection and peer influence in an online social network. Proc. Nat. Acad. Sci. **109**, 68–72 (2012)
10. Mayer, A., Puller, S.L.: The old boy (and girl) network: social network formation on university campuses. J. Public Econ. **92**, 329–347 (2008)
11. Vaquero, L.M., Cebrian, M.: The rich club phenomenon in the classroom. Scientific reports 3 (2013)
12. Venables, W.N., Smith, D.M.: The R development core team. A Programming Environment for Data Analysis and Graphics, An Introduction to R. Notes on R (2005)
13. Dokuka, S., Valeeva, D., Yudkevich, M.: Formation and evolution mechanisms in online network of students: the Vkontakte case. In: Khachay, M.Y., Konstantinova, N., Panchenko, A., Ignatov, D.I., Labunets, V.G. (eds.) AIST 2015. CCIS, vol. 542, pp. 263–274. Springer, Heidelberg (2015). doi:10.1007/978-3-319-26123-2_26
14. Newman, M.E.: Assortative mixing in networks. Phys. Rev. Lett. **89**, 208701 (2002)
15. Anderson, B.S., Butts, C., Carley, K.: The interaction of size and density with graph-level indices. Soc. Netw. **21**, 239–267 (1999)
16. Newman, M.E.: Mixing patterns in networks. Phys. Rev. E **67**, 026126 (2003)
17. Burt, R.S.: Structural holes and good ideas. Am. J. Sociol. **110**, 349–399 (2004)

# The Structure of Organization: The Coauthorship Network Case

Fedor Krasnov[1]([⊠]), Sofia Dokuka[2]([⊠]), and Rostislav Yavorskiy[3]([⊠])

[1] Gazpromneft NTC, St. Petersburg, Russia
Krasnov.FV@gazprom-neft.ru
[2] Center for Institutional Studies, Higher School of Economics, Moscow, Russia
sdokuka@hse.ru
[3] Department of Data Analysis and Artificial Intelligence,
Faculty of Computer Science, Higher School of Economics, Moscow, Russia
ryavorsky@hse.ru

**Abstract.** A balanced social structure within an organization is often considered as one of the major factors of company success. Thus the analysis of organizational networks is an important direction in network and organizational studies. In this paper we explore the mechanisms of collaboration using information about scientific paper coauthorships. We reveal the collaboration mechanisms within research departments of top Russian oil companies, Gazpromneft, Bashneft, Lukoil, and Tatneft. We examine the role of management in professional community formation.

**Keywords:** Coauthorship graph · Real structure of an organization · Professional network · Professional community · Research management

## 1 Introduction and Related Works

The productivity of the organization is a very complex and multidimensional concept. It is influenced by many factors and one of the important predictors of company's market success is the well-developed communication and cooperation between employees. Many theoretical and empirical research demonstrate the connection between organizational productivity and the structure of employees' communication, see e.g. [1,2]. The investigation of the social structure in organizations and professional communities is becoming one of the major directions in applied social network analysis. The patterns of communication within organizations are deeply investigated in the field of networks and management. Agneessens and Wittek shed light on the mechanisms on advice interactions within the organizations [3]. Ellwardt with coauthors reveal the patterns of gossip within the organization [4]. The informal structure of organizations was investigated in detail in Krackhardt's studies [5–7]. The structure and effectiveness of professional communities in Russian were investigated in [8–10].

Despite a clear understanding of person to person communication and interaction patterns within an organization, the mechanisms of collaboration are still

© Springer International Publishing AG 2017
D.I. Ignatov et al. (Eds.): AIST 2016, CCIS 661, pp. 100–107, 2017.
DOI: 10.1007/978-3-319-52920-2_10

to some degree unclear. A substantial obstacle to investigating the structure of actual collaboration within an organization is the difficulty of data collection. Information about employees' interaction can be obtained in many different ways, from corporate databases, social surveys, self-reports, etc. However such data have some limitations, because it they do not reflect all the mechanisms of professional interaction. As Wasserman and Faust outline [11], about half of what people report about their own interactions is incorrect in one way or another. Thus, people are not very good at reporting on their relations in particular situation, so the mechanism of data collection should avoid this bias.

In order to study the structure of professional communication we gathered data about coauthorship within research departments in several Russian oil companies, namely, Gazpromneft, Bashneft, Lukoil, and Tatneft. Prell [12] points out that bonuses of data collection from online databases is the avoidance of the interview effect, as well as imperfections in name recall, and other potential sources of measurement error that may acompany survey research.

The social network analysis of coauthorship has a long history [13]. There are a plenty of studies examining the structure of coauthorship ties within diverse scientific fields and reveal specific collaboration patterns for the different disciplines [1, 14–17].

Still, there are no studies which investigate the structure and mechanisms of corporate collaboration using coauthorship data. In this paper we fill this substantial gap and explore the internal structure of a research institute on the basis of coauthorship relationships. We discover the key mechanisms of coauthorship network formation and outline the specific role of management in scientific collaboration. The rest of the paper is organized as follows. In the second section we describe the data collection procedures and provide descriptive statistics. In the third section we provide the results for empirical estimation of the data. The fourth section concludes.

## 2 Data and Method

### 2.1 The Dataset

In order to investigate the structure of collaboration within research departments of top Russian oil companies, we gathered information about coauthorship ties. The data we work on is retrieved from open portal of Society of Petroleum Engineers (SPE), which is an authoritative industry organization founded in 1957, and today brings together 143 thousand members. A database of scientific articles of members of SPE is published on the Internet portal onepetro.org.

In the retrieved data each publication record includes the following information:

- title of the article;
- the list of authors and their affiliation;
- year of publication.

The most time-consuming step was to prepare the data and make the dataset clean and useful. Unfortunately, the portal does not have a single directory for authors. As a result sometimes we had up to 6 different spellings of the same name in different articles.

Finally our dataset covers 64 articles authored by researchers from Gazpromneft, 25 for Bashneft, 165 for Lukoil, and 51 for Tatneft. Some of the papers are written in Russian and some in English. It is rather common practice for the Russian academic community to publish preliminary results in Russian, and then supplemented and expanded version of the work in English.

In some cases we considered two publications as one if the authors and the topic are identical. A database can be found on https://github.com/Sdokuka/aist2016.

## 2.2    Descriptive Statistics of the Dataset

The descriptive statistics for three coauthorship networks is in Table 1. Data analysis was performed in R statistical environment [18].

**Table 1.** Descriptive statistics for coauthorship dataset

| Network index | Gazpromneft | Bashneft | Lukoil | Tatneft |
|---|---|---|---|---|
| Number of nodes | 116 | 63 | 376 | 84 |
| Number of edges | 320 | 155 | 1352 | 256 |
| Density | 0.05 | 0.08 | 0.02 | 0.07 |
| Clustering coefficient | 0.79 | 0.90 | 0.62 | 0.39 |
| Average number of papers per author | 1.3 | 1.2 | 1.4 | 1.7 |
| Average number of authors per paper | 4.1 | 4.8 | 5.0 | 4.5 |
| Average number of coauthors per author | 5.5 | 5.3 | 7.4 | 6.1 |
| Network radius | 1.7 | 1.3 | 3.9 | 2.5 |
| Assortativity by degree by number of coauthors | 0.14 | 0.43 | 0.18 | −0.17 |

It is clear from the table that the publication activity in Lukoil is much higher than in Bashneft and Gazpromneft. Such results can be influenced by the different publication strategies or by of our sample bias. It may be that scientists from Bashneft and Gazpromneft publish their research results in journals and proceedings which are not included in SPE database.

On average in both coauthorship networks each researcher has 1.2–1.7 research papers. This parameter is similar for all the companies. Still, the number of coauthors for each publication differs. In Gazprom there are 4 coauthors for each paper on average. In Bashneft and Lukoil there are 5 coauthors.

As we can see from descriptive statistics, network density for all coauthorship networks in quite low, which is one of the social network characterizing features.

However, the density of the Lukoil coauthorship network is lower than Bashneft, Gazpromneft and Tatneft. This can be the result of different publication strategies and/or collaboration patterns within different professional organizations.

The visualization of coauthorship networks for these organizations is presented on Figs. 1 and 2 below. It reveals different communication and professional patterns for these organizations.

In case of Bashneft there are several independent cohesive groups of coauthors, but we can trace the presence of big aggregated component. In case of Gazpromneft the network structure is pretty similar to Bashneft one: there are a few densely connected groups of coauthors which are isolated from each other. However, in Gazpromneft network we so not fiz the presence of giant connected component. In case of Lukoil one can notice the presence of giant component and several isolated groups of coauthors. In Tatneft case we trace the presence of giant connected component with only one isolated group of authors.

Such network structure can tell us that in Gazpromneft researchers are working on their different projects rather independently, while in other two companies (Bashneft and Gazprom) researchers are involved in more complex relationships, they have to collaborate with many more people.

Tatneft coauthorship network looks different from other social network. Such network structure shows the presence of very common scientific interests within the research department.

The assortativity index (which shows the tendency to be connected with similar others) in Bashneft, Gazpromneft and Lukoil is positive, while in Tatneft network it is negative. It means that in Tatneft actors with low degree (with little number of coauthors) tend to create connections with high degree actors (researchers with many collaborators).

According to Newman [1] such a distribution is common for coauthorship networks and scientific collaboration. It means that there are a lot of researchers are constantly working with their coauthors, but there are few active authors who are collaborating with many colleagues and participate in a lot of joint projects.

## 3   Results

In order to test Newmans idea on our sample and investigate the structure of collaboration among researchers we used Conditional Uniform Graph test (CUG-test) [19]. This test allows us to compare graph-level parameters (such as transitivity, assortativity and network radius) of real observed social networks with random networks. We compared four observed coauthorship social networks with 10000 same-size preferential attachment networks [20]. Preferential attachment mechanism implies that nodes tend to create links with 'popular' nodes, who already have a lot of connections.

It turns out that both the clustering coefficient and the assortativity index are significantly higher ($p < 0.05$) for the coauthorship networks of Bashneft, Gazpromneft and Lukoil. It means that actors tend to cluster and create cohesive

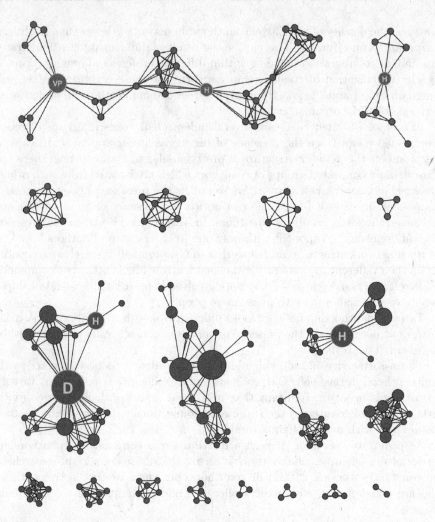

**Fig. 1.** Coauthorship social networks for Bashneft (above) and Gazpromneft (below). The node is the author of the paper. The link between two nodes means that both these persons are coauthors of the same paper. The size of the node is proportional to the number of coauthors (so-called popularity).

groups. At the same time researchers collaborate with 'stars', but they prefer to integrate them into their dense groups. In the case of Tatneft we trace totally different results. The assortativity is significantly ($p < 0.05$) lower, which means that actors do not tend to segregate based on the level of the popularity. Such network behavior is not typical for scientific collaboration. As we showed for Bashneft, Gazpromneft and Lukoil, researchers usually try to form connections with popular and successful researchers.

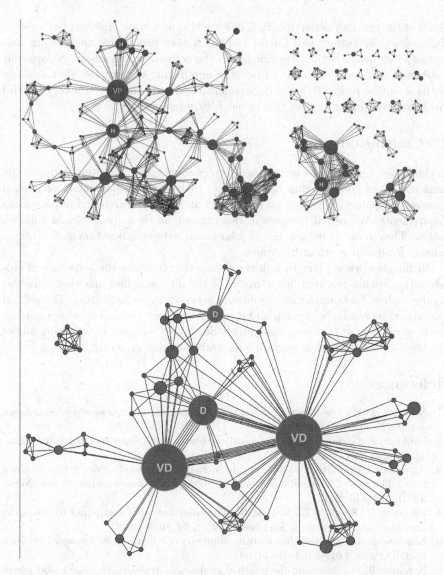

**Fig. 2.** Coauthorship social networks for Lukoil (above) and Tatneft (below). The node is the author of the paper. The link between two nodes means that both these persons are coauthors of the same paper. The size of the node is proportional to the number of coauthors (so-called popularity).

After the social network analysis we investigated the role of management and heads of the departments in scientific collaboration. We distinguished the most productive authors (the 5 persons with highest number of publications from each company) and found their affiliation and position within the company. In case of Gazpromneft and Bashneft 3 of the 5 most productive authors are the heads of the research departments. They are also part-time professors in Russian Universities. In Tatneft and Lukoil 4 of the 5 most productive authors are top-managers or heads of the departments in the company. The role of department heads and management turns to be very important. We suppose that analysis of the scientific productivity of department heads and managers is very fruitful direction. We aim to explore this in out future research.

## 4    Conclusion

In this paper we study the structure of professional community and focus on the mechanisms of coauthorship collaboration in research departments of several Russian companies. We study oil companies and their research and development departments. We reveal the role of department heads in professional collaboration. They tend to be key actors who coordinate collaboration and serve as brokers for different scientific groups.

In future work we aim to link the connection between the structure of collaboration within research departments of the oil companies and their business results such as fuel production, profit and other strategic indicators. The oil and gas sectors increasingly develop and it means that new technological approaches are in great demand for such companies. Success in research is becoming one of the key factors in business performance and economic growth.

## References

1. Newman, M.E.: The structure of scientific collaboration networks. Proc. Nat. Acad. Sci. **98**(2), 404–409 (2001)
2. Ushakov, K.: Diagnostics of an educational institution real structure. Educ. Stud. **4**, 247–260 (2013)
3. Agneessens, F., Wittek, R.: Where do intra-organizational advice relations come from? The role of informal status and social capital in social exchange. Soc. Netw. **34**(3), 333–345 (2012)
4. Ellwardt, L., Steglich, C., Wittek, R.: The co-evolution of gossip and friendship in workplace social networks. Soc. Netw. **34**(4), 623–633 (2012)
5. Krackhardt, D.: The ties that torture: Simmelian tie analysis in organizations. Res. Sociol. Organ. **16**(1), 183–210 (1999)
6. Krackhardt, D.: Assessing the political landscape: structure, cognition, and power in organizations. Adm. sci. Q. **35**, 342–369 (1990)
7. Krackhardt, D., Hanson, J.R.: Informal networks. Harvard Bus. Rev. **71**(4), 104–111 (1993)
8. Krasnov, F., Vlasova, E., Yavorskiy, R.: Connectivity analysis of computer science centers based on scientific publications datafor major russian cities. Procedia Comput. Sci. **31**, 892–899 (2014)
9. Krasnov, F., Yavorskiy, R.: Measurement of maturity level of a professional community. Bus. Inf. **1**(23), 64–67 (2013)
10. Krasnov, F., Yavorskiy, R.E., Vlasova, E.: Indicators of connectivity for urban scientific communities in Russian cities. In: Ignatov, D.I., Khachay, M.Y., Panchenko, A., Konstantinova, N., Yavorskiy, R.E. (eds.) AIST 2014. CCIS, vol. 436, pp. 111–120. Springer, Heidelberg (2014). doi:10.1007/978-3-319-12580-0_11

11. Wasserman, S., Faust, K.: Social Network Analysis: Methods and Applications, vol. 8. Cambridge University Press, Cambridge (1994)
12. Prell, C.: Social Network Analysis: History, Theory and Methodology. Sage, London (2012)
13. Hou, H., Kretschmer, H., Liu, Z.: The structure of scientific collaboration networks in scientometrics. Scientometrics **75**(2), 189–202 (2008)
14. Barabási, A.L., Jeong, H., Néda, Z., Ravasz, E., Schubert, A., Vicsek, T.: Evolution of the social network of scientific collaborations. Phys. A Stat. Mech. Appl. **311**(3), 590–614 (2002)
15. Rodriguez, M.A., Pepe, A.: On the relationship between the structural and socioacademic communities of a coauthorship network. J. Informetrics **2**(3), 195–201 (2008)
16. Ding, Y.: Scientific collaboration and endorsement: network analysis of coauthorship and citation networks. J. Informetrics **5**(1), 187–203 (2011)
17. Acedo, F.J., Barroso, C., Casanueva, C., Galán, J.L.: Co-authorship in management and organizational studies: an empirical and network analysis. J. Manage. Stud. **43**(5), 957–983 (2006)
18. R Core Team: R: a language and environment for statistical computing. R foundation for statistical computing 2013, Vienna, Austria (2014)
19. Anderson, B.S., Butts, C., Carley, K.: The interaction of size and density with graph-level indices. Soc. Netw. **21**(3), 239–267 (1999)
20. Barabási, A.L., Albert, R.: Emergence of scaling in random networks. Science **286**(5439), 509–512 (1999)

# Organizational Networks Revisited: Relational Predictors of Organizational Citizenship Behavior

Valentina Kuskova[✉], Elena Artyukhova,
Rustam Kamalov, and Daria Danilova

National Research University Higher School of Economics,
Moscow, Russian Federation
vkuskova@hse.ru

**Abstract.** Organizational citizenship behavior (OCB) is an important management construct. Despite previous investigations in relation to social capital, the role of networks in its emergence has received only limited attention. In this paper we investigate the relationship between OCB, with data collected from supervisors evaluating their subordinates; sever-al types of organizational networks (professional, friendship, support, supervisor-subordinate), and several other constructs (collected from the employees themselves), shown to affect OCB in the past. All data were collected at a large insurance company in Russia.

Outcomes of this study have several important implications. First, the impact of networks on manifestation of OCB depends not only on the strength of network ties, but on types of network. Second, interoganizational relationships are complex and consist of several levels of mediated relationships. Results of this study can impact the theoretical understanding of OCB and have practical implications for the supervisor-subordinate relationships in the workplace.

**Keywords:** Network relationships · Organizational citizenship behavior · Job characteristics

## 1 Introduction

Organizational citizenship behavior (OCB) is an important construct in the study of management, first introduced by Organ [1] to describe the behavior that was discretionary in nature, not required by the job description, performed without an expectation of a reward or payment, but contributing to organizational performance. Over the years, a large number of constructs were developed to describe the different elements of such behavior, and the multitude of studies en-compass a large variety of antecedents, covariates and consequences of OCB. The many constructs in the large domain of OCB are generally divided into five groups: altruism, courtesy, conscientiousness, civic virtue and sportsmanship [2].

The study has been funded by the Russian Academic Excellence Project '5-100'.

D.I. Ignatov et al. (Eds.): AIST 2016, CCIS 661, pp. 108–117, 2017.
DOI: 10.1007/978-3-319-52920-2_11

The importance of this construct cannot be overestimated: broadly, by some estimates, OCB im-proves organizational performance by 18–38% [3]; specifically, work quality is increased by 18%, work quantity by 19%, financial results by 25%, client satisfaction indicators by 38% [3]. In addition, OCB increases both client and employee satisfaction, and reduces turnover and absenteeism [4], among many other factors. An important element of organizational culture, this behavior is usually a part of a well-functioning organization, and once established, cannot be easily emulated or copied, providing organizations with sustained competitive advantage [5]. Over-all, this is one of the most widely studied organizational phenomena, still occupying a prominent place in organizational research despite being first introduced over 30 years ago [1].

Many theories have been put forth for the explanation of the positive effect of OCB in organizations [6], among them, norms of reciprocity and impression management. No matter the result, OCB appears to be a social behavior, which assumes that an employee is a member of a social community; multiple studies have confirmed that OCB does not manifest in isolation [7]. Moreover, as [7] have shown, the level of OCB, manifested by an employee, has a strong positive association with the level of OCB manifested by employees co-workers. While such observations seem almost intuitive, it is surprising that very little work has been done to examine the relation-ship between intraorganizational social networks and OCB. The work done so far is mostly limited to examination of reciprocity of OCB, relational correlates of interpersonal OCB, and impact of co-workers on employees manifestation of OCB [8–10]. The aspects of interorganizational network structure (e.g., type and strength of ties) remain mostly unexplored, despite obvious indications that OCB is a relational behavior.

This paper reports on a study designed to fill an important gap in our understanding of the relationship between organizational social structure and OCB. The study was conducted in a large insurance company in Moscow, and data were collected from several sources to ensure validity and reliability. Employees provided data on their networks and important organizational and job characteristics, previously shown to be related to OCB; employee supervisors evaluated the organizational citizenship behavior of their subordinates.

## 2    Relationship Between Social Networks and OCB: Theoretical Considerations

Social networks, in general, help understand the structure of social exchange [11]. For example, people with high ethical and moral norms usually tend to act ethically in most situations. However, in presence of different group norms, people tend to change their behavior. In other words, ethical behavior is explained, in large part, by the social context in which it takes place [11,12], rather than by individual preferences. So it is appears intuitive that OCB, with its strong ethical orientation, has a strong social component.

A relatively recent study explored how social factors may influence the emergence of helping behavior in the workplace. It showed that the extent to which an employee was willing to help his or her coworkers was directly related to the amount of help this employee received in the past [13], and reciprocity of helping inside a group or a team was a part of a group norm [10]. When helping behavior was expected inside the group, and employee was more likely to help, meaning social context played a large role in interpersonal relationships, even in small social groups.

In a different context, previous studies of networks in organizations have shown that employee values, as well as attitudes and perceptions, were in part a reflection of the employees relation-ship with his or her coworkers [14,15]. In another context, employees were more likely to ask their co-workers, other than supervisors, for help with understanding organizational norms and values [16]. So clearly, organizational networks emerge as an important covariate of helping behavior, a part of OCB.

Most frequently in organizations, people build friendship and advice networks [14]. A couple of studies noted that an employee who provides an advice is often perceived as more important than someone who does not do it as often, because advice information is frequently needed by others to perform work-related duties [17,18]. Results of these studies indicate that by seeking and using advice, employees build advice networks, which in turn promote advice behavior. In addition, it was shown that the number of connections that a person maintains has a positive effect on improving work-related knowledge and qualifications, as well as organizational involvement [19]. According to the social information processing theory, employees use advice received from others to form and evaluate the organizational environment, including their own work environment and other complex constructs [20], such as networks. In particular, people use social information in order to react to social signals, form their impressions of the work environment, develop attitudes towards their workplace, and evaluate demand characteristics and job expectations.

In longitudinal studies [14] has shown that advice behavior of co-workers strengthen the organizational stability, because it promotes information exchange and activity coordination inside the work group. In addition to information exchange, social influence manifests in observation and emulation of someones behavior, especially relationships and emotional reactions, including altruistic motives. So it appears that advice networks play an important role in establishment of OCB.

In addition, in work context, people experience a strong need for communication and friendship and social and emotional support [21,22]. Many emulate their friends employment and career decisions, and close ties in organization often result in convergence on views on controversial topics and promote organizational change.

Despite the fact that friendship and advice are different in their essence, both types of relation-ships are similar in tie strength. Tie strength here is defined similar to Granovetters view of time spent, emotional strength, level of intimacy and reciprocity of favors that characterize the tie. Strong ties are more intimate, allow for more self-expression, and assume deeper relationships than those required for exchange relationships in a context of a job. People who have stronger ties are more likely to have more similar views on various topics, experience and access to re-sources [23, 24]. In contrast, the exchange via weak ties assumes less intimate and less frequent contact between people, usually located further from each other in a network. Weak ties are considered important because they allow employees access to information and resources that cannot be obtained via strong ties. In the context of OCB, however, strength of ties remains a controversial construct. For example, the study by [15] has shown that strong ties in advice networks were associated with better job performance, whereas weak ties did not show such association at all. Other network characteristics, indicating the level of prestige and connectedness, may follow similar patterns.

Based on theoretical considerations above, this study has several hypotheses:

*Hypothesis 1: Strength of an employees position in a friendship network, manifested by centrality and other network characteristics, is positively associated with manifestation of organizational citizenship behavior.*

*Hypothesis 2: Strength of an employees position in an advice network, manifested by centrality and other network characteristics, is positively associated with manifestation of organizational citizenship behavior.*

*Hypothesis 3: Strength of an employees position in a supervisor-subordinate network, manifested by centrality and other network characteristics, is positively associated with manifestation of organizational citizenship behavior.*

*Hypothesis 4: Strength of an employees position in a professional network, manifested by centrality and other network characteristics, is positively associated with manifestation of organizational citizenship behavior.*

## 3    Method and Analysis

### 3.1    Sample

The study was conducted at a Moscow-based company RosGosStrakh (an abbreviation of Russian State Insurance Company), one of the largest insurance companies in Russia, 100% privately held. It offers a large line of insurance products to individuals and companies, and has 74 subsidiaries and 35,000 offices throughout Russia. The company has over 4.5 million individual clients, 240,000 corporate clients, and employs over 100,000 people, including 65,000 insurance agents. It is currently number 75 in Russias top 400 companies.

Data was collected in the HR department of the Moscow headquarters from 69 employees (out of 72 at the time), 51 of them female, average age of 29.7 years. Sixteen of them were managers at different level. Data collection was not anonymous, but respondents were assured of full confidentiality of data; surveys were collected directly by one of the investigators, without being shown to company management.

All respondents were asked to fill out a survey consisting of network data (free recall, several types of relationships friendship, advice, supervisor-subordinate, professional, with strength of ties evaluated on a scale from 1 to 7, with 7 being the strongest), as well as other constructs, shown in previous studies to covary with OCB: workload, responsibility, work difficulty, work speed, administrative problems, interpersonal conflicts, uncertainty in the future, role conflict, physical participation, emotional participation. Supervisors also filled out a separate survey, in which they evaluated each subordinate on several OCB dimensions: workplace pride, fulfillment of work demands, helping behavior, voice behavior, lack of fear. All constructs and scales were adapted from previously tested and validated OCB scales [2–4].

## 3.2  Analysis

All data were analyzed using structural equation modeling (SEM) in Lisrel 8.8 and network structure was analyzed using UCINet. Network characteristics were then used as inputs into the structural model.

The general estimation procedure for the structural model followed the standard SEM algorithm [25]. The general matrix form for model with latent variables is defined as follows:

$$\Sigma = \left[ \frac{\Sigma_{yy} | \Sigma_{yx}}{\Sigma_{xy} | \Sigma_{xx}} \right] = \Sigma(\Phi), \tag{1}$$

which can be further specified as follows:

| $\Lambda_y (I - B)^{-1} (\Gamma \Phi \Gamma' + \Psi) \left[ (I - B)^{-1} \right]' \Gamma'_y + \Theta_\epsilon$ | $\Lambda_y (I - B)^{-1} \Gamma \Phi \Lambda'_x$ |
|---|---|
| $\Lambda_x \Phi \Gamma' \left[ (I - B)^{-1} \right]' \Lambda'_y$ | $\Lambda_x \Phi \Lambda'_x + \Theta_\delta$ |

Where $\Lambda$s are the matrices of factor loadings of $x$s, or exogenous factor indicators, and $y$s, or endogenous factor indicators; $B$ and $\Gamma$ are the matrices of structural relationship parameters; $\Phi$ is the matrix of exogenous factor intercorrelations; $\Psi$ is the matrix of endogeneous factor errors; $\Theta$s are matrices of random errors on $x$-variables, or $\delta$s, and $y$-variables, or $\epsilon$s. The model is based on certain laws of variances, covariances, and means of the observed variables, which form the measurement model of the latent variables, and is subject to the following laws:

- Law 1: The covariance of a random variable $X$ with itself is equal to its variance, $Cov(X, X) = Var(X)$
- Law 2: If $a$ and $b$ are constants and $X$ and $Y$ are random variables, $Cov(aX, bY) = abCov(X, Y)$
- Law 3: If $X, Y, Z$, and $U$ are random variables and $a, b, c$, and $d$ are constants, then $Cov(aX + bY, cZ + dU) = acCov(X, Z) + adCov(X, U) + bcCov(Y, Z) + bdCov(Y, U)$

- Law 4: Using Laws 1, 2, and 3, and the knowledge that $Cov(X,Y) = Cov(Y,X)$, leads to the following: $Var(aX+bY) = Cov(aX+bY, aX+bY) = a^2Var(X) + b^2Var(Y) + 2abCov(X,Y)$
- Law 5: An important special case of Law 4 is where $Cov(X,Y) = 0$, $Var(aX + bY) = a^2Var(X) + b^2Var(Y)$

The resulting equations for the observed variables (V) can be written as follows:

$$V_i = \lambda_{ij}F_i + E_i \tag{2}$$

Or, equivalently,

$$X_i = \lambda_{ij}\Xi_i + \delta_i \tag{3}$$

where $i$ is the number of the indicator, $j$ - latent factor indicator, $F$ and $\Xi$ are factors, $E$ and $\delta$ - errors. With two indicator variables ($V_1$ and $V_2$), loading on the same factor $F_1$, their implied covariances are expressed as follows:

$$
\begin{aligned}
Cov(V_1, V_2) &= Cov(\lambda_{11}F_1 + E_1, \lambda_{21}F_1 + E_2) \\
&= Cov(\lambda_{11}F_1, \lambda_{21}F_1) + Cov(\lambda_{11}F_1, E_2 + Cov(E_1, \lambda_{21}F_1) + Cov(E_1, E2) \\
&= \lambda_{11}\lambda_{21}Cov(F_1, F_1) + \lambda_{11}Cov(F_1, F_2) + \lambda_{21}Cov(E_1, F_1) + Cov(E_1, E_2) \\
&= \lambda_{11}\lambda_{21}Cov(F_1, F_1) \\
&= \lambda_{11}\lambda_{21}Var(F_1) \\
&= \lambda_{11}\lambda_{21}
\end{aligned} \tag{4}
$$

When two indicator variables ($V_1$ and $V_4$) load on different factors ($F_1$ and $F_2$), the picture is calculated similarly, though slightly more complicated:

$$
\begin{aligned}
Cov(V_1, V_4) &= Cov(\lambda_{11}F_1 + E_1, \lambda_{4,2}F_2 + E_4) \\
&= Cov(\lambda_{11}, F_1, \lambda_{42}F_2) + Cov(\lambda_{11}F_1, E_4) + Cov(E_1, \lambda_{42}F_2) + Cov(E_1, E_4) \\
&= \lambda_{11}\lambda_{42}Cov(F_1, F_2) + \lambda_{11}Cov(F_1, E_4) + \lambda_{4,2}Cov(E_1, F_2) + Cov(E1, E4) \\
&= \lambda_{11}\lambda_{42}Cov(F_1, F_2)
\end{aligned} \tag{5}
$$

The rest of the calculations follow the same logic, though relationships become much more complicated with increased number of indicators and variables. For the model presented in this paper, we estimated the model for 15 exogenous latent factors with 39 indicators and six endogenous latent factors with 18 indicators.

With respect to network characteristics, maximum network size reported was 19 connections, though the average network size of each respondent was 8–9 people. Also, the overall network turned out to have three separate connected components, which may indicate that inside the evaluated department, there are

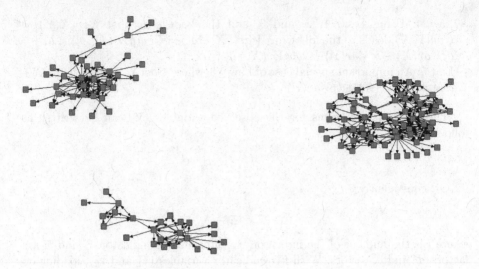

**Fig. 1.** Organizational network

three clearly distinct informal subgroups (though none were reported at the time of data collection). Resulting network is depicted in Fig. 1.

Diagram representing the overall model is presented in Fig. 2, with only the statistically significant results shown. Model was subjected to the standard tests of fit; measurement model was built before structural model, and only the significant factor loadings (above 0.7 per generally accepted standards) were retained. Then, hypothesized structural relationships were tested one at a time, and only the significant relationships were retained in the final model.

## 4    Results

Overall, study hypotheses were partially supported. Friendship network showed positive association with advice behavior, though advice network did not have any associations with OCB. Stronger supervisor-subordinate network was related to stronger manifestations of OCB, which may indirectly indicate impression management motives. Strictly professional ties did not seem to have any effect on the manifestation of OCB, but this may be due to the fact that all networks were drawn on the same nodes (people), and separating advice from friendship and professional networks have somewhat affected the variance explained by individual network type vs. the overall network.

In addition to the tested hypotheses, the study confirmed previously explored relationships, such as between higher level of responsibility and higher OCB, and higher job demands and lower levels of OCB. Confirmation of previously shown relationship was an important element of this study, because it demonstrated the overall validity of constructs used in the study.

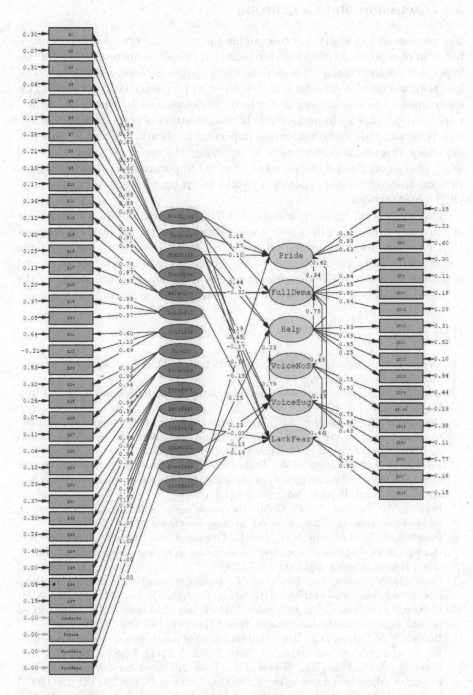

**Fig. 2.** Structural model of the hypothesized relationships

# 5   Discussion and Conclusion

The purpose of this study was to examine the role that networks play in mani-festation of organizational citizenship behavior. Despite numerous studies of this important organizational construct, and its clearly relational nature, only lim-ited work was done to examine it in the con-text of organizational networks. This study filled an important gap in the field of management by showing that net-works, indeed, play an important role in manifestation of organizational citizen-ship behavior. This study has several important implications. First, it makes an important theoretical contribution by extending the social exchange theory of OCB emergence. Second, it has some practical implications as it demonstrates that existence of certain types of networks in an organization is conducive to OCB manifestation.

This study is not without limitations. First, it was conducted on a limited set of office employees in an insurance organization; the very nature of the business with its client orientation may be conducive to OCB, so spuriousness of the found relationships cannot be ruled out. Second, the study is rather small and is limited only to the HR department of this company. Further testing of the proposed model is warranted before more generalized results could be drawn.

# References

1. Organ, D.W.: Organizational Citizenship Behavior: The Good Soldier Syndrome. Lexington Books, Lexingon (1988)
2. Organ, D.W.: The motivational basis of organizational citizenship behavior. Res. Organ. Behav. **12**(1), 43–72 (1990)
3. Podsako, P.M., Mackenzie, S.B., Pain, J.B., Bachrach, D.G.: Organizational cit-izenship behaviors: critical review of the theoretical and empirical literature and suggestions future research. J. Manag. **26**, 513–563 (2000)
4. Podsakoff, N.P., Whiting, S.W., Podsakoff, P.M., Blume, B.D.: Individual- and organizational-level consequences of organizational citizenship behaviors: a meta-analysis. J. Appl. Psychol. **94**(1), 122–141 (2009)
5. Bolino, M., Turnley, W.H.: Going the extra mile: cultivating and managing employee citizenship behavior. Acad. Manag. Executive **17**, 61 (2013)
6. Podakoff, P.M., Mackenzie, S.W., Hui, C.: Organizational citizenship behaviors and managerial evaluations of employee performance: a review and suggestion. Int. J. Hum. Resour. Manag. **18**, 1085–1097 (2007)
7. Bommer, W., Miles, W., Graver, S.N.: Does one good turn deserve another? Coworker influence on employee citizenship. J. Organ. Behav. **24**, 181–196 (2003)
8. Deckop, J.R., Circa, C.C., Andersson, L.M.: Doing unto others: the reciprocity of helping behavior in organizations. J. Bus. Ethics **47**, 101–113 (2003)
9. Bowler, W.M., Brass, D.J.: Relational correlates of interpersonal citizenship behav-ior: a social network perspective. J. Appl. Psychol. **91**(1), 78–82 (2006)
10. Bommer, W.H., Miles, W., Grover, S.L.: Does one good turn deserve another? Coworker influences on employee citizenship. J. Organ. Behav. **24**, 181–196 (2003)
11. Emerson, R.M.: Social exchange theory. Annu. Rev. Psychol. **2**, 335–362 (1976)
12. Brass, D.J., Buttereld, K.D., Skaggs, B.: Relationships and unethical behavior: a social networks perspectives. Acad. Manag. Rev. **23**, 14–31 (1998)

13. Decop, J.R., Circa, C.C., Andersson, L.M.: Doing unto others: the reciprocity of helping behavior in organizations. J. Bus. Ethics **47**, 101–113 (2003)
14. Gibbons, D.: Friendship and advice networks in the context of changing professional values. Adm. Sci. Q. **49**, 238–262 (2004)
15. Morrison, W.: Newcomer's relationships: the role of social network ties during socialization. Acad. Manag. J. **45**, 1149–1160 (2002)
16. Cross, R., Prusak, L.: The peopel who make organizations go - or stop. Harvard Bus. Rev. **80**(6), 105–112 (2002)
17. Brass, D.J., Burkhardt, M.: Potential power and power use: an investigation of structur and behavior. Acad. Manag. J. **36**, 441–470 (1993)
18. Burkhardt, M., Brass, D.J.: Changing patters or patterns of change: the effect of a change in technology on social network structure and power. Adm. Sci. Q. **35**, 104–127 (1990)
19. Zagenczyk, T.J., Murrell, J.: It is better to receive than to give: advice network effects on job and work-unit attachment. J. Bus. Psychol. **24**(2), 139–152 (2014)
20. Salanscik, G.R., Pfeffer, J.: A social information processing approach to job attitudes and task design. Adm. Sci. Q. **23**, 224–253 (1978)
21. Krakhardt, D., Stern, R.N.: Informal networks and organizational crises: an experimental simulation. Soc. Psychol. Q. **51**, 123–140 (1988)
22. Kilduff, M.: The interpersonal structure of decision-making: a social comparison approach to organizational choice. Organ. Behav. Hum. Decis. Process. **47**, 270–288 (1990)
23. Granovetter, M.S.: The strength of weak ties. Am. J. Sociol. **78**(6), 1360–1380 (1978)
24. Granovetter, M.: The strength of weak ties: a network theory revisited (1982)
25. Bollen, K.A., Long, J.S.: Testing Structural Equation Models, vol. 154. Sage, Thousand Oaks (1993)

# Natural Language Processing

# Bigram Anchor Words Topic Model

Arseniy Ashuha[1] and Natalia Loukachevitch[2(✉)]

[1] Moscow Institute of Physics and Technology, Dolgoprudny, Russia
ars.ashuha@phystech.edu
[2] Research Computing Center of Lomonosov Moscow State University,
Moscow, Russia
louk_nat@mail.ru

**Abstract.** A probabilistic topic model is a modern statistical tool for document collection analysis that allows extracting a number of topics in the collection and describes each document as a discrete probability distribution over topics. Classical approaches to statistical topic modeling can be quite effective in various tasks, but the generated topics may be too similar to each other or poorly interpretable. We supposed that it is possible to improve the interpretability and differentiation of topics by using linguistic information such as collocations while building the topic model. In this paper we offer an approach to accounting bigrams (two-word phrases) for the construction of Anchor Words Topic Model.

**Keywords:** Topic model · Anchor words · Bigram

## 1 Introduction

A probabilistic topic model is a modern statistical tool for document collection analysis that allows identifying a set of topics in the collection and describes each document as a discrete probability distribution over topics. The topic is meant a discrete probability distribution over words, considered as a thematically related set of words. Topic models are actively used for various applications such as text analysis [4,6,12], user analysis [8], information retrieval [10,14].

To recognize hidden topics, standard algorithms of topic modeling such as PLSA or LDA [3,7], take into account only the frequencies of words and do not consider the syntactic structure of sentences, the word order, or grammatical characteristics of words. Neglect of the linguistic information causes the low interpretability and the low degree of topic differentiation [14], which may hinder the use of topic models. If a topic has low interpretability then it may seem as a set of unrelated words or a mixture of several topics. It is difficult to differentiate topics when they are very similar to each other.

One of the approaches that improves the interpretability of the topics is proposed in [1,2] and is called Anchor Words. This approach is based on the assumption that in each topic there exists a unique word that describes the topic, but this approach is also built on word frequencies.

© Springer International Publishing AG 2017
D.I. Ignatov et al. (Eds.): AIST 2016, CCIS 661, pp. 121–131, 2017.
DOI: 10.1007/978-3-319-52920-2_12

In this paper we put forward a modification of the Anchor algorithm, which allows us to take into account collocations when building a topic model. The experiments were conducted on various text collections (Banks Articles, 20 Newsgroups, NIPS) and confirmed that the proposed method improved the interpretability and the uniqueness of topics without downgrading other quality measures.

The paper is organized as follows. Section 2 reviews similar work. Section 3 describes the metrics used for evaluating the quality of topic models. In Sect. 4, we propose a method that allows us to take into account collocations in the Anchor Words topic model.

## 2    Related Work

### 2.1    Notation and Basic Assumptions

Many variants of topic modeling algorithms have been proposed so far. Researchers usually suppose that a topic is a set of words that describe a subject or an event; a document is a set of topics that have generated it. A **Topic** is a discrete probability distribution over words: topic $t = \{P(w|t) : w \in W\}$ [3,7]. In this notation, each word in each topic has a certain probability, which may be equal to zero. Probabilities of words in topics are usually stored in the matrix $\Phi_{W \times T}$. A **Document** is a discrete probability distribution over topics $P(t|d)$ [3,7]. These probabilities are represented as a matrix $\Theta_{T \times D}$.

In topic modeling, the following hypotheses are usually presupposed: a **Bag of words hypothesis** is the assumption that it is possible to determine which topics have generated the document without taking into account the order of words in the document; **Hypothesis of conditional independence** is the assumption that the topic does not depend on the document, the topic is represented by the same discrete distribution in each document that contains this topic. Formally, the probability of a word in the topic is not dependent on the document – $P(w|d,t) = P(w|t)$ [3,7,14]; **Hypothesis about the thematic structure of the document** usually assumes that the probability of a word in a document depends on the hidden topics that have generated the document, as, for example, in the simplest topic model:

$$p(w|d) = \sum_{t \in T} P(w|d,t)P(t|d) = \sum_{t \in T} P(w|t)P(t|d) \tag{1}$$

### 2.2    Specific Topic Models

In this section we consider several well-known approaches to topic modeling.

**Probabilistic Latent Semantic Analysis, PLSA** was proposed by Thomas Hoffman in [7]. To build the model, he proposed to optimize the log-likelihood with the restrictions of normalization and non-negativeness:

$$log\, L(D, \Phi, \Theta) = log \prod_{d \in D} \prod_{w \in d} p(w|\, d) \rightarrow \max_{\Phi, \Theta} \qquad (2)$$

$$\phi_{wt} \geq 0; \quad \sum_{w \in W} \phi_{wt} = 1; \quad \theta_{td} \geq 0; \quad \sum_{t \in T} \phi_{td} = 1 \qquad (3)$$

To solve the optimization problem, the EM-algorithm was proposed, which is usually used to find the maximum likelihood estimate of probability model parameters when the model depends on hidden variables.

**Latent Dirichlet Allocation, LDA** was proposed by David Blei in [3]. This paper introduces the generative model that assumes that the vectors of topics and the vectors of documents are generated from the Dirichlet distribution. For training the model, it was proposed to optimize the following function:

$$log \left[ L(D, \Phi, \Theta) \prod_d Dir(\theta_d|\beta) \prod_t Dir(\phi_t|\alpha) \right] \rightarrow \max_{\Phi, \Theta} \qquad (4)$$

$$\phi_{wt} \geq 0; \quad \sum_{w \in W} \phi_{wt} = 1; \quad \theta_{td} \geq 0; \quad \sum_{t \in T} \phi_{td} = 1 \qquad (5)$$

To solve the optimization problem, the authors use the Bayesian inference, which leads to EM-algorithm similar to PLSA. Because of the factored-conditional conjugate prior distribution and the likelihood, the formula for the parameters update can be written explicitly.

**Additive Regularization Topic Model** was proposed by Konstantin Vorontsov in [14]. The "Additive Regularization Topic Model" generalizes LDA (the LDA approach can be expressed in terms of an additive regularization) and allows applying a combination of regularizers to topic modeling by optimizing the following functional:

$$log\, L(D, \Phi, \Theta) + \sum_{i=1}^{n} \tau_i R_i(\Phi, \Theta) \rightarrow \max_{\Phi, \Theta} \qquad (6)$$

$$\phi_{wt} \geq 0; \quad \sum_{w \in W} \phi_{wt} = 1; \quad \theta_{td} \geq 0; \quad \sum_{t \in T} \phi_{td} = 1 \qquad (7)$$

where, $\tau_i$ – weight of regularizer, $R_i$ – regularizer.

To introduce regularizers, the Bayesian inference is not used. On the one hand, it simplifies the process of entering regularizers because it does not require the technique of Bayesian reasoning. On the other hand, the introduction of a new regularizer is an art, which is hard to formalize.

The paper [14] shows that the use of the additive regularization allows simulating reasonable assumptions about the structure of topics, which helps to improve some properties of a topic model such as interpretability and sparseness.

**Anchor Words Topic Model** was proposed by Sanjeev Arora in [1,2]. The basic idea of this method is the assumption that for each topic $t_i$ there is an anchor word that has a nonzero probability only in the topic $t_i$. If one has the *anchor words* one can recover a topic model without EM algorithm.

The Algorithm 1 consists of two steps: the search of anchor words and recovery of a topic model with anchor words. Both procedures use the matrix $Q_{W \times W}$ that contains joint probabilities of co-occurrence of word pairs $p(w_i, w_j)$, $\sum Q_{ij} = 1$. Let us denote row-normalized matrix $Q$ as $\hat{Q}$, the matrix $\hat{Q}$ can be interpreted as $\hat{Q}_{i,j} = p(w_j|w_i)$. It should be note that Algorithm 1 does not need to keep Q matrix in memory, it can be possessed by blocks.

---

**Algorithm 1.** High Level Anchor Words

---

**Input**: collection D, number of topics |T|
**Output**: matrix $\Phi$;
  1: $Q = $ Rows_normalized Word Co-occurences(D)
  2: $\hat{Q} = Random\_projection(\hat{Q})$
  3: $S = $ FindAnchorWords($\hat{Q}$, |T|)
  4: $\Phi = $ RecoverWordTopic($\hat{Q}$, S)
  5: **return** $\Phi$

---

Let us denote indexes of anchor words $S = \{s_1, \ldots, s_T\}$. The rows indexed by elements of $S$ are special in that every other row of $\hat{Q}$ lies in the convex hull of the rows indexed by the anchor words [1]. At the next step optimization problems are solved. It's done to recover the expansion coefficients of $C_{it} = p(t|w_i)$, and then using the Bayes rule we restore matrix $(p(w|t))_{W \times T}$. The search of anchor words is equal to the search for almost convex hull in the vectors of the matrix $\hat{Q}$ [1]. The combinatorial algorithm that solves the problem of finding the anchor words is given in Algorithm 2.

---

**Algorithm 2.** The combinatorial algorithm FastAnchorWords

---

**Input**: dots $V = v_1, \ldots, v_n$, dim of convex hull $K$, parameter of error $\epsilon$;
**Output**: $\{v'_1, \ldots, v'_k\}$ – set of points which constitute the convex hull;
  1: put $v_i$ into random subspace $V$, $dimV = 4logV/\epsilon^2$
  2: S = $\{s_0\}$, $s_0$ – point that has the largest distance to origin.
  3: **for all** i **do**=1 to K-1:
  4:    denote point $\in$ V that has the largest distance to $span(S)$ as $s_i$
  5:    $S = S \cup \{s_i\}$
  6: **for all** i **do**=1 to K-1:
  7:    denote point $\in$ V that has the largest distance to $span(S \setminus \{s_i\})$ as $s'_i$
  8:    update $s_i$ on $s'_i$
     **return** S

---

## 2.3   Integration of N-Grams into Topic Models

The above-discussed topic models are based on single words (unigrams). Sometimes collocations can more exactly define a topic than individual words, therefore various approaches have been proposed to take into account word combinations while building topic models.

**Bigram Topic Model** proposed by Hanna Wallach in [15]. This model involves the introduction of the concept a hierarchical language model Dirichlet [9]. It is assumed that the appearance of a word depends on the topic and the previous word, all word pairs are collocations.

**LDA Collocation Model** proposed by M. Steyvers in [13]. The model introduces a new type of hidden variables $x$ ($x = 1$, if $w_{i-1}w_i$ is collocation 0 else). This model can take into account the bigrams and unigrams, unlike the bigram topic model, where each pair of words are collocations.

**N-Gram Topic Model** proposed by Xuerui Wang in [16]. This model adds the relation between topics and indicators of bigrams that allows us to understand the context depending on the value of the indicator [16].

**PLSA-SIM** proposed by Michail Nokel in [12]. The algorithm takes into account the relation between single words and bigrams (PLSA-SIM). Words and bigrams are considered as similar if they have the same component word. Before the start of the algorithm, sets of similar words and collocations are precalculated. The original algorithm PLSA is modified to increase the weight of similar words and phrases in case of their co-occurrence in the documents of the collection.

## 3   Methods to Estimate the Quality of Topic Models

To estimate the quality of topic models, several metrics were proposed.

**Perplexity** is a measure of inconsistency of a model towards the collection of documents. The perplexity is defined as:

$$P(D, \Phi, \Theta) = exp\left(-\frac{1}{len(D)} \ log \ L(D, \Phi, \Theta)\right) \tag{8}$$

Low perplexity means that the model correctly predicts the appearance of terms in the collection. The perplexity depends on the size of a vocabulary: usually perplexity grows with increase of the vocabulary.

**Coherence** is an automatic metric of interpretability proposed by David Newman in [11]. It was showed that the coherence measure has the high correlation with the expert estimates of topics interpretability.

$$PMI(w_i, w_j) = log\frac{p(w_1, w_2)}{p(w_1)p(w_2)} \tag{9}$$

The coherence of a topic is the median PMI of word pairs representing the topic, usually it is calculated for $n$ most probable elements in the topic. The coherence of the model is the median of the topics coherence.

**A Measure of the Kernels Uniqueness.** Human-constructed topics usually have unique kernels, that is words having high probabilities in the topic. The measure of kernel uniqueness shows to what extent topics are different to each other.

$$U(\Phi) = \frac{|\cup_t kernel(\Phi_t)|}{\sum_{t\in T}|kernal(\Phi_t)|} \tag{10}$$

If the uniqueness of the topic kernels is closer to one then we can easily distinguish topics from each other. If it is closer to zero then many topics are similar to each other, contain the same words in their kernels. In this paper the kernel of a topic means the ten most probable words in the topic.

## 4    Bigram Anchor Words Topic Modeling

The bag of words text representation does not take into account the order of words in documents, but, in fact, many words are used in phrases, which can form completely different topics.

Usually adding collocations as unique elements of the vocabulary significantly impairs the perplexity by increasing the size of the vocabulary, but the topic model interpretability is increased. The question arises: if it is possible to consider collocations in the Anchor Words algorithm without adding them to the vocabulary.

### 4.1    Extracting Collocations (Bigrams)

To extract collocations, we used the method proposed in [5]. The authors propose the following algorithm. If several words in a text mean the same entity then in this text these words should appear beside each other more often than separately. It was assumed that if a pair of words co-occurs as immediate neighbors more than half of their appearances in the same text box, it indicates that this pair of words is a collocation. For further use in topic models, we will use 1000 most frequent bigrams extracted from the source text collection.

## 4.2    Representation of Collocations in Anchor Words Model

One of the known problems in statistical topic modeling is the high fraction of repeated words in different topics. If one wants to describe topics in a collection only with unigrams there are many degrees of freedom to determine the topics. Multiword expressions such as bigrams can facilitate more diverse description of extracted topics. Typically, the addition of bigrams as unique elements of a vocabulary increases the number of model parameters and degrades the perplexity. Further in the article, we put forward the modification of Anchor Words algorithm that can use the unigrams and bigrams as anchor words and improve the perplexity of the source Anchor topic model.

In the step 3 of the Algorithm 1, each word $w_i$ is mapped to vector $\hat{Q}_i$. The problem of finding the anchor words is the allocation of the "almost convex hull" [1] in the vectors $\hat{Q}_i$. Each topic has a single anchor word with corresponding vector from the set of $\hat{Q}_i$.

The space, which contains the vector $\hat{Q}_i$, has a thematic semantics, therefore each word may become an anchor, and thus may correspond to a some topic. To search anchor words means to find vectors corresponding to the basic hidden topics, so that the remaining topics are linear combination of basic topics.

Our main assumption is that in the space of word candidates onto anchor words positions $(\hat{Q})$, bigrams $w_i w_j$, are presented as a sum of vectors $w_i + w_j$. We prepare a set of bigrams and add vectors according to this bigrams in a set of anchor word candidates. It should be noted that, after that modification, bigrams can be anchors words but are not introduced as elements of topics.

The search of the anchor words happens directly using the distance of each word on the current convex hull (Algorithm 2). Bigrams that have on their composition two vectors close to the borders of current convex hull are given the priority in the process of selection of anchor words. It is caused by the increase of the norm of the resultant vector in the direction of convex hull expansion. Therefore, while searching anchor words, we take into account bigrams and increase the probability of choosing a bigram as an anchor word that can be interpreted as a regularization.

The expansion of the convex hull helps to describe more words through the fixed basis. It is important to note that unreasonable extension of the convex hull can break the good properties of the model, such as interpretability. An algorithm for constructing the bigram anchor words model is shown at Algorithm 3. It differs from the original algorithm only in lines 4 and 5.

## 4.3    Experiments

The experiments were carried out on three collections:

1. **Banks Articles** – a collection of Russian banking articles, 2500 documents (2000 for the train, 500 for the control), 18378 words.
2. **20 Newsgroups** – a collection of short news stories, 18846 documents (11314 for the train, 7532 for the control), 19570 words http://goo.gl/6js4G5.

**Algorithm 3.** High Level Bigram Anchor Words

**Input**: collection D, number of topics |T|, set of bigrams C
**Output**: matrix $\Phi$;

1: $Q = $ Word Co-occurences(D)
2: $\hat{Q} = Rows\_normalized(Q)$
3: $\hat{Q} = Random\_projection(\hat{Q})$
4: $\hat{B} = \hat{Q}_1, \ldots, \hat{Q}_n, \hat{Q}_{C_{11}} + \hat{Q}_{C_{12}}, \ldots, \hat{Q}_{C_{n1}} + \hat{Q}_{C_{n2}}$
5: $S = $ FindAnchorWords($\hat{B}$, |T|)
6: $\Phi = $ RecoverWordTopic($\hat{Q}$, $S$)
7: **return** $\Phi$

3. **NIPS** – a collection of abstracts from the Conference on Neural Information Processing Systems (NIPS), 1738 documents (1242 for the train, 496 for the control), 21358 words https://goo.gl/EaGmT0.

All collections have been preprocessed. The characters were brought to lowercase, characters which do not belong to Cyrillic and Latin alphabet were removed, words have been normalized (or stemmed for English collections), stop words and words with the length less than four letters were removed. Also words occurring less than 5 times were rejected. Collocations have been extracted with the algorithm described in Sect. 4.1. The preprocessed collections are available on the page, github.com/ars-ashuha/tmtk. In all experiments, the number of topics was fixed $|T| = 100$.

The metrics were calculated as follows:

- To calculate perplexity, the collection was divided into train and control parts. When calculating perplexity on test samples, each document was subdivided into two parts. In the first part, the vector of topics for the document was estimated, on the second part, perplexity was calculated.
- When calculating the coherence, the conditional probabilities are calculated with a window of 10 words.
- When calculating the unique kernel, ten most probable words in a topic were considered as its kernel.

The experiments were performed on the following models: PLSA (PL), Anchor Words (AW), Bigram Anchor Words (BiAW), Anchor Words and PLSA combination (AW + PL), Bigram Anchor Words and PLSA combination (BiAW + PL). The combination was constructed as follows: the topics obtained by the Anchor Word or Bigram Anchor Word algorithm, were used as an initial approximation for PLSA algorithm. In experiments, perplexity was measured on the control sample ($P_{test}$), coherence is denoted as ($PMI$), the uniqueness of the kernel is denoted as ($U$). The results are shown in Table 1.

As in the experiments of the authors of the Anchor Words model, the perplexity grows (in two collections out of three), which is a negative phenomenon, but the uniqueness and interpretability of the topics also grows. The combination of *Anchor Words* and *PLSA* models shows the results better than *Anchor Words* or *PLSA* separately.

**Table 1.** Results of Numerical experiments

| Collection | Banks articles | | | 20 Newsgroups | | | NIPS | | |
|---|---|---|---|---|---|---|---|---|---|
| Metric | $P_{test}$ | PMI | U | $P_{test}$ | PMI | U | $P_{test}$ | PMI | U |
| PL | 2116 | 0.60 | 0.40 | 2155 | 0.31 | 0.40 | 1635 | 0.21 | 0.32 |
| AW | 2330 | 0.63 | 0.53 | 2268 | 0.38 | 0.41 | 1505 | 0.41 | 0.38 |
| BiAW | 2248 | 0.79 | 0.60 | 2183 | 0.68 | 0.54 | 1500 | 0.50 | 0.41 |
| AW+PL | 2052 | 0.78 | 0.58 | 2053 | 0.54 | 0.55 | 1434 | 0.52 | 0.46 |
| BiAW+PL | **1848** | **0.87** | **0.63** | **2027** | **0.78** | **0.64** | **1413** | **0.58** | **0.49** |

The Bigram Anchor Words model shows the results better than the original *Anchor Words*: has lower perplexity, greater interpretability and uniqueness of kernels, but is still inferior to the *PLSA* model in perplexity. The combination of *Bigram Anchor Words* and *PLSA* models shows better results than other models; this combination has higher interpretability and uniqueness of the kernels.

It can be concluded that the initial approximation, given by the *Bigram Anchor Words* model, is more optimal in terms of achieving final perplexity and other metrics of quality. This approximation improves the sensitivity of *PLSA* to the initial approximation, which, in turn, can be formed taking into account the linguistic knowledge. Tables 2 and 3 contain examples of topics for Bank and NIPS collections. Note that main achievement is that our approach allows us to use bigrams as anchors. Also we present the tables with topic examples to show that our model is not similar to others.

**Table 2.** Examples of topics for the Russian Bank collection

| PLSA | | ANW | | ANW+ PLSA + BI | |
|---|---|---|---|---|---|
| *Topic 1* | *Topic 2* | *Topic 1* | *Topic 2* | *Topic 1* | *Topic 2* |
| рынок | кредит | акция | кредит | рынок | кредит |
| российский | кредитный | рынок | кредитован | инвестор | замщик |
| инвестор | замщик | размещение | потребитель | акция | ипотечный |
| фондовый | кредитован | акционер | ипотечный | фондовый | кредитован |
| облигация | ставка | инвестор | замщик | инструмент | залог |
| бумага | банка | капитал | банк | биржа | портфель |
| инструмент | процентный | фондовый | население | облигация | задолжен |
| фонд | срок | биржа | клиент | сегмент | потребитель |

**Anchor words for unigram anchor model:** *москва* (moscow), *налоговый* (tax), *история* (history), *акция* (share), *сила* (power), *платеж* (payment)

**Anchor words for bigram anchor models:** *компания* (company), *миллионрубль* (million rubles), *страна ес* (eu country), *управление* (control), *юридическийлицо* (company), *российский федерация* (russian federation)

**Table 3.** Examples of topics for the NIPS collection

| PLSA | | ANW | | ANW+ PLSA + BI | |
|---|---|---|---|---|---|
| *Topic 1* | *Topic 2* | *Topic 1* | *Topic 2* | *Topic 1* | *Topic 2* |
| Neuron | Tree | Neuron | Tree | Synaps | Tree |
| Spike | Featur | Synaps | Decis | Synapt | Decis |
| Fire | Branch | Synapt | Branch | Neuron | Branch |
| Time | Thi | Input | Structur | Hebbian | Set |
| Synapt | Class | Pattern | Leaf | Postsynapt | Probabl |
| Synaps | Imag | Neural | Prune | Pattern | Prune |
| Rate | Object | Activ | Set | Function | Algorithm |
| Input | Decis | Connect | Probabl | Activ | Leaf |

**Anchor words for unigram anchor model:** *face, charact, fire, loss, motion, cluster, tree, circuit, trajectori, word, extra, action, mixtur*

**Anchor words for bigram anchor model:** *likelihood, network, loss, face,* **ocular domain**, **reinforc learn**, **optic flow**, **boltzmann machin**, *markov*

## 5  Conclusion

We propose a modification of the Anchor Words topic modeling algorithm that takes into account collocations. The experiments have confirmed that this approach leads to the increase of the interpretability without deteriorating perplexity.

Accounting of collocations is only the first step to add linguistic information into a topic model. Further work will focus on the study of the possibilities of using the sentence structure of a text, as well as the morphological structure of words in the construction of topic models.

**Acknowledgments.** This work was supported by grant RFFI 14-07-00383A «Research of methods of integration of linguistic knowledge into statistical topic models».

## References

1. Arora, S., Ge, R., Halpern, Y., Mimno, D., Moitra, A., Sontag, D., Wu, Y., Zhu, M.: A practical algorithm for topic modeling with provable guarantees. arXiv preprint arXiv:1212.4777 (2012)
2. Arora, S., Ge, R., Moitra, A.: Learning topic models - going beyond SVD. In: 2012 IEEE 53rd Annual Symposium on Foundations of Computer Science (FOCS), pp. 1–10. IEEE (2012)
3. Blei, D.M., Ng, A.Y., Jordan, M.I.: Latent dirichlet allocation. J. Machine Learn. Res. **3**, 993–1022 (2003)

4. Cheng, X., Yan, X., Lan, Y., Guo, J.: BTM: topic modeling over short texts. IEEE Trans. Knowl. Data Eng. **26**(12), 2928–2941 (2014)
5. Dobrov, B., Lukashevich, N., Siromytnikov, S.: Forming the base of terminological phrases in the texts of the subject area, pp. 201–210 (2003)
6. Gao, W., Li, P., Darwish, K.: Joint topic modeling for event summarization across news and social media streams. In: Proceedings of the 21st ACM International Conference on Information and Knowledge Management, pp. 1173–1182. ACM (2012)
7. Hofmann, T.: Probabilistic latent semantic indexing. In: Proceedings of the 22nd Annual International ACM SIGIR Conference on Research and Development in Information Retrieval, pp. 50–57. ACM (1999)
8. Krestel, R., Fankhauser, P., Nejdl, W.: Latent dirichlet allocation for tag recommendation. In: Proceedings of the third ACM Conference on Recommender Systems, pp. 61–68. ACM (2009)
9. MacKay, D.J., Peto, L.C.B.: A hierarchical dirichlet language model. Nat. Lang. Eng. **1**(03), 289–308 (1995)
10. Mei, Q., Cai, D., Zhang, D., Zhai, C.: Topic modeling with network regularization. In: Proceedings of the 17th International Conference on World Wide Web, pp. 101–110. ACM (2008)
11. Newman, D., Lau, J.H., Grieser, K., Baldwin, T.: Automatic evaluation of topic coherence. In: Human Language Technologies: The 2010 Annual Conference of the North American Chapter of the Association for Computational Linguistics, pp. 100–108. Association for Computational Linguistics (2010)
12. Nokel, M., Loukachevitch, N.: The method of accounting bigram structure in topical models. Computational Methods and Programming (2015)
13. Steyvers, M., Griffiths, T.: Matlab topic modeling toolbox 1.3 (2005)
14. Vorontsov, K.: Additive regularization for topic models of text collections. In: Doklady Mathematics, pp. 301–304. Pleiades Publishing (2014)
15. Wallach, H.M.: Topic modeling: beyond bag-of-words. In: Proceedings of the 23rd International Conference on Machine Learning, pp. 977–984. ACM (2006)
16. Wang, X., McCallum, A., Wei, X.: Topical n-grams: phrase and topic discovery, with an application to information retrieval. In: Seventh IEEE International Conference on Data Mining, ICDM 2007, pp. 697–702. IEEE (2007)

# Parallel Non-blocking Deterministic Algorithm for Online Topic Modeling

Oleksandr Frei[1]([⊠]) and Murat Apishev[2]([⊠])

[1] Moscow Institute of Physics and Technology, Moscow, Russia
oleksandr.frei@gmail.com
[2] National Research University Higher School of Economics, Moscow, Russia
great-mel@yandex.ru

**Abstract.** In this paper we present a new asynchronous algorithm for learning additively regularized topic models and discuss the main architectural details of our implementation. The key property of the new algorithm is that it behaves in a fully deterministic fashion, which is typically hard to achieve in a non-blocking parallel implementation. The algorithm had been recently implemented in the BigARTM library (http://bigartm. org). Our new algorithm is compatible with all features previously introduced in BigARTM library, including multimodality, regularizers and scores calculation. While the existing BigARTM implementation compares favorably with alternative packages such as Vowpal Wabbit or Gensim, the new algorithm brings further improvements in CPU utilization, memory usage, and spends even less time to achieve the same perplexity.

**Keywords:** Probabilistic topic modeling · Additive regularization of topic models · Stochastic matrix factorization · EM-algorithm · Online learning · Asynchronous and parallel computing · BigARTM

## 1 Introduction

Topic modeling [1] is a powerful machine learning tool for statistical text analysis that has been widely used in text mining, information retrieval, network analysis and other areas [2]. Today a lot of research efforts around topic models are devoted to distributed implementations of *Latent Dirichlet Allocation* (LDA) [3], a specific Bayesian topic model that uses Dirichlet conjugate prior. This lead to numerous implementations such as AD-LDA [7], PLDA [8] and PLDA+ [9], all designed to run on a big cluster. Topic models of web scale can reach millions of topics and words, yielding Big Data models with trillions of parameters [4]. Yet not all researchers and applications are dealing with so large web-scale collections. Some of them require an efficient implementation that can run on a powerful workstation or even a laptop. Such implementations are very useful, as shown by the popular open-source packages Vowpal Wabbit [13], Gensim [12] and Mallet [14], which are neither distributed nor sometimes even multi-threaded.

© Springer International Publishing AG 2017
D.I. Ignatov et al. (Eds.): AIST 2016, CCIS 661, pp. 132–144, 2017.
DOI: 10.1007/978-3-319-52920-2_13

A similar optimization problem and its distributed implementations exist in the Deep Neural Network area, as DNN and Topic modeling both use Stochastic Gradient Descent (SGD) optimization. The asynchronous SGD [11] does not directly apply to Topic modeling because it is computationally less efficient to partition a topic model across nodes. Limited parallelizability [10] of speech DNNs across nodes justify our focus on single-node optimizations. However, scaling down a distributed algorithm can be challenging. LightLDA [4] is a major step in this direction, however it focuses only on the LDA model. Our goal is to develop a flexible framework that can learn a wide variety of topic models.

BigARTM [18] is an open-source library for regularized multimodal topic modeling of large collections. BigARTM is based on a novel technique of additive regularized topic models (ARTM) [16,17,19], which gives a flexible multi-criteria approach to probabilistic topic modeling. ARTM includes all popular models such as LDA [3], PLSA [5], and many others. Key feature of ARTM is that it provides a cohesive framework that allows users to combine different topic models that previously did not fit together.

BigARTM is proven to be very fast compared to the alternative packages. According to [18], BigARTM runs approx. 10 times faster compared to Gensim [12] and twice as fast as Vowpal Wabbit [13] in a single thread. With multiple threads BigARTM wins even more as it scales linearly up to at least 16 threads. In this paper we address the remaining limitations of the library, including performance bottlenecks and non-deterministic behavior of the Online algorithm.

The rest of the paper is organized as follows. Section 2 introduces basic notation. Sections 3 and 4 summarize offline and online algorithms for learning ARTM models. Sections 5 and 6 discuss asynchronous modifications of the online algorithm. Section 7 compares BigARTM architecture between versions 0.6 and 0.7. Section 8 reports the results of our experiments on large datasets. Section 9 discusses advantages, limitations and open problems of BigARTM.

## 2  Notation

Let $D$ denote a finite set (collection) of texts and $W$ denote a finite set (vocabulary) of all words from these texts. Let $n_{dw}$ denote the number of occurrences of a word $w \in W$ in a document $d \in D$; $n_{dw}$ values form a sparse matrix of size $|W| \times |D|$, known as *bag-of-words* representation of the collection.

Given an $(n_{dw})$ matrix, a probabilistic topic model finds two matrices: $\Phi = (\phi_{wt})$ and $\Theta = (\theta_{td})$, of sizes $|W| \times |T|$ and $|T| \times |D|$ respectively, where $|T|$ is a user-defined number of *topics* in the model (typically $|T| << |W|$). Matrices $\Phi$ and $\Theta$ provide a compressed representation of the $(n_{dw})$ matrix:

$$n_{dw} \approx n_d \sum_{t \in T} \phi_{wt} \theta_{td}, \text{ for all } d \in D, w \in W,$$

where $n_d = \sum_{w \in W} n_{dw}$ denotes the total number of words in a document $d$.

To learn $\Phi$ and $\Theta$ from $(n_{dw})$ an additively-regularized topic model (ARTM) maximizes the log-likelihood, regularized via an additional penalty term $R(\Phi,\Theta)$:

$$F(\Phi,\Theta) = \sum_{d \in D} \sum_{w \in W} n_{dw} \ln \sum_{t \in T} \phi_{wt}\theta_{td} + R(\Phi,\Theta) \rightarrow \max_{\Phi,\Theta}. \tag{1}$$

Regularization penalty $R(\Phi,\Theta)$ may incorporate external knowledge of the expert about the collection. With no regularization $(R = 0)$ it corresponds to PLSA [5]. Many Bayesian topic models, including LDA [3], can be represented as special cases of ARTM with different regularizers $R$, as shown in [17].

In [16] it is shown that the local maximum $(\Phi,\Theta)$ of problem (1) satisfies the following system of equations:

$$p_{tdw} = \operatorname*{norm}_{t \in T}\left(\phi_{wt}\theta_{td}\right); \tag{2}$$

$$\phi_{wt} = \operatorname*{norm}_{w \in W}\left(n_{wt} + \phi_{wt}\frac{\partial R}{\partial \phi_{wt}}\right); \quad n_{wt} = \sum_{d \in D} n_{dw}p_{tdw}; \tag{3}$$

$$\theta_{td} = \operatorname*{norm}_{t \in T}\left(n_{td} + \theta_{td}\frac{\partial R}{\partial \theta_{td}}\right); \quad n_{td} = \sum_{w \in d} n_{dw}p_{tdw}; \tag{4}$$

where operator $\operatorname*{norm}_{i \in I} x_i = \frac{\max\{x_i,0\}}{\sum_{j \in I}\max\{x_j,0\}}$ transforms a vector $(x_i)_{i \in I}$ to a discrete distribution, $n_{wt}$ counters represent term frequency of word $w$ in topic $t$.

Learning of $\Phi$ and $\Theta$ from (2)–(4) can be done by EM-algorithm, which starts from random values in $\Phi$ and $\Theta$, and iterates E-step (2) and M-steps (3), (4) until convergence. In the sequel we discuss several variations of such EM-algorithm, which are all based on the above formulas but differ in the way how operations are ordered and grouped together.

In addition to plain text, many collections have other metadata, such as authors, class or category labels, date-time stamps, or even associated images, audio or video clips, usage data, etc. In [19] this data was represented as *modalities*, where the overall vocabulary $W$ is partitioned into $M$ subsets $W = W^1 \sqcup \cdots \sqcup W^M$, one subset per modality, and in (3) matrix $\Phi$ is normalized independently within each modality. Incorporating modalities into a topic model improves its quality and makes it applicable for classification, cross-modal retrieval, or making recommendations. In the sequel we list all algorithms for one modality, but our implementation in BigARTM supports the general case.

## 3   Offline Algorithm

Offline ARTM (Algorithm 2) relies on subroutine ProcessDocument (Algorithm 1), which corresponds to Eqs. (2) and (4) from the solution of the ARTM optimization problem (1). ProcessDocument requires a fixed $\Phi$ matrix and a vector $n_{dw}$ of term frequencies for a given document $d \in D$, and as a result it returns a topical distribution $(\theta_{td})$ for the document, and a matrix $(\hat{n}_{wt})$ of size $|d| \times |T|$, where $|d|$ denotes

---

**Algorithm 1.** ProcessDocument$(d, \Phi)$

---

**Input**: document $d \in D$, matrix $\Phi = (\phi_{wt})$;
**Output**: matrix $(\tilde{n}_{wt})$, vector $(\theta_{td})$ for the document $d$;

1  initialize $\theta_{td} := \frac{1}{|T|}$ for all $t \in T$;
2  **repeat**
3      $p_{tdw} := \underset{t \in T}{\mathrm{norm}}(\phi_{wt}\theta_{td})$ for all $w \in d$ and $t \in T$;
4      $\theta_{td} := \underset{t \in T}{\mathrm{norm}}(\sum_{w \in d} n_{dw}p_{tdw} + \theta_{td}\frac{\partial R}{\partial \theta_{td}})$ for all $t \in T$;
5  **until** $\theta_d$ converges;
6  $\tilde{n}_{wt} := n_{dw}p_{tdw}$ for all $w \in d$ and $t \in T$;

---

**Algorithm 2.** Offline ARTM

---

**Input**: collection $D$;
**Output**: matrix $\Phi = (\phi_{wt})$;

1  initialize $(\phi_{wt})$;
2  create batches $D := D_1 \sqcup D_2 \sqcup \cdots \sqcup D_B$;
3  **repeat**
4      $(n_{wt}) := \sum_{b=1,\ldots,B} \sum_{d \in D_b} \mathrm{ProcessDocument}(d, \Phi)$;
5      $(\phi_{wt}) := \underset{w \in W}{\mathrm{norm}}(n_{wt} + \phi_{wt}\frac{\partial R}{\partial \phi_{wt}})$;
6  **until** $(\phi_{wt})$ converges;

---

the number of distinct words in the document. ProcessDocument might also be useful as a separate routine which finds $(\theta_{td})$ distribution for a new document, but in the Offline algorithm it is instead used as a building block in an iterative EM-algorithm that finds the $\Phi$ matrix.

Offline algorithm performs scans over the collection, calling ProcessDocument for each document $d \in D$ from the collection, and then aggregating the resulting $(\hat{n}_{wt})$ matrices into the final $(n_{wt})$ matrix of size $|W| \times |T|$. After each scan it recalculates $\Phi$ matrix according to the Eq. (3).

At step 2 we partition collection $D$ into batches $(D_b)$. This step is not strictly necessary for Offline algorithm, but it rather reflects an internal implementation detail. For performance reasons the outer loop over batches $b = 1, \ldots, B$ is parallelized across multiple threads, and within each batch the inner loop over documents $d \in D_b$ is executed in a single thread. Each batch is stored in a separate file on disk to allow out-of-core streaming of the collection. For typical collections it is reasonable to have around 1000 documents per batch, however for ultimate performance we encourage users to experiment with this parameter. Too small batches can cause disk IO overhead due to lots of small reads, while too large batches will result in bigger tasks that will not be distributed evenly across computation threads.

Note that $\theta_{td}$ values appear only within ProcessDocument subroutine. This leads to efficient memory usage because the implementation never stores the

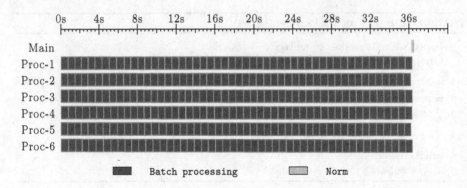

**Fig. 1.** Gantt chart for Offline ARTM (Algorithm 2)

entire theta matrix $\Theta$ at any given time. Instead, $\theta_{td}$ values are recalculated from scratch on every pass through the collection.

Figure 1 shows a Gantt chart of the Offline algorithm. Here and in the sequel Gantt charts are built for a single EM-iteration on NYTimes dataset[1] ($|D| = 300K$, $|W| = 102K$) with $|T| = 16$ topics. ProcessBatch boxes corresponds to the time spent in processing an individual batch. The final box Norm, executed on the main thread, correspond to the time spent in the step 4 in Algorithm 2 where $n_{wt}$ counters are normalized to produce a new $\Phi$ matrix.

## 4   Online Algorithm

Online ARTM (Algorithm 3) generalizes the Online variational Bayes algorithm, suggested in [6] for the LDA model. Online ARTM improves the convergence rate of the Offline ARTM by re-calculating matrix $\Phi$ each time after processing a certain number of batches. To simplify the notation we introduce a trivial subroutine

$$\text{ProcessBatches}(\{D_b\}, \Phi) = \sum_{D_b} \sum_{d \in D_b} \text{ProcessDocument}(d, \Phi)$$

that aggregates the output of ProcessDocument across a given set of batches at a constant $\Phi$ matrix. Here the partition of the collection $D := D_1 \sqcup D_2 \sqcup \cdots \sqcup D_B$ into batches plays a far more significant role than in the Offline algorithm, because different partitioning algorithmically affects the result. At step 6 the new $n_{wt}^{i+1}$ values are calculated as a convex combination of the old values $n_{wt}^i$ and the value $\hat{n}_{wt}^i$ produced on the recent batches. Old counters $n_{wt}^i$ are scaled by a factor $(1 - \rho_i)$, which depends on the iteration number. A common strategy is to use $\rho_i = (\tau_0 + i)^{-\kappa}$, where typical values for $\tau_0$ are between 64 and 1024, for $\kappa$ — between 0.5 and 0.7.

---

[1] https://archive.ics.uci.edu/ml/datasets/Bag+of+Words.

---

**Algorithm 3.** Online ARTM

---

**Input**: collection $D$, parameters $\eta, \tau_0, \kappa$;
**Output**: matrix $\Phi = (\phi_{wt})$;

1  create batches $D := D_1 \sqcup D_2 \sqcup \cdots \sqcup D_B$;
2  initialize $(\phi_{wt}^0)$;
3  **for** $i = 1, \ldots, \lfloor B/\eta \rfloor$ **do**
4  $\quad (\hat{n}_{wt}^i) := \mathsf{ProcessBatches}(\{D_{\eta(i-1)+1}, \ldots, D_{\eta i}\}, \Phi^{i-1})$;
5  $\quad \rho_i := (\tau_0 + i)^{-\kappa}$;
6  $\quad (n_{wt}^i) := (1 - \rho_i) \cdot (n_{wt}^{i-1}) + \rho_i \cdot (\hat{n}_{wt}^i)$;
7  $\quad (\phi_{wt}^i) := \underset{w \in W}{\mathrm{norm}}(n_{wt}^i + \phi_{wt}^{i-1} \frac{\partial R}{\partial \phi_{wt}})$;

---

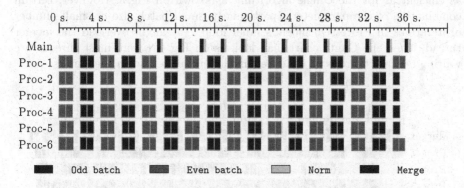

**Fig. 2.** Gantt chart for Online ARTM (Algorithm 3)

As in the Offline algorithm, the outer loop over batches $D_{\eta(i-1)+1}, \ldots, D_{\eta i}$ is executed concurrently across multiple threads. The problem with this approach is that none of the threads have any useful work to do during steps 5–7 of the Online algorithm. The threads can not start processing the next batches because a new version of $\Phi$ matrix is not ready yet. As a result the CPU utilization stays low, and the run-time Gantt chart of the Online algorithm typically looks like in Fig. 2. Boxes Even batch and Odd batch both correspond to step 4, and indicate the version of the $\Phi^i$ matrix (even $i$ or odd $i$). Merge correspond to the time spent merging $n_{wt}$ with $\hat{n}_{wt}$. Norm is, as before, the time spent normalizing $n_{wt}$ counters into the new $\Phi$ matrix.

In the next two sections we present asynchronous modifications of the online algorithm that result in better CPU utilization. The first of them (Async ARTM) has non-deterministic behavior and few performance bottlenecks. The second algorithm (DetAsync ARTM) addresses these problems.

## 5  Async: Asynchronous Online Algorithm

Async algorithm was implemented in BigARTM v0.6 as described in [18]. The idea is to trigger asynchronous execution of the Offline algorithm and store the resulting $\hat{n}_{wt}$ matrices into a queue. Then, whenever the number of elements in the queue becomes equal to $\eta$, the Async algorithm performs steps 5–7 of the Online ARTM (Algorithm 3). For performance reasons merging of the $\hat{n}_{wt}$ counters happens in a background by a dedicated *Merger* thread.

First problem of the Async algorithm is that it does not define the order in which $\hat{n}_{wt}$ are merged. This order is usually different from the original order of the batches, and typically it changes from run to run. This affects the final $\Phi$ matrix which also changes from run to run.

Another issue with Async algorithm is that queuing $\hat{n}_{wt}$ counters may considerably increase the memory usage, and also lead to performance bottlenecks in the *Merger* thread. In some cases the execution of the Async algorithm is as efficient as for the Offline algorithm, as shown on Fig. 3. However, certain combination of the parameters (particularly, small batch size or small number of iterations in ProcessDocument's inner loop 2–5) might overload the merger thread. Then the Gantt chart may look as on Fig. 4, where most threads are waiting because there is no space left in the queue to place $n_{wt}$ counters.

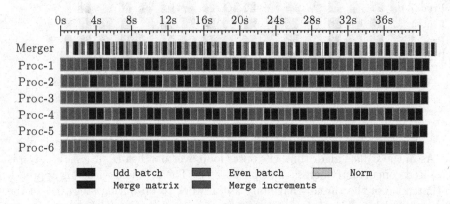

**Fig. 3.** Gantt chart for Async ARTM from BigARTM v0.6 — normal execution

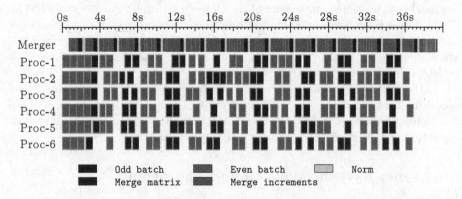

**Fig. 4.** Gantt chart for Async ARTM from BigARTM v0.6 — performance issues

---

**Algorithm 4.** DetAsync ARTM

---

**Input**: collection $D$, parameters $\eta, \tau_0, \kappa$;
**Output**: matrix $\Phi = (\phi_{wt})$;

1  create batches $D := D_1 \sqcup D_2 \sqcup \cdots \sqcup D_B$;
2  initialize $(\phi_{wt}^0)$;
3  $F^1 := \mathsf{AsyncProcessBatches}(\{D_1, \ldots, D_\eta\}, \Phi^0)$;
4  **for** $i = 1, \ldots, \lfloor B/\eta \rfloor$ **do**
5  $\quad$ **if** $i \neq \lfloor B/\eta \rfloor$ **then**
6  $\quad\quad\lfloor \; F^{i+1} := \mathsf{AsyncProcessBatches}(\{D_{\eta i+1}, \ldots, D_{\eta i+\eta}\}, \Phi^{i-1})$;
7  $\quad (\hat{n}_{wt}^i) := \mathsf{Await}(F^i)$;
8  $\quad \rho_i := (\tau_0 + i)^{-\kappa}$;
9  $\quad (n_{wt}^i) := (1 - \rho_i) \cdot (n_{wt}^{i-1}) + \rho_i \cdot (\hat{n}_{wt}^i)$;
10 $\quad (\phi_{wt}^i) := \underset{w \in W}{\mathsf{norm}}(n_{wt}^i + \phi_{wt}^{i-1} \frac{\partial R}{\partial \phi_{wt}})$;

---

In the next section we resolve the aforementioned problems by introducing a new DetAsync algorithm, which has an entirely deterministic behavior and achieves high CPU utilization without requiring user to tweak the parameters.

# 6  DetAsync: Deterministic Async Online Algorithm

DetAsync ARTM (Algorithm 4) is based on two new routines, AsyncProcess-Batches and Await. The former is equivalent to ProcessBatches, except that it just queues the task for an asynchronous execution and returns immediately. Its output is a future object (for example, an std::future from C++11 standard), which can be later passed to Await in order to get the actual result, e.g. in our case the $\hat{n}_{wt}$ values. In between calls to AsyncProcessBatches and Await the algorithm can perform some other useful work, while the background threads are calculating the $(\hat{n}_{wt})$ matrix.

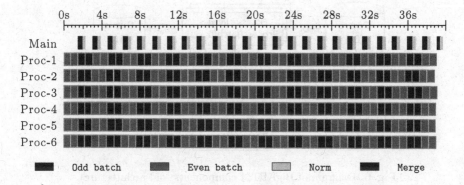

**Fig. 5.** Gantt chart for DetAsync ARTM (Algorithm 4)

To calculate $\hat{n}_{wt}^{i+1}$ it uses $\Phi^{i-1}$ matrix, which is one generation older than $\Phi^i$ matrix used by the Online algorithm. This adds an extra "offset" between the moment when $\Phi$ matrix is calculated and the moment when it is used, and as a result gives the algorithm additional flexibility to distribute more payload to computation threads. Steps 3 and 5 of the algorithm are just technical tricks to implement the "offset" idea.

Adding an "offset" should negatively impact the convergence of the DetA-sync algorithm compared to the Online algorithm. For example, in AsyncProcess-Batches the initial matrix $\Phi^0$ is used twice, and the two last matrices $\Phi^{\lfloor B/\eta \rfloor - 1}$ and $\Phi^{\lfloor B/\eta \rfloor}$ will not be used at all. On the other hand the asynchronous algorithm gives better CPU utilization, as clearly shown by the Gantt chart from Fig. 5. This tradeoff between convergence and CPU utilization will be evaluated in Sect. 8.

## 7   Implementation

The challenging part for the implementation is to aggregate the $\hat{n}_{wt}$ matrices across multiple batches, given that they are processed in different threads. The way BigARTM solves this challenge was changed between versions v0.6 (Fig. 6) and v0.7 (Fig. 7).

In the old architecture the $\hat{n}_{wt}$ matrices were stored in a queue, and then aggregated by a dedicated *Merger thread*. In the new architecture we removed Merger thread, and $\hat{n}_{wt}$ are written directly into the final $n_{wt}$ matrix concurrently from all processor threads. To synchronize the write access we require that no threads simultaneously update the same row in $\hat{n}_{wt}$ matrix, yet the data for distinct words can be written in parallel. This is enforced by spin locks $l_w$, one per each word in the vocabulary $W$. At the end of ProcessDocument we loop through all $w \in d$, acquire the corresponding lock $l_w$, append $\hat{n}_{wt}$ to $n_{wt}$ and release the lock. This approach is similar to [15], where the same pattern is used to update a shared stated in a distributed topic modeling architecture.

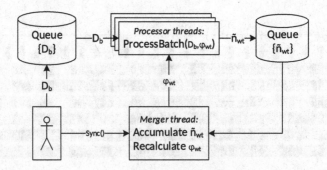

**Fig. 6.** Diagram of BigARTM components (old architecture)

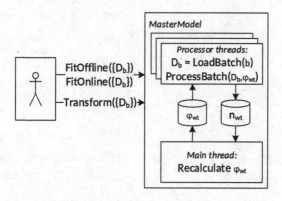

**Fig. 7.** Diagram of BigARTM components (new architecture)

In our new architecture we also removed *DataLoader* thread, which previously was loading batches from disk. Now this happens directly from processor thread, which simplified the architecture without sacrificing performance.

In addition, we provided a cleaner API so now the users may use simple FitOffline, FitOnline methods to learn the model, and Transform to apply the model to the data. Previously the users had to interact with low-level building blocks, such as ProcessBatches routine.

## 8    Experiments

In this section we compare the effectiveness of Offline (Algorithm 2), Online (Algorithm 3), Async [18] and DetAsync (Algorithm 4) algorithms. According to [18] Async algorithm runs approx. 10 times faster compared to Gensim [12], and twice as fast compared to Vowpal Wabbit (VW) [13] in a single thread; and with multiple threads BigARTM wins even more.

In the experiments we use *Wikipedia* dataset ($|D| = 3.7$M articles, $|W| = 100$K words) and *Pubmed* dataset ($|D| = 8.2$M abstracts, $|W| = 141$K words). The experiments were run on Intel Xeon CPU E5-2650 v2 system with 2 processors, 16 physical cores in total (32 with hyper-threading).

Figure 8 show the *perplexity* as a function of the time spent by the four algorithms listed above. The perplexity measure is defined as

$$\mathscr{P}(D, p) = \exp\left( -\frac{1}{n} \sum_{d \in D} \sum_{w \in d} n_{dw} \ln \sum_{t \in T} \phi_{wt} \theta_{td} \right), \tag{5}$$

where $n = \sum_d n_d$. Lower perplexity means better result. Each point on the figures corresponds to a moment when the algorithm finishes a complete scan of the collection. Each algorithm was time-boxed to run for 30 min.

Table 1 gives peak memory usage for $|T| = 1000$ and $|T| = 100$ topics model on Wikipedia and Pubmed datasets.

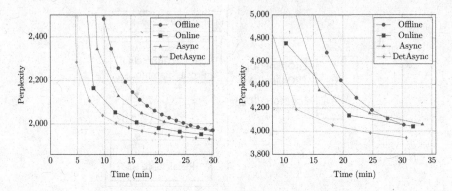

**Fig. 8.** Perplexity versus time for Pubmed (left) and Wikipedia (right), $|T| = 100$ topics

**Table 1.** BigARTM peak memory usage, GB

|        | $|T|$ | Offline | Online | DetAsync | Async (v0.6) |
|--------|-------|---------|--------|----------|--------------|
| Pubmed | 1000  | 5.17    | 4.68   | 8.18     | 13.4         |
| Pubmed | 100   | 1.86    | 1.62   | 2.17     | 3.71         |
| Wiki   | 1000  | 1.74    | 2.44   | 3.93     | 7.9          |
| Wiki   | 100   | 0.54    | 0.53   | 0.83     | 1.28         |

## 9   Conclusions

We presented a deterministic asynchronous (DetAsync) online algorithm for learning additively regularized topic models (ARTM). The algorithm supports all features of ARTM models, including multi-modality, ability to add custom regularizers and ability to combine regularizers. As a result, the algorithm allows the user to produce topic models with a rich set of desired properties. This differentiates ARTM from the existing models, such as LDA or PLSA, which give almost no control over the resulting topic model.

We provided an efficient implementation of the algorithm in BigARTM open-source library, and our solution runs an order of magnitude faster than the alternative open-source packages. Compared to the previous implementation we eliminated certain performance bottlenecks, achieving optimal CPU utilization without requiring the user to tweak batch size and the number of inner loops per document. In addition, DetAsync algorithm guarantees deterministic behavior, which makes it easier for us to unit-test our implementation and makes BigARTM ready for production use-cases.

In the future we will focus on memory efficiency to benefit from sparsity of word-topic ($\Phi$) and topic-document ($\Theta$) matrices, and extend our implementation to run on a cluster.

**Acknowledgements.** The work was supported by Russian Science Foundation (grant 15-18-00091). Also we would like to thank Prof. K. V. Vorontsov for constant attention to our research and detailed feedback to this paper.

# References

1. Blei, D.M.: Probabilistic topic models. Commun. ACM **55**(4), 77–84 (2012)
2. Daud, A., Li, J., Zhou, L., Muhammad, F.: Knowledge discovery through directed probabilistic topic models: a survey. Front. Comput. Sci. Chin **4**(2), 280–301 (2010)
3. Blei, D.M., Ng, A.Y., Jordan, M.I.: Latent Dirichlet allocation. J. Mach. Learn. Res. **3**, 993–1022 (2003)
4. Yuan, J., Gao, F., Ho, Q., Dai, W., Wei, J., Zheng, X., Xing, E.P., Liu, T.Y., Ma, W.Y.: LightLDA: big topic models on modest computer clusters. In: Proceedings of the 24th International Conference on World Wide Web, pp. 1351–1361 (2015)
5. Hofmann, T.: Probabilistic latent semantic indexing. In: Proceedings of the 22nd Annual International ACM SIGIR Conference on Research and Development in Information Retrieval, pp. 50–57 (1999)
6. Hoffman, M.D., Blei, D.M., Bach, F.R.: Online learning for latent Dirichlet allocation. In: NIPS, pp. 856–864. Curran Associates Inc. (2010)
7. Newman, D., Asuncion, A., Smyth, P., Welling, M.: Distributed algorithms for topic models. J. Mach. Learn. Res. **10**, 1801–1828 (2009)
8. Wang, Y., Bai, H., Stanton, M., Chen, W.-Y., Chang, E.Y.: PLDA: parallel latent Dirichlet allocation for large-scale applications. In: Goldberg, A.V., Zhou, Y. (eds.) AAIM 2009. LNCS, vol. 5564, pp. 301–314. Springer, Heidelberg (2009). doi:10. 1007/978-3-642-02158-9_26
9. Liu, Z., Zhang, Y., Chang, E.Y., Sun, M.: PLDA+: parallel latent Dirichlet allocation with data placement and pipeline processing. ACM Trans. Intell. Syst. Technol. **2**(3), 26:1–26:18 (2011)
10. Seide, F., Fu, H., Droppo, J., Li, G., Yu, D.: On parallelizability of stochastic gradient descent for speech DNNs. In: 2014 IEEE International Conference on Acoustics, Speech and Signal Processing (ICASSP), pp. 235–239. IEEE (2014)
11. Dean, J., Corrado, G.S., Monga, R., Chen, K., Devin, M., Le, Q.V., Mao, M.Z., Ranzato, M.-A., Senior, A., Tucker, P., Yang, K., Ng, A.Y.: Large scale distributed deep networks. In: NIPS, pp. 1223–1231 (2012)
12. Řehůřek, R., Sojka, P.: Software framework for topic modelling with large corpora. In: Proceedings of the LREC 2010 Workshop on New Challenges for NLP Frameworks, pp. 45–50, Valletta, Malta, May 2010
13. Langford, J., Li, L., Strehl, A.: Vowpal wabbit open source project. Technical report. Yahoo! (2007)
14. McCallum, A.K.: A Machine Learning for Language Toolkit (2002). http://mallet.cs.umass.edu
15. Smola, A., Narayanamurthy, S.: An architecture for parallel topic models. Proc. VLDB Endow. **3**(1–2), 703–710 (2010)
16. Vorontsov, K.V.: Additive regularization for topic models of text collections. Dokl. Math. **89**(3), 301–304 (2014)
17. Vorontsov, K.V., Potapenko, A.A.: Additive regularization of topic models. Mach. Learn. Spec. Issue Data Anal. Intell. Optim. **101**(1), 303–323 (2015)

18. Vorontsov, K., Frei, O., Apishev, M., Romov, P., Dudarenko, M.: BigARTM: open source library for regularized multimodal topic modeling of large collections. In: Khachay, M.Y., Konstantinova, N., Panchenko, A., Ignatov, D.I., Labunets, V.G. (eds.) AIST 2015. CCIS, vol. 542, pp. 370–381. Springer, Heidelberg (2015). doi:10. 1007/978-3-319-26123-2_36
19. Vorontsov, K., Frei, O., Apishev, M., Romov, P., Suvorova, M., Yanina, A.: Non-Bayesian additive regularization for multimodal topic modeling of large collections. In: Proceedings of the 2015 Workshop on Topic Models: Post-Processing and Applications, pp. 29–37. ACM, New York (2015)

# Flexible Context Extraction for Keywords in Russian Automatic Speech Recognition Results

Olga Khomitsevich[1], Kirill Boyarsky[2], Eugeny Kanevsky[3],
Anna Bulusheva[4(✉)], and Valentin Mendelev[1]

[1] Speech Technology Center Ltd, St. Petersburg, Russia
{khomitsevich,mendelev}@speechpro.com
[2] ITMO University, St. Petersburg, Russia
boyarin9@yandex.ru
[3] St. Petersburg Institute for Economics and Mathematics, RAS,
St. Petersburg, Russia
kanev@emi.nw.ru
[4] STC-Innovations Ltd, St. Petersburg, Russia
bulusheva@speechpro.com

**Abstract.** The paper deals with extracting contexts for keywords found in text, in particular in Automatic Speech Recognition (ASR) output. We propose using a syntactic parser to find contexts by analysing the sentence structure, rather than simply using a window of several words on the left and right of the keyword, or the whole sentence. This method provides concise but meaningful contexts that are easily readable by humans and can also be used in applications such as thematic clustering. We describe the Russian *SemSin* system which combines a syntactic dependency parser and elements of semantic ontology. We demonstrate the use of *SemSin* for our task both for normal text and for recognition output, and outline the suggestions for future developments of our method.

**Keywords:** Syntactic parsing · Keyword extraction · Contexts · Russian · Speech-to-text · ASR · Speech recognition

## 1 Introduction

In this paper we propose a system for extracting contexts for keywords found in text, in particular in Automatic Speech Recognition (ASR) output. The context can be displayed in keyword search results or used in thematic clustering tasks when we need to classify texts based on the occurrences of certain keywords [1,2]. The easiest way of selecting the context for a keyword is to display the whole sentence in which the keyword is found. However, this is not always desirable, since the sentence may be very long; we may also be dealing with poorly punctuated recognizer output. Another frequent solution is to use a window of n words

© Springer International Publishing AG 2017
D.I. Ignatov et al. (Eds.): AIST 2016, CCIS 661, pp. 145–154, 2017.
DOI: 10.1007/978-3-319-52920-2_14

to the right and left of the keyword, as implemented for instance in [3]. The problem with this is that sometimes the window may miss important information connected with the keyword, while superfluous words, such as parenthetical expressions, may be included. This is especially common in a free word order language like Russian.

In order to efficiently extract the information connected with the keyword, we propose to use the results of a syntactic parser and to define the context window structurally rather than in linear terms. That allows us to extract the context flexibly, depending on the sentence structure, and to keep it relatively short while at the same time retaining the relevant information about the keyword.

## 2   The SemSin System

### 2.1   Overview

SemSin [4] is a system for syntactic and semantic analysis of Russian text. It combines the functions of a Part-of-Speech tagger, ontology and syntactic parser. The system is based on a lexicon comprising 190k lemmas divided into 1680 classes, 90 of which (over 32k lemmas) are proper name classes. The lexicon was built using the dictionary constructed by V.A. Tuzov [5]. It is divided into three databases:

- Morphological database. This is the main database which contains lemmas together with their POS and other morphological features, links to inflection paradigms, semantic class numbers, and the grammatical forms that depend on them.
- Database of idioms. This database contains set expressions comprising two or more words, including uninflected phrases, such as complex prepositions, and inflected phrases such as figurative expressions or proper names. The database currently contains over 4700 items. Uninflected idioms are combined into one token during parsing. Inflected idioms are treated as separate tokens, but are combined into a phrase, which is assigned the appropriate semantic class.
- Database of prepositions. This is a separate database for storing prepositions, which lists the preposition, the case of the noun required by the preposition, the semantic class of that noun and the type of the resulting relation (for instance, "where", "when", etc.). The same preposition may be listed multiple times if it can be used in different combinations. The database comprises over 2200 items.

### 2.2   Syntactic Parsing and Semantic Analysis Using SemSin

The *SemSin* system analyses text by paragraph, involving the following steps. First, each word is processed by the morphological analyser, whereby it is assigned its lemma, POS and grammatical form, semantic class and syntactic dependents. After that, the text is tokenized and divided into sentences by the pre-syntax module. That includes resolving any ambiguities arising from the presence of periods after abbreviations, initials, digits and so on. Then syntactic parse trees are constructed for each sentence by means of the application of about 400 rules. The system attempts to apply each rule to each token consecutively, and performs certain actions in case of success. In the process, morphological and lexical ambiguity is resolved.

*SemSin* is a dependency parser that attempts to establish dependency relations between all the words in the sentence and to determine the type of each relation. The predicate (typically a finite verb) is the highest node in the clause on which all the other words ultimately depend. Dependency parsing is well suited for processing a morphologically rich, free-word-order language like Russian [6,7]. The parsing result is stored as an .xml file. A fragment of a resulting file for the sentence part *Саудовская Аравия предпочитает* "*Saudi Arabia prefers*" is given below:

```
<w Id="1" lemma="САУДОВСКИЙ" morph="ПРИЛ жр,ед,им" class="$715">
<rel id_head="2" type="Часть_Назв"/> Саудовская </w>
<w Id="2" lemma="АРАВИЯ" morph="СУЩ но,жр,ед,им" class="$1231000">
<rel id_head="3" type="Субъект"/> Аравия </w>
<w Id="3" lemma="ПРЕДПОЧИТАТЬ" morph="Г пе,нс,дст,нст,3л, ед"
class="$1241/41561"> <rel id_head="" type=""/> предпочитает </w>
```

The following features are represented in the xml fragment:

- *Id* is the unique ID of the token inside the sentence.
- *lemma* is the base form of the word.
- *morph* contains the information about the POS and grammatical features of the word, such as animacy, gender, number, case, tense, etc.
- *class* number refers to the semantic class of the word.
- *rel* is the tag containing information about relations between words in the sentence. Its parameter *id_head* contains the *Id* of the parent node, while the parameter *type* indicates the type of the dependency relation between the two words.

The results of parsing can be visualized as a dependency tree which marks the relations between words. A detailed illustration is provided below in Sect. 4.

## 3   Rules for Context Extraction

The aim of our algorithm is to provide a basic meaningful context in which a keyword occurs. The algorithm analyses the syntactic parse tree of the sentence and extracts a part of it including the target word. For the first version of our program, we developed an algorithm including the following rules:

- After the target keyword or group of keywords (e.g., a named entity) that we are looking for is detected, we find all the words immediately dependent on it.
- We extract the topmost node of the clause (normally the predicate) and all the nodes between it and the target word.

– We extract the subject of the predicate (unless it is already extracted or coincides with the target word).
– We extract the direct object of the predicate, and, for verbs of the class "speech/information/reporting", the object denoting the content of the report.
– Prepositional and other groups linked to the predicate by a "where?"-type link are also added.
– We add all the words in genitive case that depend on those already extracted.
– All the extracted words are ordered as they appear in the original sentence.

The above rules were formed empirically by analysing different test cases for keyword extraction. This basic algorithm, implemented as a prototype computer program, normally yields a short but readable context for the keyword. The input of the program is the xml output of the *SemSin* parser, and its output is a context as a string of words stored in a text file. The following section demonstrates some examples of the operation of the algorithm.

## 4    Examples of Context Extraction

### 4.1    News Articles

We first tested our system on regular (not recognized) text. The task was to extract contexts for keywords found in news articles from the web. Let us consider the example in (1), which is a sentence containing the target keyword *США* "*USA*".

Полевой командир талибов Маулави Сангин сообщил в четверг западным информационным агентствам, что военнослужащий США, пропавший в афганской провинции Пактика в конце июня, находится в руках боевиков.    (1)

(Translation: "Talib field commander Mawlawi Sangin informed Western information agencies on Thursday that the USA serviceman who went missing in the Afghan Paktika province in the end of June is in the hands of militants").

This sentence is fairly long, so we would like to avoid using the complete sentence as a context. On the other hand, if we extract a linear word window, it will likely give us a disjointed string of words (for instance, taking three words to the left and right of the target word results in the string *агентствам, что военнослужащий США, пропавший в афганской* "*to agencies that the USA serviceman who went missing in Afghan*"). To make the context both concise and coherent, we propose to extract it from the sentence structure, utilizing the information about word dependencies.

The dependency parse tree for the sentence 1 is given in Fig. 1. For the sake of clarity, English translations are given under each node, and the types of the dependency relations are given in English next to each dependency arrow. These include cases of nouns (GEN = genitive, ACC = accusative, LOC = locative), attributive relations ("what kind?"), circumstantial relations ("where?",

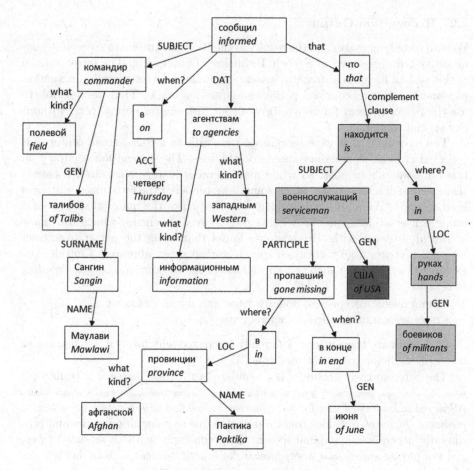

**Fig. 1.** Parse tree for sentence1.

"when?"), etc. Most words form a node of their own, however the words *в конце* "in the end" are combined into a single token as a complex preposition.

We can see that the target word *США* (highlighted dark grey in the scheme) is dependent on the word *военнослужащий* "serviceman", which in turn is dependent on the finite verb *находится* "is (located)". The verb occupies the highest position in the clause so, according to our rules, it is the highest node that we extract, even though it itself depends on the verb in the main clause. Further, we extract the expression *в руках* "in the hands", which is connected to the predicate by a "where?" link, and the genitive noun *боевиков* "of militants" dependent on it. The words extracted by using the algorithm are highlighted light grey in the figure. By combining them in the order that they appear in the original sentence we obtain the context *военнослужащий США находится в руках боевиков* "the USA serviceman is in the hands of militants", which is short but easy to comprehend and contains essential information about the keyword.

### 4.2  Recognition Output

We also tested our system on the results of the Russian Automatic Speech Recognition system developed at Speech Technology Center [8,9]. The latest version of this system includes automatic sentence boundary detection and punctuation placement, detecting commas, periods and question marks. Thus we were able to use the *SemSin* parser for recognition output just like for normal text, without any special tuning.

The recordings that were recognized were calls to a Russian call center that collected citizens' complaints about social issues. The recognition accuracy on this data was about 80%, so while many sentences were recognized correctly there were still a few errors, which sometimes rendered the recognized sentences hard to parse. An additional problem was that since the material was spontaneous conversational speech, it contained many disfluencies, which also made it difficult to work with. However, we found that using the parser to extract keyword contexts from recognized speech worked reasonably well. Consider the recognized sentence 2, which contains a form of the keyword *льготный* *"relating to benefits"*:

$$\text{меня очень интересует, почему у нас так плохо стало с с} \atop \text{лекарством бы льготным лекарствам.} \qquad (2)$$

(Approximate translation: "I'm really interested why for us it has become so bad with medicine to benefit medicines").

The    recognized    result    is    similar    to    the    original    transcript: *меня очень интересует, почему у нас так плохо стало с лекарством, льготным лекарством* (*"I'm really interested why for us it has become so bad with a medicine, a benefit medicine"*). However it does contain mistakes due to recognition errors and possibly disfluences in the original speech: the preposition "with" is repeated twice, and the phrase *льготным лекарствам* *"to benefit medicines"* is in dative case, which does not make sense in the general structure of the sentence. The parse tree that the *SemSin* parser created for this sentence is given in Fig. 2.

As can be seen, the parse tree looks relatively normal, apart from one hanging preposition. The keyword *льготным* is dependent on the noun *лекарствам* *"to medicines"*, while the latter is dependent on the complex predicate *стало плохо* *"it became bad"*. After adding the prepositional group *у нас* *"with us"* we obtain the following context: *у нас плохо стало льготным лекарствам* *"for us it has become bad to benefit medicines"*. (The keyword is highlighted dark grey in the picture, and the context words are highlighted light grey). While not completely grammatical, the resulting context is more readable than the original recognized sentence and provides the necessary information in concise form.

## 5  Experiments and Results

### 5.1  News Articles

We first performed our experiments on regular (not recognized) news articles from the web (about 500 sentences). From this data we extracted 237 contexts

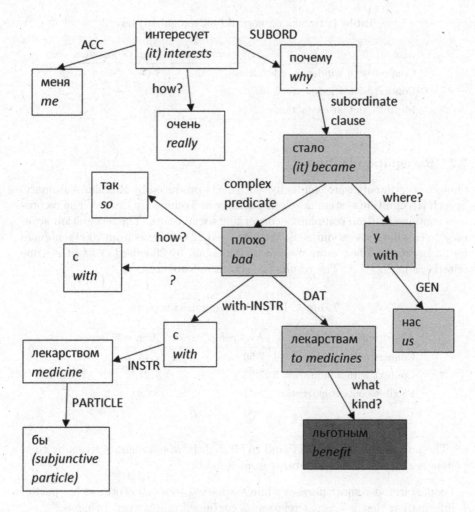

**Fig. 2.** Parse tree for sentence 2.

for 55 keywords. As the baseline performance we considered the algorithm which used a window of n words to the right and left of the keyword, as implemented for instance in [3]. Two context quality measures have been introduced: completeness and conciseness. The former reflects what part of information connected to a keyword in a text is present in a context sample. The latter indicates how much information that is not connected to the keyword are there in the context sample. Twenty human experts were asked to assess test samples by assigning a 1 to 10 value for each measure (the more the better) The results are shown in Table 1.

As one can see the algorithm proposed in this paper outperforms presented alternatives. Completeness is not much better, but probably if we expand the set of context extraction rules we can improve this indicator without losing conciseness

**Table 1.** Results on normal (not recognized) text

| Algorithm | Avg. completeness | Avg. conciseness |
|---|---|---|
| Context with window n = 4 | 6.2 | 7.64 |
| Context with window n = 5 | 6.74 | 7.3 |
| Flexible context extraction | 7.34 | 8.5 |

## 5.2 Recognition Output

Similar experiments were conducted with texts produced by Russian Automatic Speech Recognition system developed at Speech Technology Center. The recordings contained citizen complaints concerning social issues. The recognition accuracy on this data was about 80%. We extracted 223 contexts from 2000 sentences for 23 keywords. The results were evaluated again by 20 experts using the same criteria as in Sect. 5.1. The results are shown in Table 2.

**Table 2.** Results on recognized text

| Algorithm | Avg. completeness | Avg. conciseness |
|---|---|---|
| Context with window n = 4 | 7.59 | 7.44 |
| Context with window n = 5 | 7.92 | 7.24 |
| Flexible context extraction | 7.41 | 8.28 |

The proposed method was found to be slightly worse than its rivals by completeness criteria. We see several reasons for this:

- People often use short phrases while addressing to a call center so big part of information that is context relevant is contained in a n-words window.
- Spontaneous speech is different from reading or prepared speech what leads to syntactic parsing errors.
- Speech recognition errors lead to syntactic parsing errors as well.

In the future we plan to make a syntactic parser more robust in regard to aforementioned problems so it will work better with spontaneous speech recognition results.

## 6    Discussion and Future Developments

The system proposed in this paper is a pilot project that we are planning to develop in a number of ways. First, several important syntactic dependencies are not yet taken into account, such as words combined by coordinating conjunctions. An important issue to consider are preposition groups; some of them need to be included in contexts while others can safely be omitted, according to the type of relation between the group and its parent node and/or to the classes of the words involved. Furthermore, we plan to develop a flexible system where the context we extract can be made shorter or longer according to the user's needs.

Another area of development is including more advanced NLP methods into the context extraction procedure. Firstly, we plan to employ *anaphora* resolution, since it is included in the *SemSin* system. If we are dealing with simple keyword search, rather than the results of more sophisticated named entity extraction, we could utilize our knowledge of coreference and provide contexts not only for the occurrences of the keyword but also for the cases when it is replaced by a pronoun. A similar issue concerns cases when the keyword is embedded in a relative clause.

It may also be useful to indicate when the keywords are used in the context of someone's reported opinion. In the example (1) in Sect. 4.1, the context we extracted is actually embedded under the verb "informed", thus forming part of an opinion expressed by the subject of that verb. We could accompany such contexts by a special mark, or extend the context to include the subject and predicate of the main clause.

We would also like to explore the limitations of our method depending on recognition accuracy, and to make it more robust to recognition errors, especially errors in word endings, which are common in recognizer output and create problems by changing the syntactic roles of words in the sentence. Finally, we are planning to test the use of the extracted contexts in a clustering task [10], compared to using whole sentences or word strings.

## 7    Conclusion

We presented a system for extracting contexts for keywords found in text. In order to extract contexts flexibly, we propose analysing the sentence structure formed by a syntactic parser. We use the *SemSin* dependency parser for this purpose. The paper described the organization of the *SemSin* system which provides syntactic and semantic analysis of Russian text. Flexible context extraction algorithm was tested on prepared texts and speech recognition results. The algorithm outperforms n-words context window approach in terms of both completeness and conciseness on prepared texts and exhibits slightly lower completeness on spontaneous speech recognition results due to syntactic links deformation. In the future we plan to develop our system and make it more robust for use in different conditions.

**Acknowledgements.** The work was financially supported by the Ministry of Education and Science of the Russian Federation, Contract 14.579.21.0008, ID RFMEFI57914X0008.

# References

1. Beil, F., Ester, M., Xu, X.: Frequent term-based text clustering. In: Proceedings of the Eighth ACM SIGKDD International Conference on Knowledge Discovery and Data Mining, pp. 436–442. ACM (2002)
2. Mihalcea, R., Tarau, P.: A language independent algorithm for single, multiple document summarization. In: IJCNLP (2005)
3. Boyarsky, K., Kanevsky, E.: Vega - a system for text classification and analysis. LAP Lambert Academic Publishing, Saarbrücken (2011). in Russian
4. Boyarsky, K., Kanevsky, E.: The semantic-and-syntactic parser SemSin. In: Dialog 2012 (2012). http://www.dialog-21.ru/digest/2012/?type=doc. in Russian
5. Tuzov, V.A.: Computer semantics of the Russian language. Saint-Petersburg State University Publishing House, Saint-Petersburg (2004). in Russian
6. Covington, M.A.: A dependency parser for variable-word-order languages. Research Report (1990)
7. Nivre, J., Boguslavsky, I.M., Iomdin, L.L.: Parsing the SynTagRus treebank of Russian. In: Proceedings of the 22nd International Conference on Computational Linguistics, vol. 1, pp. 641–648. Association for Computational Linguistics (2008)
8. Chernykh, G., Korenevsky, M., Levin, K., Ponomareva, I., Tomashenko, N.: State level control for acoustic model training. In: Ronzhin, A., Potapova, R., Delic, V. (eds.) SPECOM 2014. LNCS (LNAI), vol. 8773, pp. 435–442. Springer, Heidelberg (2014). doi:10.1007/978-3-319-11581-8_54
9. Tomashenko, N., Khokhlov, Y.: Speaker adaptation of context dependent deep neural networks based on MAP-adaptation, GMM-derived feature processing. In: INTERSPEECH 2014 - Proceedings of the 15th Annual Conference of the International Speech Communication Association, pp. 2997–3001 (2014)
10. Popova, S., Krivosheeva, T., Korenevsky, M.: Automatic stop list generation for clustering recognition results of call center recordings. In: Ronzhin, A., Potapova, R., Delic, V. (eds.) SPECOM 2014. LNCS (LNAI), vol. 8773, pp. 137–144. Springer, Heidelberg (2014). doi:10.1007/978-3-319-11581-8_17

# WebVectors: A Toolkit for Building Web Interfaces for Vector Semantic Models

Andrey Kutuzov[1] and Elizaveta Kuzmenko[2(✉)]

[1] University of Oslo, Oslo, Norway
andreku@ifi.uio.no
[2] National Research University Higher School of Economics, Moscow, Russia
eakuzmenko_2@edu.hse.ru

**Abstract.** The paper presents a free and open source toolkit which aim is to quickly deploy web services handling distributed vector models of semantics. It fills in the gap between training such models (many tools are already available for this) and dissemination of the results to general public. Our toolkit, *WebVectors*, provides all the necessary routines for organizing online access to querying trained models via modern web interface. We also describe two demo installations of the toolkit, featuring several efficient models for English, Russian and Norwegian.

**Keywords:** Distributional semantics · Neural embeddings · Word2vec · Machine learning · Visualization

## 1 Introduction

In this paper we present *WebVectors*,[1] a free and open-source toolkit to deploy web services implementing vector semantic models, primarily word embeddings.

Vector models of distributional semantics are well established in the field of computational linguistics and have been here for decades (see [1] for an extensive review). However, recently they received substantially growing attention. The main reason for this is a possibility to employ artificial neural networks trained on large corpora to learn low-dimensional distributional vectors for words (word embeddings). The most well-known tool in this field now is *word2vec* and its Skip-Gram and CBOW algorithms, which allow fast training on huge amounts of raw linguistic data [2].

Word embeddings represent meaning of words, and can be of use in almost any linguistic task: named entity recognition [3], sentiment analysis [4], machine translation [5], corpora comparison [6], etc. Approaches implemented in *word2vec* and other similar tools are being extensively studied and tested in application to the English language: see [7] and many others. However, for many other languages the surface is barely scratched. Thus, it is important to facilitate research in this field and to provide access to relevant tools for various linguistic communities.

---

[1] https://github.com/akutuzov/webvectors.

© Springer International Publishing AG 2017
D.I. Ignatov et al. (Eds.): AIST 2016, CCIS 661, pp. 155–161, 2017.
DOI: 10.1007/978-3-319-52920-2_15

With this in mind, we release the *WebVectors* toolkit. It allows to quickly deploy a stable and robust web service for operations on vector semantic models, including querying, visualization and comparison, all available to users of any computer literacy level. It can be installed on any Linux server with a small set of standard tools as prerequisites, and generally works out-of-the-box. The administrator needs only to supply a trained model or models for one's particular language or research goal. The toolkit can be easily adapted for specific needs.

## 2 Deployment Basics

Technically, the toolkit is a web interface between distributional semantic models and a user. Under the hood, we use *Gensim* library [8] which deals with models' operations. The user interface is implemented in Python (*Flask* framework) and runs on top of a regular Apache HTTP server or as a standalone service (using, for example, *Gunicorn*). It communicates with *Gensim* (functioning as a daemon with our wrapper) via sockets, sending user queries and receiving back models' answers.

Such architecture allows fast simultaneous processing of multiple users querying multiple models over network. Models themselves are permanently stored in memory, eliminating time-consuming stage of loading them from hard drive every time there is a need to process a query.

*WebVectors* can be useful in a very common situation when one has trained a distributional semantics model for one's particular corpus or language (tools for this are now widespread and simple to use), but then there is a need to demonstrate one's results to general public. The setup process then is as follows:

1. install project and its dependencies at your Linux server according to the user guide (it is basically installing *Flask* and *Gensim* and copying *WebVectors* files to your web directory);
2. put your model(s) in '*models*' sub-directory of *WebVectors*;
3. change configuration files, stating the paths to your models;[2]
4. optionally change other settings;
5. run Python script to load models into memory and start daemon listening to queries via sockets;
6. run Apache or other web server you use, to start user interface listening to HTTP queries.

## 3 Main Features of WebVectors

Immediately after that you can interact with the loaded model via web browser. From a user's point of view, *WebVectors* is a semantic calculator which operates on relations between words in distributional models. In particular, users are able to:

---

[2] As of now, *WebVectors* supports models in generic *Word2vec* format (which is essentially a simple list of word vectors, in text or binary form) and *gensim* format (it is always binary and retains much more technical data, including output vectors).

1. find **semantic associates**: words semantically closest to the query word (results are returned as lists of words with corresponding similarity values);
2. calculate exact **semantic similarity** between pairs of words (results are returned as similarity values, in the range between −1 and 1);
3. apply simple **algebraic operations** to word vectors: addition, subtraction, finding average vector for a group of words (results are returned as lists of words nearest to the product of the operation and their corresponding similarity values); this can be used for analogical reasoning, widely known as one of the most interesting features of word embeddings [2];
4. **visualize** word vectors and their geometrical relations;
5. get the **raw vector** (array of real values) for the query word.

All these operations can optionally employ part-of-speech filters. Of course, to this end the model must be trained on a PoS-tagged corpus and must differentiate between homonyms belonging to different parts of speech. Also, in this case *WebVectors* can use an external tagger to detect PoS of the query words, if not stated explicitly by the user. By default, Freeling suite of linguistic analyzers [9] is employed for morphological processing: main reasons for choosing Freeling is that it is open-source, supports multiple languages and provides thread-safe parallel query processing, at the same time featuring sufficient accuracy (about 98% for English). However, one can easily adapt *WebVectors* to use any PoS-tagger of one's own choice.

Another feature of the toolkit is the possibility to use several models simultaneously. If several models are enumerated in the configuration file, the *Web-Vectors* daemon loads all of them. At the same time, the user interface allows to choose one of featured models or several at once. The results (for example, lists of nearest semantic associates) for different models are then presented to user side-by-side. Thus, it is convenient to conduct research related to comparing several distributional semantic models (trained on different corpora or with different hyperparameters).

For some categories of users it may be important that *WebVectors* web GUI is HTML5-compliant and fully supports mobile devices. It is also inherently multilingual: extending the interface with another language is pretty straightforward, demanding only to add an entry for the new language in the configuration file and to add new translations for text strings in the localization file (we provide English and Russian translations).

In the spirit of the Semantic Web paradigm, each word in each model has its own unique URI (Uniform Resource Identifier) explicitly stating lemma, model and PoS: for example, http://example.com/webvectors/your_model/boot_N. In response to requests for these addresses, we return a special page for this word in this model, providing lists of the nearest semantic associates which belong to the same PoS as the lemma itself (if PoS-aware model is used). Additionally, the word vector and its visualization are shown, complete with links to search for the word on the Internet or in the Wiktionary.

Web interface and neat HTML5 pages can be good for initial exploration into the data or for live demos, but real studies often demand large-scale querying of models. That is why *WebVectors* returns not only human-readable results, but also provides simple API. Using this, one can query the service from one's own application and receive results in the form of tab-separated text file or JSON.

## 3.1  Visualization Possibilities

*WebVectors* provides two kinds of visualizations: for vectors of single words and for inter-relations between several words in the model. Single word visualizations are simple plots of corresponding $n$-dimensional vectors. They can be useful for explaining how distributed semantic models work, for the audience which is not math-savvy. The Fig. 1 demonstrates such a plot for the 300-dimensional vector of the word 'лингви стика' *linguistics* from the model trained on Russian National Corpus.

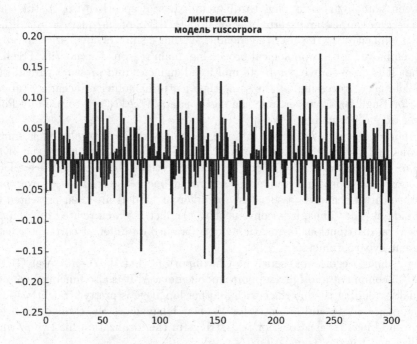

**Fig. 1.** Visualization of a single word vector

Multiple words visualizations are implemented using the well-known t-SNE algorithm [10] and project complex semantic relationships into the 2-dimensional space, possibly providing useful insights into the data structure. The algorithm tries to keep as much information about high-dimensional geometry as possible. These plots are shown for queries consisting of 7 or more words (with less words visualizations usually being not informative).

An example of such a plot for words '*mouse, keyboard, computer, laptop, aircraft, vehicle, car, tank, wine, beer, whisky*' in a model trained on Google

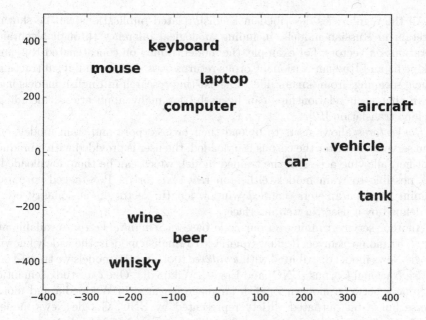

**Fig. 2.** Words relations visualization using t-SNE

News data set is shown in the Fig. 2. One can see how three groups of words are clustered in different parts of the plot: computer hardware, alcoholic beverages and transportation means.

Note that the layout of points in t-SNE plots is an approximate projection of real word layout in the multidimensional vector space of the models. Thus, distances between these points also only approximately reflect cosine similarities between high-dimensional word vectors.

## 4   Live Demos

As stated above, *WebVectors* can facilitate semantic research for languages which are less subject to computational linguists' attention. One running installation of our toolkit proves this: this is *RusVectores* service available at http://ling. go.mail.ru/dsm/. The service is already being employed in academic studies in computational linguistics [11] and digital humanities (several research projects in process as of now).

It features four models for Russian trained on different corpora. Note that prior to training, each word token in the corpora was not only lemmatized, but also augmented with a marker denoting its part of speech (for example, 'печь _V' for the verb 'to bake' which in Russian is homonymous to the noun 'furnace'). This linguistic preprocessing lends the models the ability to better handle rich morphology of Russian. In addition, it allows issuing PoS-aware queries like 'What are the most semantically similar verbs to this noun?'.

All the regularities described in English-related publications can be shown to retain in Russian models, including analogical inference through algebraic operations on vectors. For example, the model trained on concatenated Russian Wikipedia and Russian National Corpus returns быт 'daily round' if subtracting смысл 'meaning' from жизнь 'life'. The existing research in English models has shown that such relationships can be useful for many applications, including machine translation [12].

*RusVectores* allows users to upload their own corpora and train models on them server-side. After the corpus is uploaded, the user is provided with a unique identifier, allowing access to the trained model, which can be then downloaded. It is possible to train models either on raw texts or on PoS-tagged corpora. Training hyperparameters (context window length, vector dimensionality, etc.) are defined by a user via web interface.

Another service running on our code base is *Semantic Vectors* available at http://ltr.uio.no/semvec. It allows queries to 3 English models: the widely known Google News model distributed with *word2vec* tool, and the models we trained on British National Corpus (BNC) and English Wikipedia. One can study semantic differences between modern English language featured in Wikipedia and more diverse but a bit outdated variety represented by BNC. Google News model is trained on a very large corpus (100 billion words), but lacks linguistic pre-processing, which in turn leads to more ways of performing interesting comparisons. Additionally, *Semantic Vectors* features a model trained on the corpus of Norwegian news texts, *Norsk aviskorpus* [13]. To our knowledge, this is the first neural embedding model for Norwegian made available online.

One can use the aforementioned services as live demos to evaluate the *WebVectors* toolkit before actually employing it in one's own workflow.

## 5    Conclusion

The main aim of *WebVectors* is to quickly deploy web services processing queries to vector semantic models, independently of a particular language. It allows to make complex linguistic resources available to wide audience in almost no time. The authors plan to continue adding new features aiming at better understanding of embedding models, including sentence similarities, text classification and analysis of correlations between different models for different languages.

We are striving to lower the entry threshold for the distributional semantics field. Neural word embeddings seem to be a very promising approach to many NLP tasks that can be widely used. At the same time, it is important that researchers (especially linguists) with no solid programming background were able to use these powerful approaches: both in their studies and in disseminating their results. It is particularly true for a very popular discipline of digital humanities, where convenient web access to data is paramount. All these aims can be achieved with *WebVectors*.

Finally, we believe that the presented toolkit can popularize distributional semantics and computational linguistics among general public. Services based on it can promote interest among present and future students and help to make the field more compelling and attractive.

# References

1. Turney, P.D., Pantel, P., et al.: From frequency to meaning: vector space models of semantics. J. Artif. Intell. Res. **37**(1), 141–188 (2010)
2. Mikolov, T., Sutskever, I., Chen, K., Corrado, G.S., Dean, J.: Distributed representations of words and phrases and their compositionality. Adv. Neural Inf. Process. Syst. **26**, 3111–3119 (2013)
3. Siencnik, S.K.: Adapting word2vec to named entity recognition. In: Nordic Conference of Computational Linguistics, NODALIDA 2015, p. 239 (2015)
4. Maas, A.L., Daly, R.E., Pham, P.T., Huang, D., Ng, A.Y., Potts, C.: Learning word vectors for sentiment analysis. In: Proceedings of the 49th Annual Meeting of the Association for Computational Linguistics: Human Language Technologies, vol. 1, pp. 142–150. Association for Computational Linguistics (2011)
5. Zou, W.Y., Socher, R., Cer, D.M., Manning, C.D.: Bilingual word embeddings for phrase-based machine translation. In: EMNLP, pp. 1393–1398 (2013)
6. Kutuzov, A., Kuzmenko, E.: Comparing neural lexical models of a classic national corpus and a web corpus: the case for Russian. In: Gelbukh, A. (ed.) CICLing 2015. LNCS, vol. 9041, pp. 47–58. Springer, Heidelberg (2015). doi:10.1007/978-3-319-18111-0_4
7. Baroni, M., Dinu, G., Kruszewski, G.: Don't count, predict! a systematic comparison of context-counting vs. context-predicting semantic vectors. In: Proceedings of the 52nd Annual Meeting of the Association for Computational Linguistics, vol. 1 (2014)
8. Řehůřek, R., Sojka, P.: Software framework for topic modelling with large corpora. In: Proceedings of the LREC 2010 Workshop on New Challenges for NLP Frameworks, Valletta, Malta, ELRA, pp. 45–50, May 2010
9. Padró, L., Stanilovsky, E.: Freeling 3.0: towards wider multilinguality. In: Proceedings of the Eight International Conference on Language Resources and Evaluation (LREC 2012), Istanbul, Turkey, European Language Resources Association (ELRA), May 2012
10. Van der Maaten, L., Hinton, G.: Visualizing data using t-SNE. J. Mach. Learn. Res. **9**(2579–2605), 85 (2008)
11. Kutuzov, A., Andreev, I.: Texts in, meaning out: neural language models in semantic similarity task for Russian. In: Proceedings of the Dialog Conference, Moscow, RGGU (2015)
12. Mikolov, T., Le, Q., Sutskever, I.: Exploiting similarities among languages for machine translation. arXiv preprint arXiv:1309.4168 (2013)
13. Hofland, K.: A self-expanding corpus based on newspapers on the web. In: LREC (2000)

# Morphological Analysis for Russian: Integration and Comparison of Taggers

Elizaveta Kuzmenko[✉]

National Research University Higher School of Economics, Moscow, Russia
eakuzmenko_2@hse.edu.ru

**Abstract.** In this paper we present a comparison of three morphological taggers for Russian with regard to the quality of morphological disambiguation performed by these taggers. We test the quality of the analysis in three different ways: lemmatization, POS-tagging and assigning full morphological tags. We analyze the mistakes made by the taggers, outline their strengths and weaknesses, and present a possible way to improve the quality of morphological analysis for Russian.

**Keywords:** Morphological analysis · Russian · POS-tagging · Gold standard

## 1 Introduction

In this paper we present the results of testing different morphological taggerss for the Russian language. Russian is a highly inflective and morphologically rich language, and developing high-quality morphological tools for Russian presents a serious problem even for advanced researchers.

A considerable number of taggers provide morphological disambiguation while performing POS-tagging for Russian, but all of them are erroneous in some way. However, this disadvantage can be beneficial since the taggers make errors in different issues: when one analyzer fails, another may guess the correct tag. Therefore, it could be very useful to inspect the performance of each tagger and reveal the specificity of the mistakes it makes. These findings can then help to build an improved tagger for Russian that will combine in itself all the forces of other taggers. The near future of morphological analysis of Russian, as we see it, is meta-learning, in which all the cases where taggers guess tags correctly are taken and all the cases where the taggers make errors are omitted.

The question is then: do the cases where taggers make errors overlap or not? We answer this question in our paper via the experiment in which we build a gold standard corpus and compare the tags found in this corpus to those that are output by our taggers. In case of discrepancy, we analyze the cause of an error.

The structure of this paper is as follows: in Sect. 2 we describe the previous work in this field: how the standards for morphology annotation were defined and what the specific morphology problems for the Russian language are. We also give

© Springer International Publishing AG 2017
D.I. Ignatov et al. (Eds.): AIST 2016, CCIS 661, pp. 162–171, 2017.
DOI: 10.1007/978-3-319-52920-2_16

an overview of the instruments developed for Russian: taggers *Freeling*, *Pymorphy*, *MyStem* and *TreeTagger*, and describe previous attempts to compare their performance. In Sect. 3 we present an experiment in comparing the taggers: analyze the differences in the tagsets and define the rules for unification of morphological tags. In Sect. 4 we present the results of our experiment, and in Sect. 6 we discuss these results and propose the way towards organizing meta-learning of the taggers.

## 2   Background

The Russian language presents certain problems with regard to morphology annotation, because it is a highly inflectional language with many grammatical categories. There is no standard even for part-of-speech annotation, let alone subtle grammatical categories such as (im)perfectiveness and animacy. Theoretical disputes concerning Russian morphology lead to variety of solutions for morphology annotation – from positional tags following the MULTEXT-East guidelines [1] to combinations of tags employed in RNC[1]. An additional problem arises from the fact that tags in Russian can be combined and simplified in different ways. Some systems do not account for one or another grammatical category (for example, transitivity in *TreeTagger*), whereas other systems define some value of a category as the default one. Thus, the active voice of verbs is not marked in *pymorphy*, which, in its turn, follows the *OpenCorpora* guidelines [2].

The comparison of taggers for Russian is also complicated by the fact that there are different theoretical traditions for the lemmatization process. For example, some taggers count for verbs as lemmas of participles, and other taggers lemmatize participles as adjectives.

In addition to not having unified rules for morphology annotation in Russian, until recently there were no standard golden corpus of any kind. Presently, there are two corpora that could serve as models for annotation tasks: a disambiguated subcorpus of *RNC* and *Opencorpora* [2]. Moreover, there has been organized the *RU-EVAL* shared task [3], in which the participants proposed unification rules for the output of different morphological taggers and created a gold standard corpus consisting of 3 thousand word tokens.

The taggers used in our experiment are the following ones:

– *MyStem*[2] [4] is a morphological analyzer with disambiguation developed for the Russian language by Ilya Segalovich and Vitaliy Titov at "Yandex". In the core of the software lies a dictionary that helps generate morphological hypotheses for both known and unknown words.
– *Pymorphy2*[3] [5] is a morphological analyzer developed for the Russian language by Mikhail Korobov on the basis of OpenCorpora dictionaries. PyMorphy2 is written fully in the Python programming language and is able to

---

[1] http://ruscorpora.ru/.
[2] https://tech.yandex.ru/mystem/.
[3] https://pymorphy2.readthedocs.org/en/latest/.

normalize, decline and conjugate words, provide analyses or give predictions for unknown words.

- *Freeling* [6] is a set of open source linguistic analyzers for several languages. It features tokenizing, sentence splitting, morphology analyzers with disambiguation, syntax parsing, named entity recognition, etc. In this research, we use only morphological analyzer for Russian.
- *TreeTagger* [7,8] is a language independent part-of-speech tagger developed by Helmut Schmid. TreeTagger is based on decision trees and should be trained on a lexicon and a manually tagged training corpus. The program can annotate texts with part-of-speech and lemma information.

These are not all existing morphological analyzers for Russian. The choice of taggers for the comparison was motivated by their availability. For example, the *TnT* tagger, which has trained models for Russian [9], is not freely available, and we faced some problems when obtaining it from the developers. However, our work still tests all major analyzers for Russian.

Within the chosen set of analyzers, there are several issues connected to their comparability. Apart from different guidelines for lemmatization and assigning morphological categories, the taggers also feature various algorithmic designs. Thus, *pymorphy* analyzes tokens separately, without taking the context into account, whereas other analyzers determine the word characteristics from its neighboring words. However, we do not judge from the developer's point of views and do not evaluate the efficiency of various POS-tagging techniques. We take each tagger as a final product and estimate their efficiency from the user's point of view.

## 3   Experiment Design

In this work we evaluate the taggers' performance on two gold standard sets. The first set is the disambiguated subcorpus of the RNC, and the focus of evaluation is on the strict correspondence between taggers' output and the RNC data. The second set is taken from the *RU-EVAL* competition [3]. In this case we do not strictly follow the RNC guidelines and do not count the absence of some categories in the output as an error (for example, the absence of the active voice for verbs in the *pymorphy* analysis is not taken into account), and the resulting figures can be considered more objective.

The disambiguated subcorpus of Russian National Corpus contains 5.9 million tokens, annotated morphologically with the help of *MyStem* and further disambiguated and refined by hand. All tokens have only one morphological analysis, and the tagset in this corpus generally complies to the one developed for *MyStem*. An example of an annotated sentence can be found below.

```
<se>
<w><ana lex="береза" gr="S,f,inan=sg,nom"/>Берёза</w>
<w><ana lex="ждать" gr="V,ipf,tran,act=sg,praes,3p,indic"/>ждёт</w>
<w><ana lex="мороз" gr="S,m,inan=sg,gen"/>мор'оза</w>!"
</se>
```

However, if we choose only RNC as the gold standard, this leads to some limitations. First, Tretagger was trained on the disambiguated subcorpus of RNC, so it has some advantage compared to other taggers. Second, RNC has a very balanced and detailed tagset, but it is sensible to exclude some grammatical categories from the analysis, as they are highly important only for purely linguistic tasks. Thus, we also use the second gold standard set from the *RU-EVAL* task. This set contains 3300 tokens, annotated by hand. An example of an annotated sentence can be found below.

```
как как CONJ
казалось казаться V n,past,sg
раньше раньше ADV
```

One of the problems in our experiment is that all analyzers have different notations for parts of speech and morphological categories. The discrepancies between the tagsets can be of different kinds:

– **Some morphological category is present in the tagset of the gold standard but absent in the tagset of another morphological analyzer**: for example, *Mystem* distinguishes between animacy and inanimateness as it has specific dictionaries where these characteristics are defined for every word. *TreeTagger*, however, does not consider this feature to be important and does not include it in the analysis.
– **Morphological analyzers have different standards concerning part of speech identification**: for example, *Freeling* identifies participles as a separate part of speech, whereas other morphological analyzers identify participles as verbal forms.
– Consequently, **alongside with different standards towards part of speech identification, parsers assign different lemmas to tokens problematic in this aspect**: therefore, the lemma for the word '*сделанной*' would be '*сделанный*' in *Freeling* and '*сделать*' in *Mystem*.
– **If the part of speech is identified uniformly by the taggers, there still can be problems with lemmatization**: for example, *TreeTagger* assigns one and the same lemma to Russian verbs in different aspects, and so does *Freeling*. For example, the verbs '*выплывать*' and '*выплыть*' will be assigned one and the same lemma '*выплывать*', even if the aspect of a given word instance is reflected in its analysis. At the same time, other tagsets do not require the aspect to be changed in the process of lemmatization.

Due to these problems, we need to define conventions that will allow to make comparison of the taggers possible and more correct. As our gold standard is annotated by *MyStem*, we decided to convert all our tags into *MyStem* tags. The rules of conversion are presented in Table 1.

The rules for conversion into the *RU-EVAL* tagset were the same. In addition, we excluded from the analysis the following cases:

**Table 1.** Rules for conversion of the tagset into the tagset defined for *RNC*

| Gold standard tag | Tag counted as correct |
|---|---|
| A-NUM (numeric adj.) | NUM (numeral) |
| PARENTH(parenthesis) | ADV(adverb) |
| ADV-PRO (adv.-pronoun) | PRO (pronoun) |
| A-PRO (adj.-pronoun) | PRO (pronoun) |
| m-f (common gender) | both are correct |
| anim (animacy) | not important |
| inan (inanimateness) | not important |
| dat2 (the 2nd dative) | dat (dative) |
| gen2 (the 2nd genitive) | gen (genitive) |
| acc2 (the 2nd accusative) | acc (accusative) |
| loc2 (the 2nd locative) | loc (locative) |
| adnum (count form) | NUM (numeral) |
| intr (intransitiveness) | not important |
| tran (transitiveness) | not important |

– absence of voice and mood for verbs;
– confusion between predicates and other parts of speech;
– verbs which end with '*ся*';
– numerals;
– distinction between full and shortened forms for adjectives and participles.

In general, these rules mean that we accept as the right output less specific tags, for example, *dat* (the dative case) instead of *dat2* (the second dative). This leads to loss of some linguistic information, but accounts for the tagsets with less strict linguistic background. The rules for the RU-EVAL tagset in addition eliminate cases when the taggers' results differ because of tagging guidelines.

However, these are the rules only for the least problematic cases. The most problematic cases include, as it was mentioned earlier, lemmatization of participles and perfective verbs. These issues we solve by assigning lemmas given by the analyzer and taking the tag itself from another analyzer. In addition, we do not consider identifying patronyms, zoonyms and other lexical classes to be of importance for the task of morphological analysis and exclude them from our experiment.

The experimental procedure itself was as follows:

1. take the text files from the gold standard corpus and extract the tokens and their morphological characteristics;
2. analyze the tokens by the taggers in question;
3. convert the output into the RNC tagset;
4. compare token by token the output from the tagger to the morphological analysis found in the gold standard corpus.

## 4    Evaluation

For each word we compared the analyses of the three taggers and the analysis given in the gold standard corpus. In particular, we checked whether the part of speech was the same and if the set of grammatical categories contained in the tag was identical to the gold standard. There were three modes of evaluation:

1. checking the correspondence between assigned lemmas;
2. checking the correspondence between assigned parts of speech;
3. checking the correspondence between assigned morphological tags in the whole.

If the lemma, the part of speech or the tag output by the tagger agreed with the gold standard, the answer of the tagger was counted as correct for the corresponding evaluation mode. Thus, the performance of the taggers was evaluated using the accuracy metric, roughly, the proportion of correct answers given by a tagger. Table 2 presents the results for all our taggers in three modes and two sets.

**Table 2.** Evaluation of the taggers' performance

| Tagge | Mode | Accuracy | |
|---|---|---|---|
| | | RNC | RU-EVAL |
| Freeling | lemma | 0.822 | 0.816 |
| | POS | 0.907 | 0.911 |
| | full tag | 0.833 | 0.851 |
| Pymorphy | lemma | 0.882 | 0.871 |
| | POS | 0.915 | 0.904 |
| | full tag | 0.647 | 0.742 |
| TreeTagger | lemma | 0.970 | 0.869 |
| | POS | 0.952 | 0.882 |
| | full tag | 0.924 | 0.863 |

As it can be seen from the Table 2, all the taggers present decent results, but none of them perform without mistakes. *TreeTagger* was trained on the disambiguated subcorpus of RNC, and after we apply it to the *RU-EVAL* gold standard, the quality of its analysis gets worse. Other taggers perform slightly better in the *full tag* mode because of milder error criteria.

## 5    The Analysis of the Errors

After evaluating overall taggers' performance, let's have a look at the nature of the errors.

As it was said earlier, the variety of annotation guidelines makes the very notion of error in this task very ambiguous. Should a particular case of discrepancy between two taggers be attributed to the bad performance of one of them or to the differences in their guidelines? For example, if some tagger analyzes the word '*здесь*' as a predicate, and another taggers considers it as an adverb, which answer is the right one? Or, as it was described earlier, if *pymorphy* presupposes the active voice for all verbs and doesn't explicitly mark this, is this the underperformance or a tagger's feature?

There can a lot of reasoning on these grounds, and no resolution can be considered as accurate. For the purpose of our analysis we count all cases of discrepancies between the gold standard and another tagger to be errors. The tagset designed for RNC is very exhaustive and detailed, and any differences which are not taken into account by conversion rules signify either the loss of information or a proper error. Thus, the absence of active voice in the analyses of *pymorphy* is considered to be an error, as well as different representations for parts of speech (for example, analyzing a substantivized adjective '*новое*' as a noun or an adjective).

We do not claim for our definition of an error to be the ground truth. Other conventions for the correspondences between tagsets can lead to alternative figures. However, we take our decision for a balanced one and appealing to the task of morphological analyzing compliant with the RNC tagset.

Table 3 gives the figures for the taggers performance in *POS* and *lemma* modes with regard to the POS tag of a given word as determined by the tagger. The gold standard set in this task was the disambiguated subcorpus of RNC. Thus, for all words analyzed as nouns by *Freeling*, 4 % of them proved not to be nouns in the gold standard set, and almost 17 % of them did not match the gold standard lemma.

These figures allow to draw several interesting conclusion about the taggers' performance.

- The main parts of speech (such as nouns, verbs, and adjectives) are less prone to errors while tagging. The same is true in most cases for auxiliary parts of speech, as they form a closed subset.
- The heel of Achilles for the taggers are such parts of speech as pronouns of different types and categories on the border between two parts of speech. In these cases it is likely that taggers would have different tagging guidelines.
- purely erroneous tagging of a particular part of speech indicates that either this category is absent in the tagset or it is tagged as another POS. Probably, such cases should be eliminated from the analysis and evaluation.
- striking difference between error rate for POS and for lemma implies that there is a conflict between lemmatizing standards. Thus, 42% of wrong answers for verb lemmas in the *Freeling* data can be attributed mostly to the change of aspect. This is probably should be eliminated from the analysis as well, or the lemmas should be defined uniformly with the help of a dictionary of aspectual pairs.

**Table 3.** Proportion of wrong answers given in **POS** and **lemma** modes depending on the POS of a token

| POS | Freeling | | Pymorphy | | TreeTagger | |
|---|---|---|---|---|---|---|
| | POS | lemma | POS | lemma | POS | lemma |
| S | 0.039 | 0.166 | 0.043 | 0.080 | 0.027 | 0.092 |
| S-PRO | 0.144 | 0.090 | 0.075 | 0.077 | 1 | 0.114 |
| V | 0.018 | 0.422 | 0.022 | 0.029 | 0.024 | 0.377 |
| ADJ | 0.175 | 0.233 | 0.115 | 0.126 | 0.094 | 0.197 |
| ADJ-PRO | 0.085 | 0.085 | 0.195 | 0.181 | 1 | 0.117 |
| ADJ-NUM | 1 | 0.015 | 0.026 | 0.995 | 0.238 | 0.006 |
| ADV | 0.377 | 0.098 | 0.426 | 0.055 | 0.226 | 0.009 |
| ADV-PRO | 0.059 | 0.001 | 0.762 | 0.558 | 0.009 | 0.002 |
| PR | 0.004 | 0.001 | 0.008 | 0.034 | 0.002 | 0.001 |
| CONJ | 0.055 | 0.008 | 0.233 | 0.033 | 0.060 | 0.008 |
| NUM | 0.054 | 0.112 | 0.005 | 0.005 | 0.042 | 0.098 |
| PART | 0.061 | 0.004 | 0.191 | 0.014 | 0.008 | 0.001 |
| INTJ | 0.279 | 0.177 | 0.728 | 0.516 | 0.118 | 0.113 |

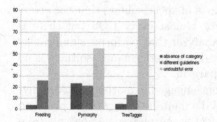

**Fig. 1.** Causes of errors among the tag-gers

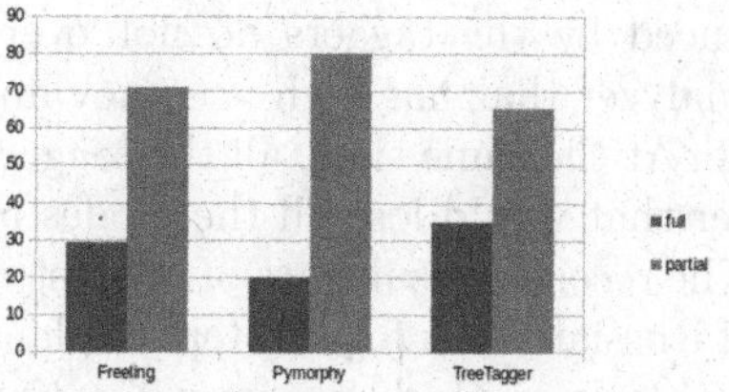

**Fig. 2.** The percentage of full or partial errors

All in all, it can be seen from Table 3 that the errors produced by taggers do not overlap in most cases. Trusting *pymorphy* on its output for interjections (INTJ) is not the best option, whereas TreeTagger shows high results in this case. On the contrary, *pymorphy* has the lowest percentage of errors for numerals (both lemma and POS) while other taggers stand down for this part of speech. This makes possible the meta-learning technique we described in the beginning of the present paper.

As for the errors made in the *full tag* mode, they are less dependent on POS. Besides, the main parts of speech (nouns, verbs, adjective) have more grammatical characteristics and thus give more space for errors. We performed the analysis of error causes in 500 erroneous cases for every tagger by hand. These results can also be of interest, though they are not formalized.

Figure 1 depicts the percentage of different error causes among the taggers. **Absence of category** accounts for cases when tags did not match because some category is not present in an analyzer's tagset. This is the case, for example, for the absence of the active voice in *pymorphy*. **Different guidelines** refers to the cases when two taggers treat the word differently because of diverse approaches to the issue. For example, it accounts for the confusion between *predicate* and *adverb* parts of speech.

Figure 2 demonstrates the percentage of 'full' and 'partial' errors in tags. Full errors are mostly represented by confusion between POS tags. If a word is tagged as a noun by TreeTagger, and in the gold standard it is an adjective, that would be the 'full' error. 'Partial' errors concern one or two categories that do not match the gold standard tag. This is the case with mismatch between assigned cases or gender.

# 6    Conclusion

In this paper we presented an analysis of the performance of three taggers for Russian. The comparison procedure was performed in three modes: assigning the POS tag, assigning lemma and assigning the full tag. Apart from evaluating the accuracy of each tagger, we analyzed the errors made by the taggers. The proportion of errors connected to different parts of speech shows that the errors produced by the taggers do not overlap. For almost every POS tag there is an analyzer that has high accuracy and an analyzer that performs significantly worse. At the same time, all the taggers show decent performance, so there is no tagger that would lose all the modes of comparison.

The received results are of interest to anyone engaged in morphological analysis of Russian. As a future step we plan to build a meta-learning system based on several taggers. Such system will take as input the morphological analyses from several taggers, identify which tagger provides the best guess for each particular case, and give as output the combination of correct variants. We expect this system to be highly accurate.

As the future work, apart from building an analyzer with meta-learning, we plan to investigate more thoroughly in which cases the taggers are more prone to errors, and what are the exact causes of these errors for every analyzer.

**Acknowledgments.** I would like to thank Elmira Mustakimova, Svetlana Toldova and Timofey Arkhangelskiy for their participation in the project. I am also grateful to Mikhail Korobov for his valuable remarks on *pymorphy* performance and explanations of error causes (and for developing *pymorphy*, of course).

This article is an output of a research project implemented as part of the Basic Research Program at the National Research University Higher School of Economics (HSE).

# References

1. Erjavec, T.: Multext-east version 3: multilingual morphosyntactic specifications, lexicons and corpora. In: LREC (2004)
2. Bocharov, V., Bichineva, S., Granovsky, D., Ostapuk, N., Stepanova, M.: Quality assurance tools in the opencorpora project. In: Proceeding of the International Conference on Computational Linguistics and Intelligent Technology, Dialog 2011, pp. 10–17 (2011)
3. Astaf'eva, I., Bonch-Osmolovskaya, A., Garejshina, A., Grishina, J., D'jachkov, V., Ionov, M., Koroleva, A., Kudrinsky, M., Lityagina, A., Luchina, E., et al.: NLP evaluation: Russian morphological parsers. In: Proceedings of Dialog Conference, Moscow, Russia (2010)
4. Segalovich, I.: A fast morphological algorithm with unknown word guessing induced by a dictionary for a web search engine. In: MLMTA, Citeseer, pp. 273–280 (2003)
5. Korobov, M.: Morphological analyzer and generator for Russian and Ukrainian languages. In: Khachay, M.Y., Konstantinova, N., Panchenko, A., Ignatov, D.I., Labunets, V.G. (eds.) Analysis of Images, Social Networks and Texts. Communications in Computer and Information Science, vol. 542, pp. 320–332. Springer, Heidelberg (2015)
6. Padró, L., Stanilovsky, E.: Freeling 3.0: towards wider multilinguality. In: LREC2012 (2012)
7. Schmid, H.: Improvements in part-of-speech tagging with an application to German. In: Proceedings of the ACL SIGDAT-Workshop, Citeseer (1995)
8. Schmid, H.: Probabilistic part-of-speech tagging using decision trees. In: Proceedings of the International Conference on New Methods in Language Processing, vol. 12, pp. 44–49. Citeseer (1994)
9. Sharoff, S., Kopotev, M., Erjavec, T., Feldman, A., Divjak, D.: Designing and evaluating a Russian tagset. In: LREC (2008)

# Anti-spoofing Methods for Automatic Speaker Verification System

Galina Lavrentyeva[1,2](✉), Sergey Novoselov[1,2], and Konstantin Simonchik[1,2]

[1] Speech Technology Center Limited, St. Petersburg, Russia
{lavrentyeva,novoselov,simonchik}@speechpro.com
[2] ITMO University, St. Petersburg, Russia
http://www.speechpro.com
http://www.ifmo.ru

**Abstract.** Growing interest in automatic speaker verification (ASV) systems has lead to significant quality improvement of spoofing attacks on them. Many research works confirm that despite the low equal error rate (EER) ASV systems are still vulnerable to spoofing attacks. In this work we overview different acoustic feature spaces and classifiers to determine reliable and robust countermeasures against spoofing attacks. We compared several spoofing detection systems, presented so far, on the development and evaluation datasets of the Automatic Speaker Verification Spoofing and Countermeasures (ASVspoof) Challenge 2015. Experimental results presented in this paper demonstrate that the use of magnitude and phase information combination provides a substantial input into the efficiency of the spoofing detection systems. Also wavelet-based features show impressive results in terms of equal error rate. In our overview we compare spoofing performance for systems based on different classifiers. Comparison results demonstrate that the linear SVM classifier outperforms the conventional GMM approach. However, many researchers inspired by the great success of deep neural networks (DNN) approaches in the automatic speech recognition, applied DNN in the spoofing detection task and obtained quite low EER for known and unknown type of spoofing attacks.

**Keywords:** Spoofing · Anti-spoofing · Spoofing detection · Speaker verification

## 1 Introduction

Biometrics technologies play an essential role in restricting access to informational resources in today's world. One of the reliable approaches of guarding access to important data is speaker recognition. Speaker recognition systems are widely used in customer identification during call to a call center, passive identification of a possible criminal using a preset "black list", Internet-banking systems and other fields of e-commerce.

© Springer International Publishing AG 2017
D.I. Ignatov et al. (Eds.): AIST 2016, CCIS 661, pp. 172–184, 2017.
DOI: 10.1007/978-3-319-52920-2_17

Automatic speaker verification systems aim to detect if the utterance belongs to the real speaker registered in the system or to the impostor. Although performance of automatic speaker verification (ASV) techniques has improved in recent years, they are still acknowledged to be vulnerable to spoofing attacks.

There are two types of spoofing attacks on the ASV systems: direct attack and indirect attacks. Indirect attacks require access permission to the system and can be applied to the inner modules (feature extraction module, voice models or classification results), while direct attacks focus only on the input data and are more likely to be used by criminals due to implementation simplicity. The most well-known spoofing attacks are "Impersonation", "Replay attack", "Cut and paste" [1]. But the most threatful are speech synthesis and voice conversion approaches. Voice conversion is the process of modifying a speech signal of the source speaker to sound like the target speaker. Speech synthesis is the computer-generated simulation of human speech.

Despite the development of new robust spoofing detection methods, most of them depend on a training dataset related to a specific spoofing attack. In real cases the nature of spoofing attack is unknown, that is why generalized spoofing detection methods are very important [2]. That was the motivation for researches from University of Eastern Finland to organize the Automatic Speaker Verification Spoofing and Countermeasures (ASVspoof) Challenge 2015 [2] in order to support the development of new spoofing detection algorithms, where we also presented our systems for spoofing detection and achieved 2nd result.

In this paper we concentrate on the investigating the most appropriate front-end features and classifiers for the spoofing detection system, which is effective in stand-alone spoofing detection task. In particular, we investigated anti-spoofing systems (ASS) introduced on the ASVspoof Challenge 2015 and compared results proposed by its authors. The aim of our research was to find the most effective method for detecting unknown spoofing attacks.

## 2   ASVspoof Challenge 2015

ASVspoof Challenge was organized by Zhizheng Wu, Tomi Kinnunen, Nicholas Evans and Junichi Yamagishi from University of Eastern Finland in 2015 to encourage the research work in spoofing detection field and stimulate the development of generalised countermeasures. According to [2] the main aim of the Challenge was to generalize the proposed spoofing detection systems on the base of their vulnerability results on one common database with varying spoofing attacks. The data set includes genuine and spoofed speech generated by 10 different spoofing algorithms using voice conversion and speech synthesis. The main purpose was to provide an opportunity to develop generalized countermeasures trained on the known type of attacks and test them on the unknown spoofing attacks.

## 2.1    Training, Development and Evaluation Data Sets

The training data set contains 3750 genuine and 12625 spoofed utterances collected from 25 speakers (10 male, 25 female). The development data set consists of 3497 genuine and 49875 spoofed trials from 35 speakers (15 male, 20 female). To generate spoofed utterances 5 spoofing methods (called known attacks) [2] were selected because of their simple implementation:

- S1 - simplified frame selection algorithm, based on voice conversion. The converted speech is generated by selecting target speech frames
- S2 - voice conversion algorithm which adjusts the first mel-cepstral coefficient to shift the slope of the source spectrum to the target
- S3, S4 - speech synthesis system based on Hidden Markov model with speaker adaptation techniques by 20 (S3) and 40 (S4) adaptation utterances
- S5 - voice conversion (using voice conversion toolkit and Festvox system)

The evaluation data set contains 9404 genuine and 184000 spoofed utterances from 46 speakers (20 male and 26 female). Spoofed trials were generated by 5 methods, used for development and training sets and additional 5 spoofing methods for unknown attacks [2]. The additional methods were:

- S6 - voice conversion algorithm based on joint density GMM and maximum likelihood parameter generation considering global variance
- S7 - voice conversion algorithm similar to S6, using line spectrum pair for spectrum representation
- S8 - tensor-based approach to voice conversion, using Japanese set for speaker space construction
- S9 - voice conversion algorithm which uses kernel-based partial least square to implement a non-linear transformation function
- S10 - speech synthesis by open-source MARY TTS.

## 3    Front-End

The main components of spoofing detection system are feature extraction and decision making modules. However some participants of the ASVspoof Challenge 2015 used additional steps in their systems, such as front-end preprocessing and high level features extraction.

### 3.1    Front-End Preprocessing

There were several signal preprocessing techniques proposed in the spoofing detection systems. The first purpose of these techniques is to enhance the impact of different features on the spoofing detection system decision, while the second is to detect simple types of spoofing attack by some enormous for natural speech artifacts and eliminate these utterances in further analysis.

**Pre-detector.** After experiments on the training part of the challenge database we decided to include pre-detection as a preliminary step in our spoof detection system [4]. The pre-detector checks whether the input speech signal has zero temporal energy values. In case of zero-sequence the signal is declared to be a spoofing attack, otherwise the speech signal is used as input data for the feature extractor. But the significant limitation of the described pre-detector is that it will be useless in case of channel effects or additive noise.

**Bandpass Filter.** Various experiments were made for acoustic features extracted from signal of different frequencies. These experiments show that different features are more informative on different ranges of frequencies. That is why we decided to check bandpass filter that rejects frequencies outside the specific for the type of features range. The results of these experiments for one of our spoofing detection systems based on phase-based features, described in next section are presented in Table 1.

**Table 1.** Bandpass filter effect on EER of TV-SVM spoofing detection system with phase-based features (%)

| Frequency range (Hz) | Spoofing type | | | | | |
|---|---|---|---|---|---|---|
|  | S1 | S2 | S3 | S4 | S5 | All |
| 0–3400 | 2.71 | 4.28 | 0.23 | 0.26 | 1.87 | 2.36 |
| 0–8000 | 2.13 | 4.3 | 0.77 | 0.74 | 3.39 | 2.51 |

**Pre-emphasis.** Pre-emphasis refers to filtering that emphasizes the higher frequencies and downplay the lower ones. Its purpose is to balance the spectrum of voiced sounds that have a steep roll-off in the high frequency region. Pre-emphasis removes some of the glottal effects from the vocal tract parameters. Comparison results of [3] demonstrates that the state-of-the-art Mel frequency cepstral coefficients (MFCC) are sensitive to pre-emphasis. They are presented in Table 2 and illustrate the usefulness of the pre-emphasis step for anti-spoofing purposes.

**Table 2.** Pre-emphasis performance on the base EER of GMM spoofing detection system on the development dataset (%)

|  | MFCC | MFCC + $\Delta$ | MFCC + $\Delta$ + $\Delta\Delta$ |
|---|---|---|---|
| No pre-emphasis | 4.00 | 2.66 | 2.80 |
| Pre-emphasis $\alpha = 0.97$ | 3.26 | 2.17 | 1.60 |

**Voice Activity Detection.** In order to discard useless information from the speech signal several participants tried to use Voice Activity Detector (VAD) as the preprocessing step for their spoofing detection systems. Authors of [5] apply DNN-based VAD and remove only first and last non-speech fragments. In [6] authors offered to use pitch based VAD on the score extraction level to discard scores of all silence patches, each of which contains 51 feature frames (with 0.025 s frame length and 0.01 s frame shift) and covers about 0.5 s of temporal context. In [7] authors remove all non-speech fragments by GMM-based VAD [14].

However experiments with our systems show that using VAD segmentation for full utterance is ineffective. According to our opinion, applying VAD may lead to throwing out informative artifacts locating between speech fragments. It is confirmed by the results of comparison for two TV-SVM systems with MFCC features performed in Table 3.

**Table 3.** Effect of VAD on the spoofing detection performance on the base of EER for TV-SVM system with MFCC features on the development dataset (%)

| Preprocessing type | Spoofing type | | | | | |
|---|---|---|---|---|---|---|
| | S1 | S2 | S3 | S4 | S5 | All |
| No VAD | 4.91 | 19.56 | 0.7 | 0.86 | 7.87 | 8.66 |
| VAD | 8.51 | 30.06 | 4.86 | 5.03 | 8.04 | 13 |

**Resampling.** Signal preprocessing in [12] includes downsampling original signal recordings from 16 kHz to 8 kHz to reduce computational load. In this case computational time greatly reduces, but our experiments show that during the downsampling process essential information is loosing which affects the performance of spoofing detection.

### 3.2 Front-End Features

Most of the participants of the ASVspoof Challenge 2015 found out the efficiency of the front-end features obtained by fusion of features appropriate for detecting specific spoofing attack. Thus, acoustic feature extractors in proposed systems are combinations of two or more different acoustic feature extraction methods. The most powerful features were attained by combining magnitude and phase information. It is hard to present full comparison of the implemented features because of the different type of classifiers used after, but we can analyse, how powerful are proposed approaches for spoofing detection task.

**Magnitude Based Features.** The magnitude spectrum contained detailed information about speech signal. Previous works has demonstrated the usefulness of magnitude information for spoofing detection task [15]. Most part of systems proposed during the ASVspoof Challenge used magnitude based features with and without their derivatives.

**Table 4.** Experiments results for the MLP-system with different features for different spoofing types obtained on the development dataset (EER %)

| Features type | Spoofing type | | | | | |
|---|---|---|---|---|---|---|
| | S1 | S2 | S3 | S4 | S5 | All |
| MS | 0.347 | 0.254 | 0.054 | 0.054 | 1.603 | 0.543 |
| RLMS | 0.000 | 0.093 | 0.039 | 0.039 | 1.456 | 0.486 |
| GD | 0.054 | 0.054 | 0.039 | 0.000 | 0.161 | 0.114 |
| MGD | 1.148 | 2.311 | 0.147 | 0.147 | 2.311 | 1.572 |
| IF | 0.161 | 0.401 | 0.147 | 0.147 | 0.948 | 0.428 |
| BPD | 2.243 | 4.955 | 0.401 | 0.347 | 5.155 | 3.431 |

Most of the successful spoofing countermeasures use *Mel-frequency Cepstral Coefficients* (MFCC) with their first and second derivatives as acoustic level features. They were used in [4,7,9–11].

In [11] authors proposed *Linear-frequency Cepstral Coefficients* (LFCC) by using linear filterbank instead of mel-filterbank.

We also use *Mel-frequency Principle Coefficients* (MFPC) coefficients, that were obtained similar to MFCC coefficients, but using principal component analysis instead of the discrete cosine transform to achieve decorrelation of the acoustic features [4]. Table 6 presents EER of spoofing detection performance for the development dataset for MFCC and MFPC-based spoofing detection systems. These results demonstrate that we achieved substantial EER improvement for all spoofing techniques by PCA basis implementation.

Another approach to use magnitude information is extraction of *Log Magnitude Spectrum* features (LMS) and *Residual Log Magnitude Spectrum* features (RLMS) [6]. Table 4 shows comparison results for systems using these features.

**Phase-Based Features.** Most approaches to detect synthetic or voice converted speech rely on processing artifacts specific to a particular synthesis or voice conversion algorithm [16] such as phase information. Phase domain features outperform magnitude related features, because spoofed speech doesn't retain the natural phase information.

The most commonly used phase-based features are related to group delay information. First of them are *Group Delay* (GD) features. Group delay is defined as derivative of the phase spectrum along the frequency axis [6]. The described way of calculation of the group delay function at frequency bins near zeros, that can occur near the unit circle, will results in high amplitude false peaks. These peaks mask out the formant structure. Due to this fact, *Modified Group Delay* (MGD) function suppress these zeros by the use of cepstrally smoothed magnitude spectrum instead of the original version. MGD features are known as more stable in speech recognition and were mostly used by participants

**Table 5.** Experiments results for the GMM-system with different features for different spoofing types obtained on the development dataset (EER %)

| Features type | Spoofing type | | |
|---|---|---|---|
| | Known attacks | Unknown attacks | All |
| MGD | 1.924 | 7.124 | 4.524 |
| PS-MFCC | 0.652 | 5.372 | 3.011 |
| WLP-GDCC | 1.436 | 8.941 | 5.188 |

of the Challenge. They were implemented in spoofing detection systems in [6–10]. However experiments from [6] presented in Table 4, demonstrate that MGD are not so effective as GD are for spoofing detection task. Another approach to solve the problem of GD were used by [7]. They implemented the *Product Spectrum* based features that were calculated as the product of power spectrum and GD function, thus combining information from amplitude and phase spectra (PC-MFCC).

The second feature type, mitigating the effect of zeros in group delay, was *All-pole Group Delay-based* features (WLP-GDCC). The main idea of this method is to keep only the vocal tract component of the speech signal and discard the contribution of the excitation source (Table 5).

As the phase changes depending on the splitting position of the input utterance it is important to normalize obtained phase information. [6,10] use *Relative Phase* extraction methods to reduce phase variation. These approaches have differences but both are based on the pitch synchronization of the slitting section instead of using fixed frame. Authors of [10] obtained impressive results. Comparing this features with MGD features on the base of one system researched obtained 0.83% EER for MGD features and 0.013% EER for Relative Phase features on the development set [10], while the performance of system based on Pitch Synchronous Phase (PSP) features [6] is slightly less than for MGD-based one. These feature type was also used in papers [11,12].

Another method to extract phase information is to use *Instantaneous frequency* (IF) estimation. While group delay is the derivative of the phase along the frequency axis, instantaneous frequency can be calculated as the derivative of the phase along the time axis. IF features are used in spoofing detection systems in [3,6].

Researchers in [6] also used *Baseband Phase Difference* (BPD) from [17] as more stable time-derivative phase based features and found out that these features contain different artifacts from the IF features. However, their results, presented below in Table 4, demonstrate that BDF features are not so efficient as IF features are, especially for voice conversion techniques S2 and S5.

In our system [4] we used *CosPhasePC* features which were extracted from unwrapped phase spectrum by applying cosine normalization and dimensionality reduction by means of principal components analysis. Results for system using CosPhasePC features on the development data set, presented in Table 6, confirms

**Table 6.** Experiments results for the TV-SVM system with different features obtained on the development dataset (EER %)

| Features type | Spoofing type | | | | | |
|---|---|---|---|---|---|---|
| | S1 | S2 | S3 | S4 | S5 | All |
| MFCC | 0.38 | 2.13 | 0.36 | 0.39 | 1.48 | 1.14 |
| MFPC | 0.13 | 0.29 | 0.09 | 0.09 | 0.37 | 0.23 |
| CosPhasePC | 0.13 | 0.20 | 0.04 | 0.05 | 0.23 | 0.15 |
| MWPC | 0.03 | 0.11 | 0.00 | 0.00 | 0.08 | 0.05 |

that CosPhasePc features are highly effective for all known types of spoofing attacks. Similar features was also used by [7,8]. Experiments in [7] confirm the power of cosine normalized phase-based features for known attacks.

**Local Binary Patterns.** Authors of [8] investigated the possibility to use spectra-temporal structure for spoofing detection task. In order to do this they used Local Binary Patterns (LBP) approach proposed for texture recognition. The spectrogram was used as acoustic representation. Authors treated it as 2D image to apply uniform LBP features extraction. Authors noticed that despite the traditional LBP algorithms for images, here they derived the histogram over each coefficient separately and used unique LPS without rotation invariance. Thus, they used the texture of the spectral magnitude as features to detect spoofed speech. By using these features they achieved 0.858% EER for all spoofing attacks from the development set.

**Wavelet Transform.** In our work, in order to include detailed time-frequency analysis of the speech signal in spoofing detection countermeasures, we proposed features based on applying the multiresolution wavelet transform [18], that was adapted to the mel scale, called *Mel Wavelet Packet Coefficients*. We used Daubechies wavelets db4 in the wavelet-decomposition. Using Teager Keiser Energy Operator instead of classical energy of the frequency sub-band makes these features more informative and noise-robust than classical sample energy. We also applied projection on the eigenvector basis for features decorrelation. Our experiments on the development set (Table 6) demonstrated that described features showed the best results in terms of performance of individual system based on concrete feature type.

Authors of [3] proposed auditory-based cepstral coefficients called *Cochlear Filter Cepstral Coefficients* (CFCC). They can be extracted by applying the cochlear filterbank based on auditory transform, hair cell function, nonlinearity and discrete cosine transform. A brief description of the feature extraction procedure is presented in [3]. Authors use CFCC features together with IF features, described above, to combine both envelope structure and IF information (CFCCIF). Framewise IF features are multiplied with the corresponding nerve spike density envelope, obtained during the CFCC extraction operation. Thus, IF obtained in silence regions will be suppressed. The derivative operation is used to capture the changing information in envelope and IF for consecutive frames. In [3] researchers obtained 2.6% EER for CFCC-based and 1.4% EER for CFCCIF-based individual systems. Comparison with 2.66% EER for MFCC-based system shows that CFCCIF features are highly effective for spoofing detection task.

**Phonetic Level.** Based on the achievements of [19] authors of [9] proposed to use combination of MFCC with the *Phonetic Level Phoneme Posterior Probability* (PPP) tandem features for spoofing detection task. They used multilayer perceptron based phoneme recognizer with a English acoustic model trained on the TIMIT database for phoneme decoding and obtained 1.72% EER on their SVM based spoofing detection system. That is expressive improvement in comparison with 8.46% EER for MFCC features on the similar system.

## 4    High Level Features Extraction

In our work for the acoustic space modelling we used the standard Total Variability approach, which is widely used in speaker verification systems [20]. The main idea of this approach consists in finding a low dimensional subspace of the GMM supervector space, named the total variability space that represents both speaker and channel variability. The vectors in the low-dimensional space called super-vectors or i-vectors. These i-vectors were extracted by means of Gaussian factor analyser defined on mean supervectors of the Universal Background Model (UBM) and Total Variability matrix T. UBM was represented by the diagonal covariance Gaussian mixture models of the used features. This approach was also used by [8,9], while [3] used two simple GMM models for natural and spoofed speech.

Systems from [5] used Deep Neural Network (DDN) models. The mean values of outputs of last hidden layer from the trained neural network were used as a final representation of the signal s, which are new robust representations, called spoofing vectors (s-vector).

## 5    Back-End

**GMM.** Most part of the participants used standard GMM-classifiers in their systems. These are [3,7,8,10,12].

**SVM.** Support Vectors Machine (SVM) was the second popular classifier in the ASVspoof challenge. We used SVM with linear kernel in our primary system as it presented the best performance in our experiments. To train SVN we used the efficient LIBLINEAR [13] library with default C-values equal to 1. Authors of [8,9,11] also chose SVM as classifiers in their submitted systems.

**Table 7.** EER for spoofing detection systems based on different classifiers for system from [9] (%)

| Linear kernel SVM | Polynomial kernel SVM | Cosine scoring | KNN | Simplified PLDA | Two stage PLDA |
|---|---|---|---|---|---|
| 1.86 | 1.06 | 2.86 | 2.46 | 1.89 | 10.18 |

**DNN.** System performed in [6] combine all 6 proposed features types. Moreover the feature vectors were concatenating within a window to incorporate long term temporal information. In order to handle the high demensional feature vectors authors used Deep Neural Network with one hidden layer. This system achieved 0.001% EER on all spoofing types from the development set.

Authors in [5] investigated Deep Neural Network classifier for two types of constructions: 6 classes (individual for each type of spoofing attack) and 2 classes (1 class for all spoofing attacks). They obtained better results for DNN than for GMM classifier. And although 2 classes DNN classification performed better that 6 classes configuration on the development set, their small scaled experiments convinced that 6 classes classification has better performance on the unknown spoofing attacks.

**DBN.** In our system we used classifier based on Deep Belief Network with softmax output units and stochastic binary hidden units. We used layer-wise pretreating of the layers by means of Restricted Boltzmann Machines (RBMs) and then applied back-propagation to train the DBN in a supervised way to perform classification. However our experiments on the development set demonstrated that linear SVM classifier works better on the proposed features. In this system we probably failed to avoid the effects of the stronger overfitting on these training dataset, in comparison with SVM.

**K-Nearest.** Authors of [9] compared several classification approaches: K-nearest neighbor classification (KNN) with 2 classes for human and spoofed speech, cosine similarity scoring, simplified PLDA classifier with 6 classes (individual class for each spoofing type), two stage LDA with 2 subspace (speaker subspace and spoofing subspace) and SVM as 2 class classification. Table 7 demonstrates results obtained for their system based on MFCC and PPP features described in Sect. 3.2 with score-fusion. According to these results SVM classifiers outperform the others.

## 6   Evaluation Results

A comparison of all final systems of the participants is possible only on the evaluation data set. These experiments results were presented in [2] and described below in Table 8. The best results in terms of unknown attacks and average was

**Table 8.** Evaluation results of ASVspoof Challenge 2015 (EER %)

| System ID | Equal Error Rates (EERs) | | |
|---|---|---|---|
| | Known attacks | Unknown attacks | Average |
| A [3] | 0.408 | **2.013** | 1.211 |
| B [4] | 0.008 | 3.922 | 1.965 |
| C [5] | 0.058 | 4.998 | 2.528 |
| D [6] | **0.003** | 5.231 | 2.617 |
| E [7] | 0.041 | 5.347 | 2.694 |
| F [8] | 0.358 | 6.078 | 3.218 |
| G [9] | 0.405 | 6.247 | 3.326 |
| H | 0.670 | 6.041 | 3.355 |
| I [10] | 0.005 | 7.447 | 3.726 |
| J [11] | 0.025 | 8.168 | 4.097 |
| K [12] | 0.210 | 8.883 | 4.547 |

obtained by system based on score-level fusion of MFCC and CFCCIF features, GMM modeling and log-likelihood scoring. Many systems used total variability modelling for high level features extraction, which also improve the performance of spoofing detection. Talking about classifiers, it should be mentioned that it is highly complicated to define the best classifiers based on the evaluation results of the ASVspoof Challenge because it depends also on the pre-processing effect, type of features, modelling type and on the details of the classification task (2 class classification or 6 class classification with each class for each spoofing attack type). Nevertheless we can figure out the strong success of SVM classifiers, that was confirmed by several researches, and high performance on neural networks for spoofing detection task. Probably, further study in this field will lead to more significant results.

Anti-spoofing system, presented in [6], that used 6 different types of features, including magnitude based, GD and MDG, IF and PSP features and used MLP classification was the best system for known types of attack, while it achieved only 4th result in terms of unknown types of attacks. Our primary system, based on the MFCC, MFPC and CosPhasePC feature-level fusion, TV modelling and SVM classifier achieved the second place with a stable 2nd results for known and unknown spoofing attacks.

All proposed systems perform poor performance for S10 type of spoofing attack. This fact leaves the problem of efficient spoofing detection countermeasures to be actual for further investigations.

## 7    Conclusion

In this paper we investigated modern tendencies in spoofing detection on the base of ASVspoof Challenge 2015 results. Experimental results of the participants, confirm that the most efficient systems use several types of features, responsible for different information and artifacts of the speech signal. Because these systems can catch complementary information that is not evident for individual feature-based systems. Most often these features contain magnitude and phase information. However, phoneme features also were highly effective.

Several preprocessing techniques were found out to be crucial for concrete features type. For example, MFCC features are sensitive to pre-emphasis step, and it can be helpful with fine tuned parameters. According to our experiments VAD may throw out informative artifacts locating between speech fragments.

Classification comparisson show that SVM is highly efficient for spoofing detection task, as well as neural network approaches.

**Acknowledgements.** This work was financially supported by the Ministry of Education and Science of the Russian Federation, Contract 14.578.21.0126 (ID RFMEFI57815X0126).

# References

1. Villalba, E., Lleida, E.: Speaker verification performance degradation against spoofing and tampering attacks. In: Proceedings of the FALA Workshop, pp. 131–134 (2010)
2. Wu, Z., Kinnunen, T., Evans, N., Yamagishi, J., Hanilc, C., Sahidullah, M., Sizov, A.: ASVspoof 2015: the First Automatic Speaker Verification Spoofing and Countermeasures Challenge (2015). http://www.spoofingchallenge.org/is2015_asvspoof.pdf
3. Patel, T.B., Patil, H.A.: Combining Evidences from Mel Cepstral, Cochlear Filter Cepstral and Instantaneous Frequency Features for Detection of Natural vs. Spoofed Speech, Interspeech (2015)
4. Novoselov, S., Kozlov, A., Lavrentyeva, G., Simonchik, K., Shchemelinin, V.: STC Anti-spoofing systems for the ASVspoof Challenge arXiv:1507.08074 (2015)
5. Chen, N., Qian, Y., Dinkel, H., Chen, B., Kai, Y.: Robust Deep Feature for Spoofing Detection - The SJTU System for ASVspoof Challenge, Interspeech (2015)
6. Xiao, X., Tian, X., Steven, D., Haihua, X., Chng, E.S., Li, H.: Spoofing Speech Detection Using High Dimensional Magnitude and Phase Features: the NTU Approach for ASVspoof Challenge, Interspeech (2015)
7. Alam, M.J., Kenny, P., Bhattacharya, G., Stafylakis, T.: Development of CRIM System for the Automatic Speaker Verification Spoofing and Countermeasures Challenge, Interspeech (2015)
8. Liu, Y., Tian, Y., He, L., Liu, J., Johnson, M.T.: Simultaneous Utilization of Spectral Magnitude and Phase Information to Extract Supervectors for Speaker Verification Anti-spoofing, Interspeech (2015)
9. Weng, S., Chen, S., Lei, Y., Xuewei, W., Cai, W., Liu, Z., Li, M.: The SYSU System for the Interspeech Automatic Speaker Verification Spoofing and Countermeasures Challenge arXiv:1507.06711 (2015)
10. Wang, L., Yoshida, Y., Kawakami, Y., Nakagawa, S.: Relative phaseinformation for detecting human speech and spoofed speech, Interspeech (2015)
11. Villalba, J., Miguel, A., Ortega, A., Lleida, E.: Spoofing Detection with DNN and One-class SVM for the ASVspoof Challenge, Interspeech (2015)

12. Sanchez, J., Saratxaga, I., Hernaez, I., Navas, E., Erro, D.: The AHOLAB RPS SSD Spoofing Challenge submission, Interspeech (2015)
13. LIBLINEAR: A library for Large Linear Classification. https://www.csie.ntu.edu.tw/cjlin/liblinear/
14. Kinnunen, T., Rajan, P.: A practical, self adaptive voice activity detector for speaker verification with noisy telephone and microphone data. In: Proceedings of ICASSP, pp. 7229–7233 (2013)
15. Marcel, S., Nixon, M.S., Li, S.Z.: Handbook of Biometric Anti-spoofing: Trusted Biometrics Under Spoofing Attacks. Springer, London (2014)
16. Wu, Z., Evans, N., Kinnunen, T., Yamagishid, J., Alegreb, F., Lia, H.: Spoofing and countermeasures for speaker verification: a survey. Speech Commun. **66**, 130–153 (2015)
17. Krawczyk, M., Gerkmann, T.: Shift phase reconstruction in voiced speech for an improved single-channel speech enhancement. IEEE/ACM Trans. Audio Speech Lang. Process. (TASLP) **22**(12), 1931–1940 (2014)
18. Mallat, S.: A Wavelett Tour of Signal Processing, 3rd edn. Academic Press, New York (2008)
19. D'Haro, L., Cordoba, R., Salamea, C., Echeverry, J.: Extended phone log-likelihood ratio features and acoustic-based i-vectors for languages recognition. In: Proceedings of ICASSP, pp. 5379–5383. IEEE (2014)
20. Novoselov, S., Pekhovsky, T., Simonchik, K.: STC speaker recognition system for the NIST i-vector challenge. In: Proceedings of Odyssey - The Speaker and Language Recognition Workshop (2014)
21. Hinton, G.E., Osindero, S., Teh, Y.: A fast learning algorithm for deep beliefnets. Neural Comput. **18**, 1527–1554 (2006)

# Combining Knowledge and CRF-Based Approach to Named Entity Recognition in Russian

V.A. Mozharova[✉] and N.V. Loukachevitch[✉]

Lomonosov. Moscow State University, Moscow, Russia
valerie.mozharova@gmail.com, louk_nat@mail.ru

**Abstract.** Current machine-learning approaches for information extraction often include features based on large volumes of knowledge in form of gazetteers, word clusters, etc. In this paper we consider a CRF-based approach for Russian named entity recognition based on multiple lexicons. We test our system on the open Russian collections "Persons-1000" and "Persons-1111" labeled with personal names. We additionally annotated the collection "Persons-1000" with names of organizations, media, locations, and geo-political entities and present the results of our experiments for one type of names (Persons) for comparison purposes, for three types (Persons, Organizations, and Locations), and five types of names. We also compare two types of labeling schemes for Russian: IO-scheme and BIO-scheme.

**Keywords:** CRF · Named entity recognition

## 1 Introduction

Information extraction is one of the most important tasks in natural language processing. There are several basic types of information to extract. The first type is named entities, such as person names, company names, or locations. The second type is relationships between named entities, for example, a person post in an organization. The third type is events that occur with named entities, for example, company merging, stock purchasing, or business meetings. All this information is used in information retrieval tasks, document annotation tasks, business analytics, and many other areas.

Most papers devoted to named entity recognition [1,2] present studies for English. For Russian, such experiments were carried out, mainly, on proprietary text collections, and the issues on comparison of approaches, the best feature sets still exist. In this paper we present our experiments on named entity extraction using the open Russian text collections "Persons-1000" and "Persons-1111". In our approach we combine statistical and knowledge-based methods. Also we compare two labeling schemes for Russian named entity recognition: IO-scheme and BIO-scheme. We use the CRF method as a machine learning method for this task.

© Springer International Publishing AG 2017
D.I. Ignatov et al. (Eds.): AIST 2016, CCIS 661, pp. 185–195, 2017.
DOI: 10.1007/978-3-319-52920-2_18

## 2    Named Entity Recognition Task

A named entity is a word or phrase that means a specific object or an event and distinguishes it from other similar objects [1]. Named entities must have a referent and they are usually written with a capital letter, for example:

– Президент Владимир Путин    17    декабря провел традиционную пресс-конференцию перед Новым годом. (The President Vladimir Putin organized the traditional press conference on December 17 before the New Year.)
– Студенты и Татьяны получат эксклюзивный пропуск на Главный каток страны. (Students and Tatyanas will get the exclusive permit to the Main Country Ice Rink.)

In the first sentence, the phrase "Владимир Путин" ("Vladimir Putin") is a named entity because it means a specific person. In the second sentence, the word "Татьяны" ("Tatyanas") is not a named entity because it does not have a specific referent.

There are many types of named entities, such as persons, organizations, locations, events, and time.

## 3    Related Work

Machine learning methods, such as CRF, maximum-entropy, or SVM, are very popular in the named entity recognition task in many languages, including Slavic languages.

In [3] the authors carried out experiments in Czech with forty-two named entity types. To recognize named entities, they used a maximum-entropy based recognizer. Two-stage prediction was implemented: the second stage used the results of the first stage. To extract the features, the authors used the large number of gazetteers and corpus-based word clusters (Brown clusters [4]).

For the open Polish collections CZER, CEN, and CPR, the authors of [5] applied the CRF method for five-type named entity recognition (first names, surnames, countries, cities, roads). They used such features as orthographic features, wordnet-based features, morphological features, and gazetteer-based features.

There are several works applying CRF in the Russian named-entity recognition.

In [6] the authors presented the results of the CRF method on various tasks, including the named entity recognition. The experiments were carried out on their own Russian text corpus, which contained 71,000 sentences. They used only n-grams and orthographic features of tokens without utilizing any knowledge-based features. They achieved 89.89% of F-score on three named entity types: names (93.15%), geographical objects (92.7%), and organizations (83.83%).

In [7] the experiments were based on the open Russian text collection "Persons-600"[1] for the person name recognition task. The authors also chose the

---

[1] http://ai-center.botik.ru/Airec/index.php?option=com_content&amp;view=article &amp;id=27:persons-600&amp;catid=15&amp;Itemid=40.

CRF method for recognition. Such features as token features, context features, and the features based on knowledge about persons (roles, professions, posts, and other) were utilized. They achieved 88.32% of F-score on person names.

In [8] the experiments were carried out on the Russian text collection, which contained 97 documents. The authors used two approaches for the named entity recognition: knowledge-based and CRF-based approach. In the machine learning framework they utilized such features as the token features and the knowledge features based on word clustering (LDA topics [10], Brown clusters [4], Clark clusters [11]). They achieved 75.05% of F-score on two named entity types: persons (84.84%) and organizations (71.31%).

To extract Russian personal names, in [9] the author used the knowledge-based approach without any machine learning method. This approach was based on regular expressions and gazetteers. The system was tested on the open collection "Persons-1000". Initially, the system achieved 81.36% of F-score, but after adding the global context feature, it achieved 96.62% of F-score on person names.

## 4   Text Collections and Labeling Rules

To extract Russian entities, we experimented on the two open Russian text collections. The first collection is "Persons-1000"[2], which contains 1000 news documents with person labels. This collection was annotated by Research Center of Artificial Intellegence [12] in a similar way to MUC-7 labeling [13].

We additionally labeled this collection with other named entities:

- Organizations (ORG)
- Media organizations having a specific function of information providers (MEDIA)
- Locations (LOC)
- States and capitals in the role of a state (GEOPOLIT), for example, "Москва аннонсировала..." ("Moscow announced that ...")

The second collection "Persons-1111"[3] is a collection containing 1111 news documents mentioning Eastern names including Arabian, Indian, Chinese, Japanese, etc., which are usually more difficult for correct named entity recognition.

### 4.1   Labeling Rules

Originally, only named entities, related to persons (PER), were labeled in the "Person-1000" text collection. According to the guidelines, only proper personal names were annotated. Roles and posts (for example, "Президент" ["The President"]), and persons, which names were not explicitly declared in the text (for example, "его отец" ["his father"]), were not labeled as named entities [12].

---

[2] http://ai-center.botik.ru/Airec/index.php/ru/collections/28-persons-1000.

[3] http://ai-center.botik.ru/Airec/index.php/ru/collections/29-persons-1111-f.

We additionaly labeled the collection with names of organizations, media organizations, locations, and geopolitical entities. We employed the following rules:

1. A descriptor is a word or a phrase indicating a generic type of a named entity. A descriptor is a part of a named entity:
   (a) If it is an abbreviation
       – [ОАО "Газпром"] ORG ([JSC "Gazprom"] ORG)
   (b) If it is the head of a noun group, but it is not the supplement
       – [Санкт-Петербургский государственный университет] ORG ([Saint-Petersburg State University] ORG)
       – город [Тула] LOC (town [Tula] LOC)
2. A person name inside a proper name is not labeled separately
   – [Библиотека имени Ленина] ORG ([State Lenin Library] ORG)
3. A geographical object inside a named entity is labeled separately, if a named entity is not in quotes
   – [Правительство] ORG [РФ] GEOPOLIT ([The Government] ORG of the [Russian Federation] GEOPOLIT)
   – гостиница ["Москва"] ORG (hotel ["Moscow"] ORG)

Our text collection labeling is similar to the markup standard accepted in MUC7 [13] and CoNLL [14]. Similar to these conferences, the labeled named entities are non-nested, non-overlapping and annotated with exactly one label. Table 1 presents the quantitative characteristics of the labeled named entities in the collection "Persons-1000".

**Table 1.** The quantitative characteristics of the labeled named entities in the collection "Persons-1000"

| | |
|---|---|
| PER | 10623 |
| ORG | 7032 |
| MEDIA | 1509 |
| LOC | 3141 |
| GEOPOLIT | 4103 |

## 4.2 Labeling Schemes

To represent labeled text segments as features for machine learning, several labeling shemes can be used. In our experiments we considered two schemes: IO-scheme and BIO-scheme.

**IO-scheme (Inside-Outside).** In IO-scheme, every token can be labeled by only two types: "it belongs to named entity" (I), "it does not belong to named entity" (O). On the one hand, the classifier can learn easier when the number of label types is smaller, but, on the other hand, it is necessary to introduce additional rules for finding named entity boundaries. Table 2 presents an example of this labeling.

Further, for this labeling scheme, the prefix "I-" is omitted, and the label "O" is replaced by the label "NO". This scheme presupposes the prediction of $|C| + 1$ classes, where $|C|$ is the number of the named entitiy categories.

**Table 2.** IO-scheme example

| Владимир (Vladimir) | I-PER |
|---|---|
| Путин (Putin) | I-PER |
| посетил (visited) | O |
| Англию (England) | I-GEOPOLIT |

**BIO-scheme (Begin-Inside-Outside).** In BIO-scheme, every text token can be associated with the label from one of three types: "named entity beginning" (B), "named entity continuation" (I), or "not named entity" (O). Thus, the classifier determines the boundaries of named entities by itself, and it should make named entity recognition much easier. Table 3 shows an example of this labeling scheme. The BIO-scheme requires to predict $|C| + 1$ classes, where $|C|$ is the number of named entity categories.

**Table 3.** BIO-scheme example

| Владимир (Vladimir) | B-PER |
|---|---|
| Путин (Putin) | I-PER |
| посетил (visited) | O |
| Англию (England) | B-GEOPOLIT |

In the well-known Stanford named entity recognizer [15], the authors gave preference to the IO-scheme because, in English, named entities of the same type rarely locate beside each other in texts, therefore it is not necessary to use the complicated scheme of labeling.

We found that in Russian, there are a lot of examples of the same type named entities locating beside each other in texts. Table 4 shows the statistics of the co-occurrence of the same-type objects for three types of named entities. It can negatively influence on further named entity token aggregation because of the problem with named entity boundaries. To compare labeling shemes, we carry out the experiments using both schemes of labeling (see Sect. 6).

**Table 4.** Statistics of the co-occurrence of the same type named entities

| NE type | Statistics |
|---------|------------|
| PER | 72 (1.3%) |
| ORG | 261 (6.1%) |
| LOC | 58 (1.6%) |

## 5  Features and Rules

To extract Russian named entities, we utilize the CRF classifier as a machine learning method because it showed good results in many works devoted to the named entity recognition. We used CRF++[4], which is an open source implementation for labeling sequential data. This implementation is fast and easy to tune.

### 5.1  Preprocessing

Before the feature extraction, the text collection was processed with a morphological analyzer. As a result, for each token such features as a part of speech, gender, number, and case were extracted.

### 5.2  Features

The fixed set of features was computed for every token. We used token features, context features, and features based on lexicons. Below the basic features are listed.

**Token Features**

- Token initial form (lemma)
- Number of symbols in a token
- Letter case. If a token begins with a capital letter, and other letters are small then the value of this feature is "BigSmall". If all letters are capital then the value is "BigBig". If all letters are small then the value is "SmallSmall". In other cases the value is "Fence"
- Token type. The value of this feature for lexemes is the part of speech, for punctuation marks the value is the type of punctuation
- The presence of a vowel (a binary feature)
- If a token ends a sentence (a binary feature)
- If a token contains a known letter n-gram from a pre-defined set:
  - If the last letters match one of the typical last name ends (-енко, -швили, -ова, -ов etc.)
  - If the first letters match one of the typical first name beginnings
  - If in a token there is a letter n-gram that usually appears in organization names (-ком-, -орг-, -деп-, etc.)

---

[4] https://taku910.github.io/crfpp/.

**Features Based on Lexicons.** To improve the results of named entity recognition, we used vocabularies that store lists of useful objects. An object can be expressed with a word or a phrase.

For every token, our system determines if a token is a known word or a token is included in a known phrase. The phrase length was also taken into account. Table 5 presents basic vocabularies and their sizes. The overall size of all vocabularies is more than 335 thousand entities. The lexicons were extracted from several sources: phonebooks, Russian Wikipedia, RuThes[5] thesaurus [16], etc.

**Table 5.** Vocabulary sizes

| Vocabulary | Size, objects | Clarification | Examples |
|---|---|---|---|
| Famous persons | 31482 | Famous people | Владимр Путин, Ангела Меркель |
| First names | 2773 | First names | Василий, Анна, Том |
| Surnames | 66108 | Surnames | Кузнецов, Грибоедов |
| Person roles | 9935 | Roles, posts | министр, китаевед |
| Verbs of informing | 1729 | Verbs that usually occur with persons | высказать, отпроситься, признаться |
| Companies | 33380 | Organization names | Сбербанк |
| Company types | 6774 | Organization types | организация, авиафирма |
| Media | 3909 | Media | РИА Новости, Первый канал |
| Geography | 8969 | Geographical objects | Балтийское море, Владивосток |
| Geographical adjectives | 1739 | Geographical adjectives | финский, томский, югославский |
| Usual words | 58432 | Frequent Russian words (nouns, verbs, adjectives) | автомобиль, падать, желтый |
| Equipment | 44094 | Devices, equipment, tools | устройство, телефон |

**Features Based on Context.** The values of the above listed features were also calculated for neighbor tokens in two-word window for every token to the left and to the right.

As a result, the same number of features were computed for every token. Table 6 presents a feature set example.

### 5.3   IO-labeling: Aggregation of Tokens into Named Entities

In result of the classifier work, each token obtains a specific tag. Tokens corresponding to the same named entity should be aggregated. In case of BIO-labeling,

---

[5] http://www.labinform.ru/pub/ruthes/.

**Table 6.** Features

| Token | Lemma | Register | Token Type | Second Name | Geo | Label |
|-------|-------|----------|------------|-------------|-----|-------|
| В | В | Small | Auxiliary | False | False | NO |
| России | РОССИЯ | BigSmall | Noun | False | Geo1 | GEOPOLIT |
| Алиев | АЛИЕВ | BigSmall | Noun | True | False | PER |
| третий | ТРЕТИЙ | Small | Numeral | False | False | NO |
| раз | РАЗ | Small | Auxiliary | False | False | NO |

named entities are distinguished by label boundaries (Begin-Inside-Outside). To determine named entity boundaries in case of IO-scheme, the special rules were used.

The main rule is that the same type tokens located beside each other are considered as a single multiword entity. Exceptions from the main rule are as follows:

- If a punctuation mark is in the same type token sequence and it is a quote, open bracket or dot that is not a sentence ending, then a named entity is not separated by this punctuation mark.
- Otherwise, a named entity is separated by punctuation marks
- Punctuation marks are not included in named entities
- If there are more than two words in a person named entity, and two first names with different grammatical cases are met in this named entity, then the named entity is separated. The boundary is the second name.

We also use additional specialized rules:

1. If the template "<ORG> имени <PER>" is met, the word fragment that matched with this template is joined together into the same organization named entity.
2. If a token sequence $(p_1, ..., p_n)$, where $p_i$ is a token labeled as a person, contains a known first name $p_j$, then all tokens $p_k$ $(k \neq j)$ are memorized as possible persons. In cases of missed person labels for $p_k$, these labels can be restored. For example, if in a text personal name "Анатолий Котляр" ("Anatoliy Kotlyar") was recognized then token "Котляр" ("Kotlyar") will be labeled with the person tag even if the CRF classifier missed it. This is an attempt to utilize the global context of the text (see discussion about global features in [17]).

## 6     Experiments

The experiments were carried out using two labeling schemes: IO-scheme and BIO-scheme. For the IO-scheme, two runs were performed: with and without rules.

The *Fscore* was used as a target metric. It was calculated as follows:

$$Precision = \frac{intersectionCount}{classifierCount}$$

$$Recall = \frac{intersectionCount}{expertCount}$$

$$Fscore = 2 \cdot \frac{Precision \cdot Recall}{Precision + Recall}$$

where *intersectionCount* is the number of named entities labeled by both: the classifier and the expert; *classifierCount* is the number of named entities labeled by only the classifier; *expertCount* is the number of named entities labeled by only the expert.

To calculate the target metric, we used the 3:1 cross-validation technique. The collection was divided into four parts, and each part was iteratively utilized as a test part and others as train parts. The final value of the target metric was calculated as the average value of the intermediate results.

Table 7 presents the results of the experiments with three types of named entities for all sorts of text labeling are shown. Table 8 shows the results for five named entity types.

**Table 7.** Results for three types of NE

| NE type | F-score, % | | |
|---------|------|-----------|------|
|         | IO   | IO + rules | BIO |
| PER     | 94.95 | 95.09    | **96.08** |
| ORG     | 80.03 | 80.23    | **83.84** |
| LOC     | 92.60 | 92.60    | **94.57** |
| Average | 89.54 | 89.67    | **91.71** |

Taking into account the results from tables, we can conclude that, during work with the IO-scheme, the rules of token aggregation positively influence on the target metric, but, on an average, the BIO-scheme gives more significant contribution especially for recognizing organization names. It means that for the Russian texts BIO-scheme plays an important role because named entities can locate beside each other in texts, and we need to separate them.

The text collection "Persons-1000" is an extension of the collection "Persons-600", used in [7]. In that work, the F-score achieved 88.32% on the person entity type. Our experiments showed the improvement of this metric to 96.08% on the extension of the collection. We consider that it happened because in the previous work the authors used the small number of vocabularies oriented only to personal names. In our approach, we utilize much more various types of knowledge.

**Table 8.** Results for five types of NE

| NE type | F-score, % | | |
|---------|------|-----------|-------|
|         | IO   | IO + rules | BIO  |
| PER     | 94.81 | 95.01 | **95.63** |
| ORG     | 75.90 | 76.16 | **80.06** |
| MEDIA   | 87.95 | 87.95 | **87.99** |
| LOC     | 84.53 | 84.53 | **86.91** |
| GEOPOLIT | **94.65** | **94.65** | 94.50 |
| Average | 88.21 | 88.37 | **89.93** |

The rule-based system described in [9] was specially tuned on the collection "Persons-1000" and achieved 96.62% of F-sore, which can be considered as the maximum for this collection, and we almost reached this result. To check robustness of our system for processing complicated names, we apply our named entity recognizer trained on the collection "Person-1000" to the collection "Person-1111" containing Eastern names. Our system achieved 81.68% of F-measure (Table 9). The rule-based system obtained only 64.43%. It demonstrates the robustness of our system.

**Table 9.** Comparison of the rule-based system [9] and our system on two collections

| Collection | F-score, % | |
|------------|---------------------|------------|
|            | Rule-based system [9] | Our system |
| Persons-1000 | **96.62** | 96.08 |
| Persons-1111-F | 64.43 | **81.68** |

## 7   Conclusion

In this work we present our experiments aimed at the Russian named entity recognition task on the open text collection "Person-1000". We used the knowledge-based approach together with the CRF classifier. The knowledge was expressed in gazetteers and rules. We described our results for three types (persons, organizations, and locations) and five types of names (persons, organizations, media, locations, and geopolitical objects). We compared our study with previous works on the same or similar collections.

**Acknowledgments.** This work is partially supported by RFBR grant No. 15-07-09306.

# References

1. Nadeau, D., Sekine, S.: A survey of named entity recognition and classification. Lingvisticae Investigationes **30**(1), 3–26 (2007)
2. Tkachenko, M., Simanovsky, A.: Named entity recognition: exploring features. In: 11th Conference on Natural Language Processing, KONVENS 2012, pp. 118–127. Eigenverlag ÖGAI (2012)
3. Straková, J., Straka, M., Hajič, J.: A new state-of-the-art Czech named entity recognizer. In: Habernal, I., Matoušek, V. (eds.) TSD 2013. LNCS (LNAI), vol. 8082, pp. 68–75. Springer, Heidelberg (2013). doi:10.1007/978-3-642-40585-3_10
4. Brown, P.F., Della Pietra, V.J., Desouza, P.V., Lai, J.C., Mercer, R.L.: Class-based n-gram models of natural language. Comput. Linguist. **18**(4), 467–479 (1992)
5. Marcińczuk, M., Stanek, M., Piasecki, M., Musiał, A.: Rich set of features for proper name recognition in polish texts. In: Bouvry, P., Kłopotek, M.A., Leprévost, F., Marciniak, M., Mykowiecka, A., Rybiński, H. (eds.) SIIS 2011. LNCS, vol. 7053, pp. 332–344. Springer, Heidelberg (2012). doi:10.1007/978-3-642-25261-7_26
6. Antonova, A.Y., Soloviev, A.N.: Conditional random field models for the processing of Russian. In: International Conference "Dialog 2013", pp. 27–44. RGGU (2013)
7. Podobryaev, A.V.: Persons recognition using CRF model. In: 15th All-Russian Scientific Conference "Digital Libraries: Advanced Methods and Technologies, Digital Collection", RCDL-2013, pp. 255–258. Demidov Yaroslavl State University (2013)
8. Gareev, R., Tkachenko, M., Solovyev, V., Simanovsky, A., Ivanov, V.: Introducing baselines for russian named entity recognition. In: Gelbukh, A. (ed.) CICLing 2013. LNCS, vol. 7816, pp. 329–342. Springer, Heidelberg (2013). doi:10.1007/978-3-642-37247-6_27
9. Trofimov, I.V.: Person name recognition in news articles based on the persons-1000/1111-F collections. In: 16th All-Russian Scientific Conference "Digital Libraries: Advanced Methods and Technologies, Digital Collections", RCDL 2014, pp. 217–221 (2014)
10. Chrupala, G.: Efficient induction of probabilistic word classes with LDA. In: 5th International Joint Conference on Natural Language Processing, IJCNLP 2011, pp. 363–372. Asian Federation of Natural Language Processing (2011)
11. Clark, A.: Combining distributional and morphological information for part of speech induction. In: 10th Conference on European Chapter of the Association for Computational Linguistics, EACL 2003, vol. 1, pp. 59–66. ACL (2003)
12. Vlasova, N.A., Suleimanova, E.A., Trofimov, I.V.: The message about Russian collection for named entity recognition task. In: TEL 2014, pp. 36–40 (2014)
13. Chinchor, N., Robinson, P.: MUC-7 named entity task definition. In: 7th Conference on Message Understanding, p. 29 (1997)
14. Sang, T.K., Erik, F., De Meulder, F.: Introduction to the CoNLL-2003 shared task: language-independent named entity recognition. In: 7th conference on Natural language learning at HLT-NAACL 2003, vol. 4, pp. 142–147. ACL (2003)
15. Finkel, J.R., Grenager, T., Manning, C.: Incorporating non-local information into information extraction systems by gibbs sampling. In: 43rd Annual Meeting of the Association for Computational Linguistics, pp. 363–370. ACL (2005)
16. Loukachevitch, N., Dobrov, B.: RuThes linguistic ontology vs. Russian wordnets. In: Global WordNet Conference GWC-2014. Tartu (2014)
17. Ratinov, L., Roth, D.: Design challenges and misconceptions in named entity recognition. In: 13th Conference on Computational Natural Language Learning, CoNLL, pp. 147–155. ACL (2009)

# User Profiling in Text-Based Recommender Systems Based on Distributed Word Representations

Anton Alekseev[1] and Sergey Nikolenko[1,2,3(✉)]

[1] Steklov Mathematical Institute at St. Petersburg, St. Petersburg, Russia
sergey@logic.pdmi.ras.ru
[2] National Research University Higher School of Economics, St. Petersburg, Russia
[3] Deloitte Analytics Institute, Moscow, Russia

**Abstract.** We introduce a novel approach to constructing user profiles for recommender systems based on full-text items such as posts in a social network and implicit ratings (in the form of likes) that users give them. The profiles measure a user's interest in various topics mined from the full texts of the items. As a result, we get a user profile that can be used for cold start recommendations for items, targeted advertisement, and other purposes. Our experiments show that the method performs on a level comparable with classical collaborative filtering algorithms while at the same time being a cold start approach, i.e., it does not use the likes of an item being recommended.

**Keywords:** User profiling · Recommender systems · Distributed word representations

## 1 Introduction

In the modern Web, with the advance of social interactions between users and full-scale data mining of all information related to users, user profiling has become a very important problem. In this context, user profiling means converting the recorded user behaviour into a certain set of labels or probability distributions that capture the most important aspects of the user that can be further used for making new recommendations, providing targeted advertisement, and various other purposes. User profiles can incorporate real-life knowledge such as demographic information (age, gender, location etc.) or attempt to infer it from user behaviour. However, the holy grail of user profiling is to concisely represent a user's topical interests, preferably as narrowly as possible, with obvious applications to targeted advertising and new recommendations. The methods of summarizing information about users lie at the core of many personalized search and advertisement engines and various recommender systems. Being able to make predictions based on appropriately summarized prior user-system interaction allows, among other things, to alleviate the so-called *cold start* problem,

© Springer International Publishing AG 2017
D.I. Ignatov et al. (Eds.): AIST 2016, CCIS 661, pp. 196–207, 2017.
DOI: 10.1007/978-3-319-52920-2_19

which is one of the main problems of recommender systems: how do you recommend a new item that has not been rated before or has had very few ratings? Given user profiles and a way to match the new item to these profiles, one can make recommendations when collaborative filtering is inapplicable.

This motivation ties in well with full-text recommendations. When users interact with items that have actual text associated with them, this allows for a possibility to infer topical user profiles based on automated mining of the texts they interact with. This problem has become especially relevant in recent years due to the growth of the social Web, where users interact with various texts all the time, not only reading but actively rating them. And as for possible solutions, recent advances in natural language understanding, especially in distributional semantics, provide many promising new methods for this problem. This is precisely the path that we take in this work, using topical clusters based on distributed word representations to construct user profiles.

The paper is organized as follows. In Sect. 2, we describe the problem setting in detail and survey related work. In Sect. 3, we describe in detail the novel approach we present in this work. Section 4 shows experimental results that validate our approach, and Sect. 5 concludes the paper.

## 2 Related Work

### 2.1 User Profiling in Recommender Systems

User profiling is a special case of user modeling. For general reviews of the field and key papers, we refer to [1–5]. Specific techniques that have been applied to represent user interests in content-based and hybrid recommender systems include, for example, relevance feedback and Rocchios algorithm [6,7], where a user profile is represented as a set of words and their weights, penalized if the retrieved textual item is uninteresting, as in [8]. Ontologies and encyclopaedic knowledge sources have been used, e.g., in Quickstep and Foxtrot systems [9] that recommend papers based on browsing history, automatically classify the topics of a paper and make use of relations between the topics in ontology to obtain their similarity; rank is computed based on the correlation of user profile topics and estimated paper topics. Nearest neighbours are often used in such systems; e.g., DailyLearner [10] stores tf-idf representations of recently liked stories in a short-term memory component, using it to recommend new stories [6,7]. Decision rules have been used, e.g., in the RIPPER system [11,12], where rules are a conjunction of several tests against items features. Interpretable predicted user characteristics are also often utilized in practice; cf., e.g., Yandex.Crypta.

### 2.2 Distributed Word Representations

Distributed word representations have become important for natural language processing with the rise of NLP systems based on neural networks; models for individual words, known as word embeddings or distributed word representations, map the words into a low-dimensional semantic space, trying to map semantic relations to geometric ones in that space. To train word representations, a model with one hidden layer attempts to predict the next word based on a window of several preceding words or words that surround it (skip-gram); this approach has been applied, for instance, in the Polyglot system [13], GloVe [14] Recent studies on the performance of various vector space models for word semantic similarity evaluation [15] demonstrated that compositions of models such as GloVe and Word2Vec as well as unsupervised one-model approaches show reasonable results for the Russian language.

# 3   User Profiling with Distributed Word Representations

## 3.1   Problem Statement and General Outline

In this work, we propose a novel method for user profiling in full-text recommender systems, constructing a user profile as an interpretable summary of the user's interests that can also be utilized for recommending new items solely based on the prior state of the system. We begin with a brief outline of our approach.

First, we cluster all word representations trained on an external corpus (see Sect. 3.2). We have obtained high-quality clusters that are easy to interpret, so they were chosen to serve as a basis for user profiling; a user would be characterized by his or her affinity to these clusters.

For the recommender system, we used a large dataset from the "Odnoklassniki" online social network; we used group posts (texts in online communities written by their members) and individual user posts (texts published by a user on his/her profile page) as full-text items and user likes for these posts as ratings. There are two important obstacles along this way.

1. First, the dataset contains only positive signals from the users (likes), which is common in real life recommender systems but which makes it hard to train. While recommender systems based on such implicit information do exist, e.g., recommender systems based on max-margin nonnegative matrix factorization [16], it is unclear how to adapt them to full-text recommendation and user profiles in the semantic space.
2. Whatever technique one tries for the problem, user profiles always tend to be dominated by clusters/topics consisting of common words that occur often in the texts of various topics, but are useless for recommendations.

The second problem was especially hard to solve; we solved it with a novel approach to user profiling based on logistic regression trained multiple times on random subsets of the dataset; this approach is described in Sect. 3.3.

## 3.2   Clustering Word Vectors

In our experiments, we use a skip-gram *word2vec* model of dimension 500 trained on a large Russian language corpus [15,17]; the corpus consisted of:

- Russian Wikipedia: 1.15 M documents, 238 M tokens;
- automated web crawl data: 890 K documents, 568 M tokens;
- texts from the *lib.rus.ec* library: 234 K documents, 12.9 G tokens.

A large collection of general-purpose texts ensured good resulting distributed representations; for an in-depth description of the model we refer to [15,17].

To get a finite set of possible user interests or document topics, we clustered the word vectors directly. Note that while for some other applications topic modeling [18,19] might prove to be more useful, but in our case the basic underlying texts were too short and of too poor quality to hope for a good topic model, decisions regarding the topics would often have to be made on the basis of one or two keywords. Besides, we wanted to develop a top-down general approach that would be applicable directly even without a large and all-encompassing collection of texts available directly in the recommender system.

The embeddings of terms that occurred in our social network posts dataset resulted in 111281 vectors to be clustered in the $\mathbb{R}^{500}$ space. We have tried several methods for large-scale data clusterization, including Birch [20], DBSCAN [21], and mean shift clustering [22], but despite being generally able to process 100K+ items, these methods have proven to be not fast enough for high-dimensional data (for dimension 500 in our case), coupled with a large number of clusters (several thousand). The best option was still provided by classical $k$-means clustering. We applied mini-batch $k$-means that samples subsets of data (mini-batches) and then applies standard $k$-means to then: they are assigned to centroids, and centroids are "moved" to actual centers; the updates are done stochastically, after every mini-batch [23]. For initialization, we used the $k$-means++ approach that initializes cluster centers as far from each other as possible and then applies standard $k$-means to a random data subset to refine initialization [24].

Table 1 shows sample clusters together with their *idf* (inverse document frequency) values. It is clear that the most frequent clusters largely consist of common words that do not represent any specific topic that could be used for recommendations; they will be our major problem in the next section.

**Table 1.** Sample clusters.

| IDF | Terms |
| --- | --- |
| 3.276 | Decide family leave buy parent read case week ... |
| 4.469 | Smile work answer appreciate state goal given inside brain remind ... |
| 5.703 | Comment Quran culture union Kim German note interview East forum historical ... |
| 5.902 | Salt pepper garlic sour-cream greens vegetables carrot cucumber ... |
| 6.126 | Pain disease shock depression abortion cardiac dense muscular insomnia ... |
| 7.608 | Stick axe thunder arrow sword boomerang shield spike steel armor paddle ... |
| 8.239 | Lead fly move once drive run walk ... |
| 9.650 | Bacteria molecule spermatozoid leukocyte chromosome mitochondria amoeba ... |
| 11.004 | Scaffold gallows pardon torture quartering hanging beheading ... |

## 3.3   Estimating Cluster Affinity with Resampling Logistic Regression

We begin with the following notation:

- $D$ is the set of documents in concern,
- $C$ is the set of clusters,
- $T$ is the set of all words in concern,
- $T_c$ is the set of words in a cluster $c \in C$,
- word2vec : $T \to \mathbb{R}^d$ is the function assigning each word its embedding,
- $\mathrm{df}(t)$ is the number of documents the word $t \in T$ occurs in,
- clust : $T \to C$ is a function returning the cluster of a word,
- $I_u^{\mathrm{like}}$ is the set of all items user $u$ liked.

To produce user profiles, we first constructed fixed-dimensional vector representations of documents $v_d \in \mathbb{R}^d$ for each document $d \in D$, representations of clusters of documents $v_c \in \mathbb{R}^d$ for each cluster $c \in C$ based on the representations of documents, and finally representations of users $v_u \in \mathbb{R}^d$ for each user $u \in U$ based on representations of the documents they liked and the corresponding clusters; in our experiments, $d = 500$. To build vector representations, we used a straightforward approach based on averaging and tf-idf weighting. Suppose that we know word2vec word embeddings for a large proportion of words in our data (not all due to typos, proper names and the like), $v_c = \frac{1}{|T_c|} \sum_{t \in T_c} \mathrm{word2vec}(t)$ for each $c \in C$. Then we define

$$\mathrm{df}_c = \sum_{t \in T_c} \mathrm{df}(t), \quad IDF_c = \log\left(\frac{\sum_{c^* \in C} \mathrm{df}_{c^*}}{\mathrm{df}_c}\right), \quad v_d = \sum_{t \in d} \frac{IDF_{\mathrm{clust}(t)} \cdot v_{\mathrm{clust}(t)}}{IDF_{sum}^d},$$

and $IDF_{sum}^d = \sum_{t\ ind} IDF_{\mathrm{clust}(t)}$. Finally, the user representation is

$$v_u = \sum_{d \in I_{liked}^u} \frac{\sum_{t \in d} IDF_{\mathrm{clust}(t)} \cdot v_{\mathrm{clust}(t)}}{Z},$$

where $Z$ is a corresponding normalization value.

Then we constructed a new representation of a document, designed as a vector of cluster likelihoods $p(c|d)$; namely, for every document $d \in D$ and every cluster $c \in C$ we computed

$$L(d|c) = e^{\frac{-\|v_d - v_c\|}{2\sigma^2}}, \quad p(c_i|d_j) = \frac{L(d_j|c_i)}{\sum_d L(d|c_i)}.$$

Then, to construct the profile of a user $u$ by his or her set of liked items $I_u^{\mathrm{like}}$, we repeated the following procedure $N$ times independently (in the experiments below, we used $N = 100$):

(i) on step $k$, draw a random sample from the documents the user $u$ didn't like, taking the size of the sample equal to the number of documents the user actually liked; we denote it by $I_{u,k}^{\mathrm{dislike}}$;

(ii) train logistic regression with the following data: $I_u^{like}$ are the positive examples, $I_{u,k}^{dislike}$ are the negative examples, and the features are document affinities to clusters $p(c|d)$, $d \in I_u^{like} \cup I_{u,k}^{dislike}$;

(iii) as a result of this logistic regression, we get a set of weights $w_{u,c,k}$ for each cluster $c$.

Then, for every user $u$ his or her profile is defined as the parameters of the normal distribution for every weight $\{(c, \mu_{u,c}, \sigma_{u,c})|c \in C\}$, each $(\mu_{u,c}, \sigma_{u,c})$ trained on the set $\{w_{u,c,k}|c \in C\}$.

In other words, logistic regression here is used to approximate the probability of a like; it trains a hyperplane separating liked items from items that have not been liked in the semantic feature space. This simple approach would be sure to fail if we simply trained liked documents against non-liked documents since the dataset is vastly imbalanced (a single user can be expected to view but a tiny fraction of all items); hence the random sampling of non-liked documents.

However, there is one more purpose to the random sampling apart from balancing the problem. We would like to solve the problem of common-word clusters, clusters that contain common words, are ubiquitous in the dataset, and tend to dominate all user profiles simply because by random chance, a user will like more than their fair share of some of common word clusters. In randomly

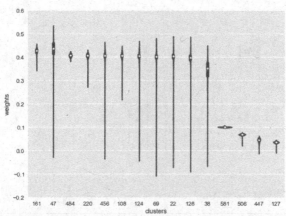

| # | $\mu$ | $\sigma$ | Words |
|---|---|---|---|
| 161 | 0.422 | 0.017 | uni din tel tine adam riga van eden loc publ etc judo art professor polis... |
| 47 | 0.421 | 0.074 | Nido Josep Jordi Victor Fbio Paulinho Avito Oswaldo Oliveira Julio... |
| 484 | 0.407 | 0.007 | Virgo Aries Taurus Dey hill branch statue strand Hon patron figure Chakhon... |
| 220 | 0.405 | 0.014 | Christ soul spirit God pray purify verse forgiven salvation devote Saturday... |
| 108 | 0.396 | 0.035 | Daniel Vladimir chronicle Tver Svyatoslav Ryazan Novgorod Pskov veche... |
| 124 | 0.391 | 0.062 | ask talk call listen report try calm assuring prompt pleading convince... |
| 69 | 0.386 | 0.081 | minute managed approach this autumn song male children ages anxiety... |

**Fig. 1.** Sample user profile produced by resampling logistic regression.

sampling the negative examples, we get a certain distribution of "concentrations" of different clusters in the negative example. Note that:

- "topical" clusters that contain rare words will seldom occur in negative examples and thus the variance of the resulting weight distribution $\sigma_{u,c}$ with respect to these clusters will be low;
- "common word" clusters that contain words that are widely distributed across the entire dataset will sometimes appear more often and sometimes less often in the negative examples, and thus the variance of the resulting weight distribution $\sigma_{u,c}$ with respect to these clusters will be high.

Hence, this approach lets us distinguish between common word clusters and topical clusters by the value of $\sigma_{u,c}$. The higher the standard deviation, the more likely it is that the cluster consists of common words. As the final scoring metric for the user profile, we propose to use the mean weight penalized by its variance; we used $\mu - 2\sigma$ as the final score in the examples below. This scoring metric can also be thought of as the lower bound of a confidence interval for the cluster affinity. Figures 1 and 2 show sample results of the users. Note how common-words clusters have high average affinity but also high standard deviation that drags them down in the final scoring and lets topical clusters come out on top.

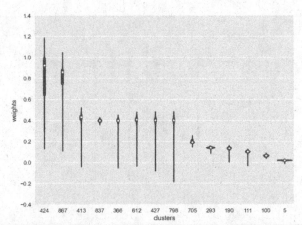

| # | $\mu$ | $\sigma$ | Words |
|---|---|---|---|
| 867 | 0.772 | 0.165 | hours two-hour break minute half-hour five-minute two-hour ten-hour... |
| 424 | 0.833 | 0.202 | kissing call cry silent scream laughing nod dare restrain angry slam... |
| 837 | 0.399 | 0.010 | youtube blog net mail facebook player online yandex user tor ado... |
| 366 | 0.396 | 0.042 | associate attitude seems quite horoscope ideal religious face era... |
| 413 | 0.406 | 0.080 | feel glad remember worrying offended jealous inhale pity envy suffer autumn... |
| 427 | 0.385 | 0.073 | hijack bombing raid to steal loot bomb |
| 798 | 0.385 | 0.080 | uro missile air defense mine RL submarine Vaenga Red Banner Pacific Fleet... |

**Fig. 2.** Sample user profile produced by resampling logistic regression.

### 3.4   Recommender Algorithm

Here, we present an actual recommender algorithm based on the user profiles mined as in Sect. 3.3. This serves as both a sample application for our user profiling system and as a way to evaluate our results numerically, by comparing it to baseline recommender algorithms.

We propose the following item-based algorithm to make recommendations based on a user profile in the form $\{(c, \mu_{u,c}, \sigma_{u,c}) | c \in C\}$:

(1) penalize the mean of a cluster's weight distribution with its variance, $w_{u,c}^{\star} = \mu_{u,c} - \alpha\sigma_{u,c}$, where $\alpha$ is a coefficient to be tuned for a specific system;
(2) predict the probabilities of likes according to the logistic regression model with modified weights, $p(\text{Like} | d, \boldsymbol{w}_u') = \left(1 + \exp(\boldsymbol{v}_d^{\top} \boldsymbol{w}_u^{\star})\right)^{-1}$;
(3) rank the items according to the predicted probability.

Note that this is a cold start algorithm for the items: it does not use an item's likes at all, only the likes of a user to construct his or her profile.

## 4   Evaluation

### 4.1   Experimental Setup

We have conducted experimental evaluation with a large dataset provided by the "Odnoklassniki" social network. For the experiment, we have chosen to use posts in groups (online communities) and likes provided by the users for these posts since a post in a group, as opposed to a post in a user's profile, is likely to be evaluated by many users with different backgrounds, and the users are more likely to like it based on its topic and content rather than the person who wrote it.

Thus, the dataset consists of texts of posts in the communities (documents) and lists of users who liked the posts. The basic dataset statistics are as follows:

- 286 K words in the vocabulary (after stemming and stop words removal);
- 14.3 M documents (group posts);
- 284.6 M total tokens in these documents;

As the user set $U$, we chose top 2000 users with most likes from a randomly sampled subset of users (so that we get users with a lot of likes but not outliers with huge number of likes that are most probably bots or very uncharacteristic users). We divided their likes into disjoint training and test sets; there were 16000 likes by these users in the training set and 4797 likes in the test set.

### 4.2   Experiments

We carried out our evaluation procedures on three algorithms: two baseline collaborative filtering algorithms and the new algorithm described above in Sect. 3.4.

1. User-based collaborative filtering: find $k$ nearest neighbors for a user and recommend documents according to users' likes. Specifically, for each user $u$ we build a list of $k$ nearest neighbours $N(u)$ by cosine distance (via LSHForest) in the space of their vector representations in $\mathbb{R}^d$. Then we set the affinity between users as cosine distance between their vector representations:

$$w(u_1, u_2) = v_{u_1}^\top v_{u_2}, \quad \forall u_1, u_2 \in U.$$

Documents are ordered by the following ranking function:

$$\text{rank}(u, d) = \frac{\sum_{u' \in N(u) \cap \text{Liked}(d)} w(u, u')}{\sum_{u' \in N(u)} w(u, u')},$$

where $\text{Liked}(d)$ is the set of users who liked document $d$. Thus, we rank documents according to the weighted sum of representations of users who liked it.

2. Item-based collaborative filtering: find $k$ nearest neighbors for a document and recommend documents similar to the ones a user liked. Specifically, for each $d \in D$ we build the set of $k$ nearest neighbours $N(d)$ by cosine distance in the space of vector representations for the documents, compute similarities between the documents, $w(d_1, d_2) = v_{d_1}^\top v_{d_2}$, and rank documents as

$$\text{rank}(u, d) = \frac{\sum_{d' \in N(d) \cap \text{Like}(u)} w(d, d')}{\sum_{d' \in N(d)} w(d, d')},$$

where $\text{Like}(u)$ is the set of documents user $u$ liked.

3. Regression-based algorithm: recommend according to the negative-biased posterior distribution (Sect. 3.4); specifically, given the profile of each user $\{(\mu_{u,c}, \sigma_{u,c} | c \in C\}$, we rank documents according to

$$\text{rank}(u, d) = \frac{1}{1 + e^{-\sum_{c \in C} p(c|d) w_{u,c}^*}},$$

where $w_{u,c}^* = \mu_{u,c} - \alpha \sigma_{u,c}$.

All users and documents vector representations are normalized before applying each of the algorithms above. Each of the evaluated recommender algorithms provides the ranking of documents for a given user. We build a set of all likes from test set and the same number of unliked documents for each user. It is expected that the liked documents will be ranked higher than others on average, which is a common ranking task. Hence, we used standard ranking evaluation metrics to evaluate the algorithms:

– NDCG (Normalized Discounted Cumulative Gain) is a unified metric of ranking quality [25]; the discounted cumulative gain is defined as

$$\text{DCG}_p = \sum_{i=1}^p \frac{\text{liked}_i}{\log_2(i+1)},$$

where liked$_i$ = 1 iff item $i$ in the ranked list is recommended correctly, and NDCG normalizes this value to the maximal possible:

$$NDCG_p = \frac{DCG_p}{IDCG_p},$$

where the ideal DCG IDCG$_p$ is the DCG of a ranked list with all correct items on top;

– Top1, Top5, and Top10 metrics show the share of liked documents at the first place, among the top five, and among the top ten recommendations respectively; these metrics are important for real-life recommender systems since an average user commonly views only a very small number of recommendations.

Results of our experimental evaluation are shown in Table 2. We see that the simple cold start recommender algorithm based on our user profiles performs virtually on par with collaborative filtering algorithms that actually take into account the likes already assigned to this item. These are very good results for a cold start algorithm; note, however, that actual recommendations of full-text items in the same system are not the only or even the main purpose of our approach: the ultimate goal would be to employ user profiles to make outside recommendations for other items with textual content or tags that could be related to the interest profile, such as targeted advertising.

Another way to demonstrate that the regression method learns new things about the users and items being recommended is to show its contribution into the performance of ensembles of rankers. We used the following blending method: first, we normalized the scores obtained by the methods in the blend ($Score_m$ for each ranking method $m$):

$$Score^m_{norm}(d) = \frac{Score^m(d) - \min_{doc} Score^m(doc)}{\max_{doc} Score^m(doc) - \min_{doc} Score^m(doc)}$$

for every document $d$ and ranking method $m$, and then constructed the final scoring function as

$$Score_{blended}(d) = e^{\sum_m e^{Score^m_{norm}(d)} \alpha_m},$$

where $\alpha_m$ are blending weights to be found. We use hill climbing to tune parameters $\alpha_m \in [-1, 1]$, maximizing average NDCG for a separate validation set and finally

**Table 2.** Experimental results for the recommender algorithms.

|  | Algorithm | NDCG | Top1 | Top5 | Top10 |
|---|---|---|---|---|---|
| 1 | User-based CF | 0.7817 | 0.4557 | 1.9440 | 2.6557 |
| 2 | Item-based CF | 0.7904 | 0.4934 | 1.9636 | 2.6589 |
| 3 | Regression-based cold start | 0.7777 | 0.4852 | 1.8741 | 2.5960 |
| 4 | User-based CF + regr | 0.8089 | 0.5508 | 2.0130 | 2.6920 |
| 5 | Item-based CF + regr | 0.8043 | 0.5364 | 1.9834 | 2.6589 |

testing performance on a production set that constituted 20% of the values. Rows 4 and 5 of Table 2 show that the blends noticeably improved upon the performance of both our regression-based approach and classical collaborative filtering.

## 5  Conclusion

In this work, we have presented a new approach to user profiling based on logistic regression on randomly resampled subsets of items. This approach leads to readily interpretable user profiles in case of full-text recommender systems. Our experiments have shown that the simple cold start recommender algorithm based on user profiles produces results comparable to collaborative filtering approaches and can be blended with them for further improvement.

As future work, we envision further applications of this method with other methods for constructing the basic vectors. First, the proposed model may be improved by moving from word vectors to document vectors, e.g., the ones provided by the paragraph2vec models [26]. Full-text recommendations based on distributed word representations may be further improved, especially for the Russian language, by better training and design of custom deep models. On the other hand, for applications where a complete enough text dataset is available it might make more sense to use topic models rather than simple clustering of word vectors. The user profiling approach we have introduced remains applicable in all of these cases.

**Acknowledgements.** This work was supported by the "Recommendation Systems with Automated User Profiling" project sponsored by Samsung and the Government of the Russian Federation grant 14.Z50.31.0030. We thank Dmitry Bugaichenko and the "Odnoklassniki" social network for providing us with the social network dataset with texts of posts and user likes and Alexander Panchenko and Nikolay Arefyev for the trained *word2vec* model along with its Russian-language training data.

## References

1. Webb, G.I., Pazzani, M.J., Billsus, D.: Machine learning for user modeling. User Model. User-Adap. Inter. **11**(1–2), 19–29 (2001)
2. Johnson, A., Taatgen, N.: User modeling. In: Handbook of Human Factors in Web Design. Lawrence Erlbaum, pp. 424–439 (2005)
3. Fischer, G.: User modeling in human–computer interaction. User Model. User-Adap. Inter. **11**(1–2), 65–86 (2001)
4. Brusilovsky, P., Kobsa, A., Nejdl, W. (eds.): The Adaptive Web: Methods and Strategies of Web Personalization. Springer, Heidelberg (2007)
5. Bjorkoy, O.: User modeling on the web: an exploratory review of recommendation systems. PhD thesis, NTNU Trondheim (2010)
6. Lops, P., Gemmis, M.D., Semeraro, G., Lops, P., Gemmis, M.D., Semeraro, G.: Chapter 3 content-based recommender systems: state of the art and trends
7. Pazzani, M.J., Billsus, D.: The Adaptive Web. Springer, Heidelberg (2007)
8. Pazzani, M., Billsus, D.: Learning and revising user profiles: the identification ofinteresting web sites. Mach. Learn. **27**(3), 313–331 (1997)
9. Middleton, S.E., Shadbolt, N.R., De Roure, D.C.: Ontological user profiling in recommender systems. ACM Trans. Inf. Syst. **22**(1), 54–88 (2004)

10. Billsus, D., Pazzani, M.J.: User modeling for adaptive news access. User Model. User-Adap. Inter. **10**(2–3), 147–180 (2000)
11. Cohen, W.W.: Fast effective rule induction. In: 12th International Conference on Machine Learning (ML95), pp. 115–123 (1995)
12. Basu, C., Hirsh, H., Cohen, W.: Recommendation as classification: using social and content-based information in recommendation. In: Proceedings of the 15th National/10th Conference on Artificial Intelligence/Innovative Applications of Artificial Intelligence, AAAI 1998/IAAI 1998, pp. 714–720, Menlo Park, CA, USA. AAAI (1998)
13. Al-Rfou, R., Perozzi, B., Skiena, S.: Polyglot: distributed word representations for multilingual NLP. In: Proceedings of the 17th Conference on Computational Natural Language Learning, Sofia, Bulgaria, ACL, pp. 183–192, August 2013
14. Pennington, J., Socher, R., Manning, C.: Glove: global vectors for word representation. In: Proceedings of the 2014 Conference on Empirical Methods in Natural Language Processing (EMNLP), Doha, Qatar, Association for Computational Linguistics, pp. 1532–1543, October 2014
15. Panchenko, A., Loukachevitch, N., Ustalov, D., Paperno, D., Meyer, C.M., Konstantinova, N.: RUSSE: the first workshop on Russian semantic similarity. In: Proceedings of the International Conference on Computational Linguistics and Intellectual Technologies (Dialogue), pp. 89–105, May 2015
16. Kumar, B.V., Kotsia, I., Patras, I.: Max-margin non-negative matrix factorization. Image Vision Comput. **30**(45), 279–291 (2012)
17. Arefyev, N., Panchenko, A., Lukanin, A., Lesota, O., Romanov, P.: Evaluating three corpus-based semantic similarity systems for Russian. In: Proceedings of International Conference on Computational Linguistics Dialogue (2015, to appear)
18. Vorontsov, K., Frei, O., Apishev, M., Romov, P., Suvorova, M., Yanina, A.: Non-Bayesian additive regularization for multimodal topic modeling of large collections. In: Proceedings of the 2015 Workshop on Topic Models: Post-Processing and Applications, TM 2015, pp. 29–37, New York, NY, USA. ACM (2015)
19. Blei, D.M., Ng, A.Y., Jordan, M.I.: Latent Dirichlet allocation. J. Mach. Learn. Res. **3**(4–5), 993–1022 (2003)
20. Zhang, T., Ramakrishnan, R., Livny, M.: Birch: an efficient data clustering method for very large databases. SIGMOD Rec. **25**(2), 103–114 (1996)
21. Sander, J., Ester, M., Kriegel, H.P., Xu, X.: Density-based clustering in spatial databases: the algorithm GDBSCAN and its applications. Data Min. Knowl. Discov. **2**(2), 169–194 (1998)
22. Comaniciu, D., Meer, P.: Mean shift: a robust approach toward feature space analysis. IEEE Trans. Pattern Anal. Mach. Intell. **24**(5), 603–619 (2002)
23. Sculley, D.: Web-scale k-means clustering. In: Proceedings of the 19th International Conference on World Wide Web, WWW 2010, pp. 1177–1178, New York, NY, USA. ACM(2010)
24. Arthur, D., Vassilvitskii, S.: K-means++: the advantages of careful seeding. In: Proceedings of the Eighteenth Annual ACM-SIAM Symposium on Discrete Algorithms, SODA 2007, pp. 1027–1035, Philadelphia, PA, USA. Society for Industrial and Applied Mathematics (2007)
25. Jarvelin, K., Kekalainen, J.: Cumulated gain-based evaluation of IR techniques. ACM Trans. Inf. Syst. **20**(4), 422–446 (2002)
26. Le, Q., Mikolov, T.: Distributed representations of sentences and documents. In: Jebara, T., Xing, E.P., (eds.) Proceedings of the 31st International Conference on Machine Learning, JMLR Workshop and Conference Proceedings, pp. 1188–1196 (2014)

# Constructing Aspect-Based Sentiment Lexicons with Topic Modeling

Elena Tutubalina[1] and Sergey Nikolenko[1,2,3,4(✉)]

[1] Kazan (Volga Region) Federal University, Kazan, Russia
sergey@logic.pdmi.ras.ru
[2] Steklov Institute of Mathematics at St. Petersburg, St. Petersburg, Russia
[3] Laboratory for Internet Studies, NRU Higher School of Economics,
St. Petersburg, Russia
[4] Deloitte Analytics Institute, Moscow, Russia

**Abstract.** We study topic models designed to be used for sentiment analysis, i.e., models that extract certain topics (aspects) from a corpus of documents and mine sentiment-related labels related to individual aspects. For both direct applications in sentiment analysis and other uses, it is desirable to have a good lexicon of sentiment words, preferably related to different aspects in the words. We have previously developed a modification for several popular sentiment-related LDA extensions that trains prior hyperparameters $\beta$ for specific words. We continue this work and show how this approach leads to new aspect-specific lexicons of sentiment words based on a small set of "seed" sentiment words; the lexicons are useful by themselves and lead to improved sentiment classification.

## 1 Introduction

Over the last decade, opinion mining tasks have attracted a lot of attention from both industry and academia. Modern approaches have mostly emphasized the importance of sentiment lexicons to classify reviews and aspects of a product or a service. Traditional methods determine positive or negative labels of the extracted adjectives and adverbs manually [1], while recent studies use machine learning models to train a classifier or neural network to assign sentiment scores for most words in the corpora [2,3]. However, the efficiency of these supervised methods depends on the domain and size of a labeled data set, and high-quality manual annotation of large corpora and/or vocabularies is very expensive and time-consuming. Moreover, some of recent studies [4] reported that large lexicons created by classifiers are more difficult to use than manually annotated lexicons since rare words added to automatically created lexicons might be irrelevant to the major topics of the text. Hence, it has become an important problem to automatically mine sentiment lexicons for specific aspects, and this is exactly the problem that we deal with in this work.

Recently, topic models have become the method of choice for aspect-based opinion mining: topic models are able to learn to identify latent topical aspects

D.I. Ignatov et al. (Eds.): AIST 2016, CCIS 661, pp. 208–220, 2017.
DOI: 10.1007/978-3-319-52920-2_20

with sentiments towards them in reviews in an unsupervised way. As an example, a standard review about a restaurant will discuss several topical aspects like food, interior, service and price. Therefore, diners are often interested not only in overall star rating of a restaurant, but also in a fine-grained sentiment analysis for a certain aspect. In this study, we focus on applying topic modeling techniques to learn sentiment lexicons for individual aspects. Using an extension of latent Dirichlet allocation (LDA) for sentiment analysis, we identify sentiment scores for semantically related words that have high probability in topics. Then we extend the list of sentiment words by applying distributed word representations trained on a large Russian language dataset. To evaluate the resulting aspect-based lexicons, we plug them back into sentiment classifiers and evaluate the resulting classifiers on a benchmark dataset from the SentiRuEval-2015 competition; we show an improvement in sentiment classification quality with the new aspect-based lexicons. We also perform a qualitative evaluation with sample excerpts from the lexicons.

The paper is organized as follows. In Sect. 2, we introduce the basic LDA model and briefly survey some LDA extensions, concentrating on sentiment-related ones; we also survey the current state of the art in learning sentiment lexicons. In Sect. 3, we introduce our approach to training aspect-based lexicons based on our previous work in improving sentiment-related LDA extensions. In Sect. 4, we show the results of an extensive evaluation on the SentiRuEval-2015 dataset that shows improvement in sentiment classification with the trained lexicons and shows examples of new sentiment words mined from the dataset with our approach. Section 5 concludes the paper.

## 2    LDA and Its Extensions for Sentiment Analysis

In this work, we consider extensions of the latent Dirichlet allocation (LDA) model. We should mention, though, that LDA is not the only modern topic model; for instance, a recently developed Additive Regularization of Topic Models (ARTM) has shown how to construct topic models with many important properties with additive regularization of pLSA [5]. Topic models in sentiment analysis fall into the area of aspect-based sentiment analysis.

### 2.1    Aspect-Based Sentiment Analysis

Over the last decade, a number of methods have been proposed to identify aspect terms with corresponding sentiment polarities; for a good up-to-date overview of the field see [6].

To extract aspect terms, traditional studies applied frequency-based approaches and machine learning. Many approaches extract phrases that contain words from predefined and usually manually constructed lexicons or words that have been shown by trained classifiers to predict a sentiment polarity. In [7], Hu and Liu identified nouns and noun phrases and reduced low-frequency expressions to extract only relevant aspect terms. In [8], the authors treated aspect term extraction as a sequence labeling problem and proposed a scheme for Conditional Random Fields (CRF). In [9], the authors proposed supervised methods based on a Maximum Entropy model with a set of term frequency features in a context of the aspect term and lexicon-based features. Recent studies have applied distributed representations of words due to its ability to cluster semantically similar words [10]. In [11], Blinov and Kotelnikov proposed an method for explicit aspect extraction based on cosine similarity between two vectors of a single word and a seed word. A set of seed words was collected from a train collection. They also applied a set of rules to extract multiword terms. An skip-gram model was trained on unlabeled text data in Russian about for restaurant domain (19,034 reviews). The method achieved best results in aspect term extraction in a sentiment analysis evaluation task called SentiRuEval'15. In [12], Tarasov trained deep recurrent neural networks without manual feature engineering on 300,000 user reviews in Russian. The method achieved second result in prediction of aspect term polarity in SentiRuEval'15.

A sentiment lexicon and manually created dictionaries play a central role in most methods. Many works usually distinguish *affective* words that express feelings ("happy", "disappointed") and *evaluative* words that express sentiment about a specific thing or aspect ("perfect", "awful"); these words as seed words come from a known dictionary, and the model is supposed to combine the sentiments of individual words into a total estimate of the entire text and individual evaluations of specific aspects.

Recently, several topic models, i.e., several LDA extensions have been proposed and successfully used for sentiment analysis [13–16]. These models assume that there is a document-specific distribution over sentiments since sentiment depends on document, and the models' priors are based on the lexicon. In [13], the authors proposed sentiment modifications of LDA, called Joint Sentiment-Topic (JST) and Reverse Joint Sentiment-Topic (Reverse-JST) models, which we will discuss in detail below. The basic assumption was that in the JST model, topics depend on sentiments from a document's sentiment distribution $\pi_d$ and words are generated conditional on sentiment-topic pairs, while in the Reverse-JST model sentiments are generated conditional on the document's topic distribution $\theta_d$.

Similar to the JST, Jo and Oh proposed the Aspect and Sentiment Unification Model (ASUM) [14], where all words in a sentence are generated from one topic with same sentiment. Topics (aspects from reviews) are generated from a sentence distribution over sentiments. ASUM achieved an improvement over supervised classifiers and other generative models, including JST. We also note a nonparametric hierarchical extension of ASUM called HASM [17]; we do not consider nonparametric topic models further in this paper but note them as a possible direction for further work. In [13], a domain-independent sentiment lexicon MPQA was used to incorporate prior knowledge into the models as the word prior sentiment polarity.

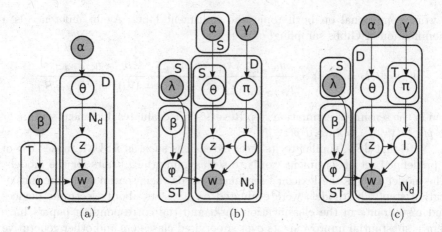

**Fig. 1.** LDA and its sentiment extensions: (a) LDA; (b) JST; (c) Reverse-JST.

## 2.2 JST and Reverse-JST

The JST graphical model is shown on Fig. 1(b). Formally speaking, for $S$ different possible sentiment labels (usually $S$ is small, e.g., for $S = 3$ we have "positive", "neutral", and "negative" labels) it extends the $\phi$ word-topic distributions, generating separate distributions $\phi_{lt}$ for each sentiment label $l \in \{1, \ldots, S\}$ (with a corresponding Dirichlet prior $\lambda$) and a multinomial distribution on sentiment labels $\pi_d$ (with a corresponding Dirichlet prior $\gamma$) for each document $d \in D$.

Each document now has $S$ separate topic distributions for each sentiment label, and the generative process operates as follows: for each word position $j$, (1) sample a sentiment label $l_j \sim \text{Mult}(\pi_d)$; (2) sample a topic $z_j \sim \text{Mult}(\theta_{d,l_j})$; (3) sample a word $w \sim \text{Mult}(\phi_{l_j,z_j})$. The work [13] derives Gibbs sampling distributions for JST by marginalizing out $\pi_d$; denoting by $n_{w,k,t,d}$ the number of words $w$ generated with topic $t$ and sentiment label $k$ in document $d$ and extending the notation accordingly, a Gibbs sampling step can be written as

$$p(z_j = t, l_j = k \mid \nu) \propto \frac{n_{*,k,t,d}^{\neg j} + \alpha_{tk}}{n_{*,k,*,d}^{\neg j} + \sum_t \alpha_{tk}} \cdot \frac{n_{w,k,t,*}^{\neg j} + \beta_{kw}}{n_{*,k,t,*}^{\neg j} + \sum_w \beta_{kw}} \cdot \frac{n_{*,k,*,d}^{\neg j} + \gamma}{n_{*,*,*,d}^{\neg j} + S\gamma},$$

where $\alpha_{tk}$ is the Dirichlet prior for topic $t$ with sentiment label $k$, and $\nu = (\boldsymbol{z}_{-j}, \boldsymbol{w}, \alpha, \beta, \gamma, \lambda)$ is the set of all other variables and model hyperparameters.

The Reverse-JST graphical model is shown on Fig. 1(c). In JST, topics were generated conditional on sentiment labels drawn from $\pi_d$; in Reverse-JST, it works in the opposite direction: for each word $w$, first we draw a topic label $z_j \sim \text{Mult}(\theta_d)$ and then draw a sentiment label $k$ conditional on the topic and a word conditional on both topic and sentiment label. Again, inference is a modification of Gibbs sampling:

$$p(z_j = t, l_j = k \mid \nu) \propto \frac{n_{*,*,t,d}^{\neg j} + \alpha_t}{n_{*,*,*,d}^{\neg j} + \sum_t \alpha_t} \cdot \frac{n_{w,k,t,*}^{\neg j} + \beta}{n_{*,k,t,*}^{\neg j} + W\beta} \cdot \frac{n_{*,k,t,d}^{\neg j} + \gamma}{n_{*,*,t,d}^{\neg j} + S\gamma},$$

and in the sentiment estimation step Reverse-JST evaluates document sentiment as $p(k \mid d) = \sum_z p(k \mid z, d)p(z \mid d)$.

Note that JST, similar to its later extensions such as ASUM, makes use of a predefined set of sentiment words to set asymmetric $\beta$ priors for the models. The model assigns the lexicon's sentiment words their sentiment in the initialization step. The models were evaluated on datasets about electronic devices and restaurants in the classification task, and the corresponding papers have shown substantial improvements over supervised classifiers and other generative models.

## 2.3   Learning Sentiment Lexicons

Most of the commonly used supervised approaches for classifying aspects rely on a predefined list of positive or negative words or a manually generated lexicon. However, there is no one multi-purpose sentiment dictionary which is suitable for each product domain. There has been a lot of research on extraction of sentiment words from text, although in a different way from this work. Specifically for the Russian language, we can distinguish methods that use linguistic resources [1], statistical approaches based on term frequencies [18], and machine learning methods [2,3,19].

The work [1] presents a method for finding new sentiment words based on synonyms and morphologic modifications of known words. In [20], Turney associated each phrase with a sentiment score by calculating it as the mutual information between the phrase and two seed words. In [19], Chetviorkin and Loukachevitch studied domain-specific sentiment lexicon extraction in Russian. They applied a classifier with a set of statistical, morphological, and rating-based features to assign sentiment for words from movie reviews. In [2], Tang et al. presented sentiment-specific phrase embedding based on a neural network and created a classifier to predict the sentiment score of each phrase in the phrase embedding. Severyn and Moschitti automatically classified tweets by positive or negative emoticons and trained a classier on weakly supervised corpora to compute the sentiment association scores as weights from an SVM model [3]. Combining created machine-learned lexicon with the state-of-the-art classifiers gave an improvement in F-measure (from 70.06% to 71.32%). Note that these works have studied document-level classification for domain-specific extraction of sentiment words while we apply probabilistic models to extract topics of semantically related words. To the best of our knowledge, topic models have not been applied to extract sentiment scores by using similarity between words in topics.

# 3   Learning and Extending Aspect-Based Lexicons

The first part of our approach follows our previous work [21]. As we have seen above, existing topic models for aspect-based sentiment analysis almost invariably assume a predefined dictionary of sentiment words, usually incorporating this information into the $\beta$ priors for the word-topic distributions in the LDA model. In [21], we proposed a novel approach for automatic updates of sentiment labels for individual words in a semi-supervised fashion, starting from a small seed dictionary. This modification works as an expectation-maximization generalization of the topic models, learning word-topic priors $\beta$ on E-steps and doing regular inference based on Gibbs sampling on M-steps.

This approach has been applied to JST, Reverse-JST, ASUM, and USTM and led to improved results in every case. Formally speaking, we start with some initial approximation of the sentiments $\beta_s^w$, obtained from a small seed dictionary and/or some simpler learning method used for initialization and then smoothed. Then, iteratively,

(1) at the E-step of iteration $i$, update $\beta_{kw}$ as $\beta_{kw} = \frac{1}{\tau(i)} n_{w,k,*,*}$ with, e.g., $\tau(i) = \max(1, 200/i)$;
(2) at the M-step, perform several iterations of Gibbs sampling for the corresponding model with fixed values of $\beta_{kw}$.

We have shown that this approach leads to improved results in terms of sentiment prediction quality [21]. In this work, we take the next step and use the results of improved sentiment-topic models to learn new aspect-based sentiment dictionaries. We began by extracting the top words from the distribution $\phi$ in sentiment topics in the JST and Reverse-JST models, trying to construct aspect-based lexicons from the topics that define these aspects. However, these were only the words that most often occur in the dataset and are most characteristic for an aspect (and not necessarily carry any sentiment), they did not have sentiment labels trained word-per-word automatically (only topic sentiments), and they did not take into account available general-purpose sentiment dictionaries, so the results were limited at best. Hence, we decided to expand aspect-based lexicons with an approach based on distributed word representations.

Distributed word representations are models that map each word occurring in the dictionary to an Euclidean space with the attempt to capture semantic relationships between the words as geometric relationships in the Euclidean space. In usual model, one first constructs a vocabulary with one-hot representations of individual words, where each word corresponds to its own dimension, and then trains representations for individual words starting from there, basically as a dimensionality reduction problem. For this purpose, researchers have usually employed a model with one hidden layer that attempts to predict the next word based on a window of several preceding words. Then representations learned at the hidden layer are taken to be the word's features; this approach has been applied, for instance in the Polyglot system developed in 2013 [22] and in other methods for learning distributed word representations [23].

For our experiments, we used the skip-gram model trained on a huge Russian-language corpus [24, 25]; it is the first model for the Russian language trained not only on Russian Wikipedia (1.15 M documents, 238 M tokens) and automated Web crawl data (890 K documents, 568 M tokens) but also on the huge lib.rus.ec library corpus (234 K documents, 12.9 G tokens), providing unprecedented quality of the resulting representations; the representations were constructed in the space $\mathbb{R}^{500}$. The intuition behind our use of distributed word representations is that words similar in some aspects of their meaning, such as sentiment, will be expected to be close in the semantic Euclidean space. To expand the top words of resulting topics, we:

- extracted word vectors for all top words from the distribution $\phi$ in topics and all words in available general-purpose sentiment lexicons;
- for every top word in the topics, we constructed a list of its nearest neighbors according to the cosine similarity measure in the $\mathbb{R}^{500}$ space among the sentiment words from the lexicons; our experiments showed that 20 neighbors almost always suffice to cover semantically similar words.

We have also experimented with other similarity metrics ($L_1$, $L_2$, variations on $L_\infty$) but they led to either worse or very similar results so we concentrated on the cosine similarity. The list of nearest neighbors was then used to construct sentiment scores for each word; we have evaluated three different approaches for this, see Sect. 4 below for details and experimental results.

## 4    Experiments

We conducted experiments over a data set with reviews about restaurants in Russian released on SentiRuEval-2015 task that dealt with aspect-based sentiment classification [26]. We used 17,132 unlabeled reviews for training the Reverse Joint-Sentiment Topic model (Reverse-JST). During a preprocessing step, we removed all stopwords and punctuation, converted word tokens to lowercase, and applied the morpho-syntactic analyzer Mystem[1] for text normalization. We also removed the words that occurred less than three times in the dataset.

Following recent studies, we incorporate knowledge from a manually constructed sentiment lexicon into asymmetric Dirichlet priors $\beta_{lw}$. We first set the neutral, positive and negative priors for all words to $\beta_{*w} = 0.01$. Then for a positive word presented in the lexicon we set $\beta_{lw} = (0.01, 2.0, 0.001)$ (2.0 for *pos*, 0.001 for *neg*, and 0.01 for *neut*); for a negative word, $\beta_{lw} = (0.01, 0.001, 2.0)$. Inference for Reverse-JST was done with 1000 Gibbs iterations with priors $\alpha = 50/K$ and $\gamma = 0.01$, where $K$ is the number of topics.

To identify aspect-based sentiment words, we first evaluated Reverse-JST with different number of topics, choosing 20 topics. Then we selected top 500 words from the topic-sentiment distribution $\phi$. Table 1 shows sample sentiment-specific topics; sentiment subtopics reflect different aspects discussed in reviews.

---

[1] https://tech.yandex.ru/mystem/.

For example, the first neutral, positive, and negative topics are related to food, atmosphere/interior, and service respectively.

Then we applied the approach described in the previous section. The created aspect-based lexicons consist of 726 topical aspects that commonly divided into three types in sentiment analysis:

(i) explicit aspects that denote parts of a product (e.g., *сотрудник* [*worker*], *баранина* [*lamb*], *овощ* [*vegetable*], *мексиканский* [*mexican*]);

(ii) implicit aspects that refer indirectly to a product (e.g., *чисто* [*clean*], *ароматный* [*aromatic*], *сытно* [*filling*], *шумно* [*noisy*]);

(iii) narrative words which related to major topics in the text and indirectly refer to sentiment polarity of the text (e.g., *пересолить* [*to oversalt*], *пожелать* [*to wish*], *почувствовать* [*to sense*], *отсутствовать* [*be missing*]).

All aspects are presented with (sentiment word, word-aspect similarity score) pairs. Below we present quantitative results of applying the lexicons to a classification task and discuss examples of discovered words associated with various aspects.

In this work, we focus on predicting the sentiment of aspects about restaurants, aiming to classify whether an aspect expresses positive or negative sentiment. We applied a classifier from [9] based on a maximum entropy model. This classifier relies on a set of term frequency features in the context of an aspect

**Table 1.** Sample topics discovered by Reverse-JST.

| # | sent. | sentiment words |
|---|---|---|
| 1 | neu | соус [sauce], салат [salad], кусочек [slice], сыр [cheese], тарелка [plate], овощ [vegetable], масло [oil], лук [onions], перец [pepper] |
| | pos | приятный [pleasant], атмосфера [atmosphere], уютный [cozy], вечер [evening], музыка [music], ужин [dinner], романтический [romantic] |
| | neg | ресторан [restaurant], официант [waiter], внимание [attention], сервис [service], обращать [to notice], обслуживать [to serve], уровень [level] |
| 2 | neu | столик [table], заказывать [to order], вечер [evening], стол [table], приходить [to come], место [place], заранее [in advance], встречать [to meet] |
| | pos | место [place], хороший [good], вкус [taste], самый [most], приятный [pleasant], вполне [quite], отличный [excellent], интересный [interesting] |
| | neg | еда [food], вообще [in general], никакой [none], заказывать [to order], оказываться [appear], вкус [taste], ужасный [awful], ничто [nothing] |
| 3 | neu | девушка [girl], спрашивать [to ask], вопрос [question], подходить [to come], официантка [waitress], официант [waiter], говорить [to speak] |
| | pos | большой [big], место [place], выбор [choice], хороший [good], блюдо [dish], цена [price], порция [portion], небольшой [small], плюс [plus] |
| | neg | цена [price], обслуживание [service], качество [quality], уровень [level], кухня [kitten], средний [average], ценник [price tag], высоко [high] |
| 4 | neu | пицца [pizza], паста [paste], итальянский [italian], заказывать [to order], вкусный [tasty], италия [Italy], цезарь [caesar], большой [big], сыр [cheese] |
| | pos | кухня [kitchen], место [place], блюдо [dish], высоко [highly], прекрасный [fine], похвала [praise], настоящий [real], удовольствие [pleasure] |
| | neg | коктейль [cocktail], бар [bar], бармен [bartender], напиток [drink], алкоголь [alcohol], выпивать [to drink], мохито [mohito], народ [people] |

term and lexicon-based features commonly used in opinion mining classifiers [27]. We extract the following features from an aspect's context window of 4 words:

(i) lowercased *character n-grams* with document frequency greater than two;
(ii) *lexicon-based unigrams* and *context unigrams and bigrams*;
(iii) *aspect-based bigrams* as a combination of the aspect term itself and words;
(iv) *lexicon-based features*: the maximal sentiment score, the minimum sentiment score, the total and averaged sums of the words' sentiment scores.

We compare the performance of classifiers with lexicon-based features computed on a manually constructed general-purpose lexicon (baseline classifier) and computed on a general-purpose lexicon for all words and aspect-based lexicons for individual aspects. The resulting aspect-based lexicon consists of (sentiment word, similarity score) pairs for each aspect, the score being the semantic similarity between the aspect and the word. We evaluated three different versions of sentiment scores:

- *scoresDict*: take sentiment score from the manually created lexicon if the word occurs in the lexicon with a positive or negative label; otherwise, set the score to 0;
- *scoresMult*: set the sentiment score of a word as a product of the dictionary score and the similarity score;
- *scoresCos*: set the sentiment score to the similarity score if the similarity between the word in question and the word *хороший* [*good*] is higher than the similarity between the word in question and *плохой* [*bad*]; otherwise, shift sentiment score towards the opposite polarity; similarity between words is measured using cosine measure.

For the evaluation, we use training and testing datasets released for the SentiRuEval-2015 task. In order to demonstrate the impact of aspect-based lexicons, we selected only those labeled aspects from datasets that are present in our lexicon. Dataset statistics are shown in Table 2; we see that aspects with positive sentiments dominate in the corpora, which makes uncovering negative information challenging.

We applied two manually constructed lexicons to evaluate classifiers. The first lexicon (denoted *Lexicon1*) from [9] consists of 1079 positive words and 1474 negative words without scores; we set sentiment score to 1 and −1 for positive and negative words respectively. The second lexicon (denoted *Lexicon2*) consists of 1241 positive words and 431 negative words with real-valued scores.

**Table 2.** Summary statistics of review datasets.

| Dataset | # of aspects | # of positive | # of negative | # of unique asp. |
|---|---|---|---|---|
| Training dataset | 1796 | 1526 | 270 | 173 |
| Testing dataset | 1779 | 1489 | 290 | 171 |

**Table 3.** Classification results

| Maximum entropy classifier | Micro-averaged | | | Macro-averaged | | |
|---|---|---|---|---|---|---|
| | P | R | F1 | P | R | F1 |
| baseline features - Lexicon1 | 0.595 | 0.344 | 0.436 | 0.738 | 0.649 | 0.676 |
| With features based on scoresDict | 0.592 | 0.344 | 0.436 | 0.737 | 0.649 | 0.676 |
| With features based on scoresMult | 0.600 | 0.351 | 0.442 | 0.740 | 0.653 | 0.680 |
| With features based on scoresCos | **0.610** | 0.372 | 0.462 | **0.748** | 0.663 | **0.691** |
| baseline features - Lexicon2 | 0.572 | 0.341 | 0.427 | 0.727 | 0.646 | 0.671 |
| With features based on scoresDict | 0.568 | 0.345 | 0.430 | 0.725 | 0.647 | 0.672 |
| With features based on scoresMult | 0.556 | 0.338 | 0.420 | 0.719 | 0.643 | 0.667 |
| With features based on scoresCos | 0.566 | 0.368 | 0.447 | 0.725 | 0.657 | 0.680 |
| baseline - Lexicon1 - Lexicon2 | 0.594 | 0.348 | 0.439 | 0.738 | 0.651 | 0.679 |
| With features based on scoresDict | 0.595 | 0.376 | 0.461 | 0.741 | 0.663 | 0.689 |
| With features based on scoresMult | 0.590 | 0.372 | 0.457 | 0.738 | 0.661 | 0.687 |
| With features based on scoresCos | 0.602 | **0.376** | **0.463** | 0.744 | **0.664** | 0.690 |

We used precision, recall, and F1-measure to evaluate classifiers and computed micro-averaged scores based only on the negative class and macro-averaged scores based on both classes. Table 3 presents classification results on the test dataset. It is clear sentiment features based on aspect-based lexicons can improve classification results over the baseline if sentiment scores are based on similarity scores for all words in the lexicon (scoresCos).

Several factors could contribute to this improvement. First, Lexicon1 and Lexicon2 contain a limited number of words while an automatically generated lexicon could contain new words. Second, the similarity scores multiplied by real-valued scores decrease the importance of aspect-based sentiment words (macro F1 with scoresMult is 0.667 vs the baseline's F1 is 0.671). A classifier with features based on two manually created lexicons gives better results than classifiers with features based on only one of the lexicons.

To illustrate how our model associates aspects and corresponding sentiments, we present some aspect-sentiment words in Table 4. Sentiment words reflect a particular set of topical aspects, e.g., *развлекательный* [entertaining] of the aspect *караоке* [karaoke] or *неисполнительный* [careless] of the aspect *администратор* [manager]. People tend to use *вкусный* [tasty] and *сладкий* [sweet] to express positive sentiments about food (*баранина* [lamb] and *пирог* [pie]), while *уютный* [cozy] is related to *интерьер* [interior]. These examples show sentiment lexicons specific to aspects.

**Table 4.** Examples of discovered sentiment words.

| aspect | sentiment words |
| --- | --- |
| баранина [lamb] | вкусный [tasty], сытный [filling], аппетитный [delicious], душистый [sweet smelling], деликатесный [speciality], сладкий [sweet] |
| караоке [karaoke] | музыкальный [musical], попсовый [pop], классно [awesome], развлекательный [entertaining], улетный [mind-blowing] |
| пирог [pie] | вкусный [tasty], аппетитный [delicious],обсыпной [bulk], сытный [filling], черствый [stale], ароматный [aromatic], сладкий [sweet] |
| ресторан [restaurant] | шикарный [upscale], фешенебельный [fashionable], уютный [cozy], люкс [luxe], роскошный [luxurious], недорогой [affordable] |
| вывеска [sign] | обветшалый [decayed], выцветший [faded], аляповатый [flashy], фешенебельный [fashionable], фанерный [veneer] |
| администратор [manager] | люкс [luxe], неисполнительный [careless], ответственный [responsible], компетентный [competent], толстяк [fatty] |
| интерьер [interior] | уют [comfort], уютный[cozy], стильный [stylish], просторный [spacious], помпезный [magnific], роскошный [luxurious], шикарный [upscale] |
| вежливый [delicate] | вежливый [delicate], учтивый[polite], обходительный [affable], доброжелательный [good-minded], тактичный [diplomatic] |

## 5  Conclusion

In this work, we have presented a method for automatically extracting aspect-based sentiment lexicons based on an extension of sentiment-related topic models augmented with similarity search with distributed word representations. Our experiments show that this approach lets one extract important new sentiment words for aspect-specific lexicons that further improve sentiment classification on standard benchmarks. In further work, we plan to continue elaborating upon the interplay between sentiment priors in LDA extensions and distributed word representations; we hope it will be possible to incorporate distributed word representations directly into the priors, designing a new model that will improve sentiment classification further and provide even better aspect-based lexicons.

**Acknowledgements.** This work was supported by the Russian Science Foundation grant no. 15-11-10019. We thank Alexander Panchenko and Nikolay Arefyev for providing us the *word2vec* model and its Russian-language training data.

## References

1. Neviarouskaya, A., Prendinger, H., Ishizuka, M.: Sentiful: generating a reliable lexicon for sentiment analysis. In: 3rd International Conference on Affective Computing and Intelligent Interaction and Workshops, 2009, ACII 2009, pp. 1–6. IEEE (2009)
2. Tang, D., Wei, F., Qin, B., Zhou, M., Liu, T.: Building large-scale twitter-specific sentiment lexicon: a representation learning approach. In: Proceedings of COLING, pp. 172–182 (2014)
3. Severyn, A., Moschitti, A.: On the automatic learning of sentiment lexicons. In: Proceedings of the Conference of the North American Chapter of the Association for Computational Linguistics (NAACL HLT 2015) (2015)
4. Kiritchenko, S., Zhu, X., Mohammad, S.M.: Sentiment analysis of short informal texts. J. Artif. Intell. Res. **50**, 723–762 (2014)

5. Vorontsov, K., Frei, O., Apishev, M., Romov, P., Suvorova, M., Yanina, A.: Non-bayesian additive regularization for multimodal topic modeling of large collections. In: Proceedings of the 2015 Workshop on Topic Models: Post-Processing and Applications, TM 2015, New York, NY, USA, pp. 29–37. ACM (2015)
6. Liu, B.: Sentiment Analysis: Mining Opinions, Sentiments, and Emotions. Cambridge University Press, Cambridge (2015)
7. Hu, M., Liu, B.: Mining and summarizing customer reviews. In: Proceedings of the Tenth ACM SIGKDD International Conference on Knowledge Discovery and Data Mining, pp. 168–177. ACM (2004)
8. Chernyshevich, M.: Ihs r&d belarus: cross-domain extraction of product features using conditional random fields. In: Proceedings of the 8th International Workshop on Semantic Evaluation (SemEval 2014), pp. 309–313 (2014)
9. Ivanov, V., Tutubalina, E., Mingazov, N., Alimova, I.: Extracting aspects, sentiment and categories of aspects in user reviews about restaurants and cars. In: Proceedings of International Conference Dialog, pp. 22–34 (2015)
10. Mikolov, T., Sutskever, I., Chen, K., Corrado, G.S., Dean, J.: Distributed representations of words and phrases and their compositionality. In: Advances in Neural Information Processing Systems, pp. 3111–3119 (2013)
11. Blinov, P.D., Kotelnikov, E.V.: Semantic similarity for aspect-based sentiment analysis. In: Proceedings of the 21st International Conference on Computational Linguistics Dialog-2015, vol. 2, pp. 36–45 (2015)
12. Tarasov, D.S.: Deep recurrent neural networks for multiple language aspect-based sentiment analysis of user reviews. In: Proceedings of the 21st International Conference on Computational Linguistics Dialog-2015, vol. 2 (2015)
13. Lin, C., He, Y., Everson, R., Ruger, S.: Weakly supervised joint sentiment-topic detection from text. IEEE Trans. Knowl. Data Eng. **24**, 1134–1145 (2012)
14. Jo, Y., Oh, A.H.: Aspect and sentiment unification model for online review analysis. In: Proceedings of the Fourth ACM International Conference on Web Search and Data Mining, WSDM 2011, New York, NY, USA, pp. 815–824. ACM (2011)
15. Yang, Z., Kotov, A., Mohan, A., Lu, S.: Parametric and non-parametric user-aware sentiment topic models. In: Proceedings of the 38th ACM SIGIR (2015)
16. Lu, B., Ott, M., Cardie, C., Tsou, B.: Multi-aspect sentiment analysis with topic models. In: 2011 IEEE 11th International Conference Data Mining Workshops (ICDMW), pp. 81–88 (2011)
17. Kim, S., Zhang, J., Chen, Z., Oh, A.H., Liu, S.: A hierarchical aspect-sentiment model for online reviews. In: Proceedings of the Twenty-Seventh AAAI Conference on Artificial Intelligence, Bellevue, Washington, USA, 14–18 July 2013 (2013)
18. Hatzivassiloglou, V., McKeown, K.R.: Predicting the semantic orientation of adjectives. In: Proceedings of the 35th Annual Meeting of the Association for Computational Linguistics and Eighth Conference of the European Chapter of the Association for Computational Linguistics, pp. 174–181. ACL (1997)
19. Chetviorkin, I., Loukachevitch, N.V.: Extraction of Russian sentiment lexicon for product meta-domain. In: COLING, pp. 593–610. Citeseer (2012)
20. Turney, P.D.: Thumbs up or thumbs down?: semantic orientation applied to unsupervised classification of reviews. In: Proceedings of the 40th Annual Meeting on Association for Computational Linguistics, pp. 417–424. Association for Computational Linguistics (2002)
21. Tutubalina, E., Nikolenko, S.: Inferring sentiment-based priors in topic models. In: Lagunas, O.P., Alcántara, O.H., Figueroa, G.A. (eds.) MICAI 2015. LNCS (LNAI), vol. 9414, pp. 92–104. Springer, Heidelberg (2015). doi:10.1007/978-3-319-27101-9_7

22. Al-Rfou, R., Perozzi, B., Skiena, S.: Polyglot: distributed word representations for multilingual NLP. In: Proceedings of the Seventeenth Conference on Computational Natural Language Learning, Sofia, Bulgaria, pp. 183–192. Association for Computational Linguistics (2013)
23. Pennington, J., Socher, R., Manning, C.: Glove: global vectors for word representation. In: Proceedings of the 2014 Conference on Empirical Methods in Natural Language Processing (EMNLP), Doha, Qatar, pp. 1532–1543. Association for Computational Linguistics (2014)
24. Arefyev, N., Panchenko, A., Lukanin, A., Lesota, O., Romanov, P.: Evaluating three corpus-based semantic similarity systems for Russian. In: Proceedings of International Conference on Computational Linguistics Dialogue (2015)
25. Panchenko, A., Loukachevitch, N., Ustalov, D., Paperno, D., Meyer, C.M., Konstantinova, N.: Russe: the first workshop on russian semantic similarity. In: Proceedings of the International Conference on Computational Linguistics and Intellectual Technologies (Dialogue), pp. 89–105 (2015)
26. Loukachevitch, N., Blinov, P., Kotelnikov, E., Rubtsova Yu, V., Ivanov, V., Tutubalina, E.: Sentirueval: testing object-oriented sentiment analysis systems in Russian. In: Proceedings of International Conference Dialog, pp. 3–9 (2015)
27. Hagen, M., Potthast, M., Büchner, M., Stein, B.: Twitter sentiment detection via ensemble classification using averaged confidence scores. In: Hanbury, A., Kazai, G., Rauber, A., Fuhr, N. (eds.) ECIR 2015. LNCS, vol. 9022, pp. 741–754. Springer, Heidelberg (2015). doi:10.1007/978-3-319-16354-3_81

# Human and Machine Judgements for Russian Semantic Relatedness

Alexander Panchenko[1]([✉]), Dmitry Ustalov[2], Nikolay Arefyev[3], Denis Paperno[4], Natalia Konstantinova[5], Natalia Loukachevitch[3], and Chris Biemann[1]

[1] TU Darmstadt, Darmstadt, Germany
{panchenko,biem}@lt.informatik.tu-darmstadt.de
[2] Ural Federal University, Yekaterinburg, Russia
dmitry.ustalov@urfu.ru
[3] Moscow State University, Moscow, Russia
louk_nat@mail.ru
[4] University of Trento, Rovereto, Italy
denis.paperno@unitn.it
[5] University of Wolverhampton, Wolverhampton, UK
n.konstantinova@wlv.ac.uk

**Abstract.** Semantic relatedness of terms represents similarity of meaning by a numerical score. On the one hand, humans easily make judgements about semantic relatedness. On the other hand, this kind of information is useful in language processing systems. While semantic relatedness has been extensively studied for English using numerous language resources, such as associative norms, human judgements and datasets generated from lexical databases, no evaluation resources of this kind have been available for Russian to date. Our contribution addresses this problem. We present five language resources of different scale and purpose for Russian semantic relatedness, each being a list of triples $(word_i, word_j, similarity_{ij})$. Four of them are designed for evaluation of systems for computing semantic relatedness, complementing each other in terms of the semantic relation type they represent. These benchmarks were used to organise a shared task on Russian semantic relatedness, which attracted 19 teams. We use one of the best approaches identified in this competition to generate the fifth high-coverage resource, the first open distributional thesaurus of Russian. Multiple evaluations of this thesaurus, including a large-scale crowdsourcing study involving native speakers, indicate its high accuracy.

**Keywords:** Semantic similarity · Semantic relatedness · Evaluation · Distributional thesaurus · Crowdsourcing · Language resources

## 1 Introduction

Semantic relatedness numerically quantifies the degree of semantic alikeness of two lexical units, such as words and multiword expressions. The relatedness score

© Springer International Publishing AG 2017
D.I. Ignatov et al. (Eds.): AIST 2016, CCIS 661, pp. 221–235, 2017.
DOI: 10.1007/978-3-319-52920-2_21

is high for pairs of words in a semantic relation (e.g., synonyms, hyponyms, free associations) and low for semantically unrelated pairs. *Semantic relatedness* and *semantic similarity* have been extensively studied in psychology and computational linguistics, see [1–4] inter alia. While both concepts are vaguely defined, similarity is a more restricted notion than relatedness, e.g. "apple" and "tree" would be related but not similar. Semantically similar word pairs are usually synonyms or hypernyms, while relatedness also can also refer to meronyms, co-hyponyms, associations and other types of relations. Semantic relatedness is an important building block of NLP techniques, such as text similarity [5,6], word sense disambiguation [7], query expansion [8] and some others [9].

While semantic relatedness was extensively studied in the context of the English language, NLP researchers working with Russian language could not conduct such studies due to the lack of publicly available relatedness resources. The datasets presented in this paper are meant to fill this gap. Each of them is a collection of weighted word pairs in the format $(w_i, w_j, s_{ij})$, e.g. (*book, proceedings*, 0.87). Here, the $w_i$ is the source word, $w_j$ is the destination word and $s_{ij} \in [0; 1]$ is the semantic relatedness score (see Table 1).

More specifically, we present (1) four resources for evaluation and training of semantic relatedness systems varying in size and relation type and (2) the first open distributional thesaurus for the Russian language (see Table 2). All datasets contain relations between single words.

The paper is organized as follows. Sect. 2 describes approaches to evaluation of semantic relatedness in English. Section 3 presents three datasets where semantic relatedness of word was established manually. The HJ dataset, further described in Sect. 3.1, is based on Human Judgements about semantic relatedness; the RuThes (RT) dataset is based on synonyms and hypernyms from a handcrafted thesaurus (see Sect. 3.2); the Associative Experiment (AE) dataset, introduced in Sect. 3.3, represents cognitive associations between words. Section 4 describes datasets where semantic relatedness between words is established automatically: the Machine Judgements (MJ) dataset, presented in Sect. 4.1, is based on a combination of submissions from a shared task on Russian semantic similarity; Sect. 4.2 describes the construction and evaluation of the Russian Distributional Thesaurus (RDT).

## 2   Related Work

There are three main approaches to evaluating semantic relatedness: using human judgements about word pairs, using semantic relations from lexical-semantic resources, such as WordNet [10], and using data from cognitive word association experiments. We built three evaluation datasets for Russian each based on one of these principles to enable a comprehensive comparison of relatedness models.

**Table 1.** Example of semantic relations from the datasets described in this paper: the five most and least similar terms to the word "книга" (book) in the MJ dataset

| Source Word, $w_j$ | Destination Word, $w_j$ | Semantic Relatedness, $s_{ij}$ |
|---|---|---|
| книга (book) | книжка (book, little book) | 0.719 |
| книга (book) | книжечка (little book) | 0.646 |
| книга (book) | сборник (proceedings) | 0.643 |
| книга (book) | монография (monograph) | 0.574 |
| книга (book) | том (volume) | 0.554 |
| книга (book) | трест (trust as organization) | 0.151 |
| книга (book) | одобрение (approval) | 0.150 |
| книга (book) | киль (keel) | 0.130 |
| книга (book) | Марокко (Marocco) | 0.124 |
| книга (book) | Уругвай (Uruguay) | 0.092 |

**Table 2.** Language resources presented in this paper. The pipe (|) separates the sizes of two dataset versions: one with manual filtering of negative examples and the other version, marked by an asterix (*), where negative relations were generated automatically, i.e. without manual filtering

| Dataset | HJ | RT | AE | MJ | RDT |
|---|---|---|---|---|---|
| # relations | 398 | 9 548 \| 114 066* | 3 002 \| 86 772* | 12 886 | 193 909 130 |
| # source words, $w_i$ | 222 | 1,008 \| 6 832* | 340 \| 5 796* | 1 519 | 931 896 |
| # destination words, $w_j$ | 306 | 7 737 \| 71 309* | 2 498 \| 56 686* | 9 044 | 4 456 444 |
| Types of relations | Relatedness | Synonyms, hypernyms | Associations | Relatedness | Relatedness |
| Similarity score, $s_{ij}$ | From 0 to 1 | 0 or 1 | 0 or 1 | From 0 to 1 | From 0 to 1 |
| Part of speech | Nouns | Nouns | Nouns | Nouns | Any |

## 2.1 Datasets Based on Human Judgements About Word Pairs

Word pairs labeled manually on a categorical scale by human subjects is the basis of this group of benchmarks. High scores of subjects indicate that words are semantically related, low scores indicate that they are unrelated. The HJ dataset presented in Sect. 3.1 belongs to this group of evaluation datasets.

Research on relatedness starts from the pioneering work of Rubenstein and Goodenough [11], where they aggregated human judgments on the relatedness of 65 noun pairs into the RG dataset. 51 human subjects rated the pairs on a scale from 0 to 4 according to their similarity. Later, Miller and Charles [12] replicated the experiment of Rubenstein and Goodenough, obtaining similar results on a subset of 30 noun pairs. They used 10 words from the high level (between 3 and 4), 10 from the intermediate level (between 1 and 3), and 10 from the low level (0 to 1) of semantic relatedness, and then obtained similarity judgments from 38 subjects, given the RG annotation guidelines, on those 30 pairs. This dataset is known as the MC dataset.

A larger set of 353 word pairs was put forward by Filkenstein et al. [13] as the WordSim353 dataset. The dataset contains 353 word pairs, each associated with 13 or 16 human judgements. In this case, the subjects were asked to rate word pairs for relatedness, although many of the pairs also exemplify semantic similarity. That is why, Agirre et al. [14] subdivided the WordSim353 dataset into two subsets: the WordSim353 similarity set and the WordSim353 relatedness set. The former set consists of word pairs classified as synonyms, antonyms, identical, or hyponym-hypernym and unrelated pairs. The relatedness set contains word pairs connected with other relations and unrelated pairs. The similarity set contains 204 pairs and the relatedness set includes 252 pairs.

The three abovementioned datasets were created for English. There have been several attempts to translate those datasets into other languages. Gurevych translated the RG and MC datasets into German [15]; Hassan and Mihalcea translated them into Spanish, Arabic and Romanian [16]; Postma and Vossen [17] translated the datasets into Dutch; Jin and Wu [18] presented a shared task for Chinese semantic similarity, where the authors translated the WordSim353 dataset. Yang and Powers [19] proposed a dataset specifically for measuring verb similarity, which was later translated into German by Meyer and Gurevych [20].

Hassan and Mihalcea [16] and Postma and Vossen [17] used three stages to translation pairs: (1) disambiguation of the English word forms; (2) translation for each word; (3) ensuring that translations are in the same class of relative frequency as the English source word.

More recently, SimLex-999 was released by Hill et al. [21], focusing specifically on similarity and not relatedness. While most datasets are only available in English, SimLex-999 became a notable exception and has been translated into German, Russian and Italian. The Russian version of SimLex-999 is similar to the HJ dataset presented in our paper. In fact, these Russian datasets were created in parallel almost at the same time.[1] SimLex-999 contains 999 word pairs, which is considerably larger than the classical MC, RG and WordSim353 datasets.

The creators of the MEN dataset [22] went even further, annotating via crowdsourcing 3 000 word pairs sampled from the ukWaC corpus [23]. However, this dataset is also available only for English. A comprehensive list of datasets for evaluation of English semantic relatedness, featuring 12 collections, was gathered by Faruqui and Dyer [24]. This set of benchmarks was used to build a web application for evaluation and visualization of word vectors.[2]

## 2.2   Datasets Based on Lexical-Semantic Resources

Another group of evaluation datasets evaluates semantic relatedness scores with respect to relations described in lexical-semantic resources such as WordNet. The RT dataset presented in Sect. 3.2 belongs to this group of evaluation datasets.

Baroni and Lenci [25] stressed that semantically related words differ in the type of relation between them, so they generated the BLESS dataset containing tuples of the form $(w_j, w_j, type)$. Types of relations included co-hyponyms, hypernyms, meronyms, attributes (relation between a noun and an adjective expressing its attribute), event (relation between a noun and a verb referring to

---

[1] The HJ dataset was first released in November 2014 and first published in June 2015, while the SimLex-999 was first published December 2015.

[2] http://wordvectors.org/suite.php.

actions or events). BLESS also contains, for each target word, a number of random words that were checked to be semantically unrelated to this word. BLESS includes 200 English concrete single-word nouns having reasonably high frequency that are not very polysemous. The destination words of the non-random relations are English nouns, verbs and adjectives selected and validated manually using several sources including WordNet, and collocations from the Wikipedia and the ukWaC corpora.

Van de Cruys [26] used Dutch WordNet to evaluate distributional similarity measures. His approach uses the structure of the lexical resource, whereby distributional similarity is compared to shortest-path-based distance. Biemann and Riedl [27] follow a similar approach based on the English WordNet to assess quality of their distributional semantics framework.

Finally, Sahlgren [28] evaluated distributional lexical similarity measures comparing them to manually-crafted thesauri, but also associative norms, such as those described in the following section.

### 2.3  Datasets Based on Human Word Association Experiments

The third strain of research evaluates the ability of current automated systems to simulate the results of human word association experiments. Evaluation tasks based on associative relations originally captured attention of psychologists, such as Griffiths and Steyvers [29]. One such task was organized in the framework of the Cogalex workshop [30]. The participants received lists of five words (e.g. "circus", "funny", "nose", "fool", and "Coco") and were supposed to select the word most closely associated to all of them. In this specific case, the word "clown" is the expected response. 2 000 sets of five input words, together with the expected target words (associative responses) were provided as a training set to participants. The test dataset contained another 2 000 sets of five input words. The training and the test datasets were both derived from the Edinburgh Associative Thesaurus (EAT) [31]. For each stimulus word, only the top five associations, i.e. the associations produced by the largest number of respondents, were retained, and all other associations were discarded. The AE dataset presented in Sect. 3.3 belongs to this group of evaluation datasets.

## 3  Human Judgements About Semantic Relatedness

In this section, we describe three datasets designed for evaluation of Russian semantic relatedness measures. The datasets were tested in the framework of the shared task on RUssian Semantic Similarity Evaluation (RUSSE) [32].[3] Each participant had to calculate similarities between a collection of word pairs. Then, each submission was assessed using the three benchmarks presented below, each being a subset of the input word pairs.

---

[3] http://russe.nlpub.ru.

### 3.1    HJ: Human Judgements of Word Pairs

**Description of the Dataset.** The HJ dataset is a union of three widely used benchmarks for English: RG, MC and WordSim353, see [14,33–35,35,36] inter alia. The dataset contains 398 word pairs translated to Russian and re-annotated by native speakers. In addition to the complete dataset, we also provide separate parts that correspond to MC, RG and WordSim353.

To collect human judgements, an in-house crowdsourcing system was used. We set up a special section on the RUSSE website and asked volunteers on Facebook and Twitter to participate in the experiment. Each annotator received an assignment consisting of 15 word pairs randomly selected from the 398 pairs, and has been asked to assess the relatedness of each pair on the following scale: 0 – not related at all, 1 – weak relatedness, 2 – moderate relatedness, and 3 – high relatedness. We provided annotators with simple instructions explaining the procedure and goals of the study.[4]

A pair of words was added to the annotation task with the probability inversely proportional to the number of current annotations. We obtained a total of 4 200 answers, i.e. 280 submissions of 15 judgements. Ordinal Krippendorff's alpha of 0.49 indicates a moderate agreement of annotators. The scores included in the HJ dataset are average human ratings scaled to the $[0, 1]$ range.

**Using the Dataset.** To evaluate a relatedness measure using this dataset one should (1) calculate relatedness scores for each pair in the dataset; (2) calculate Spearman's rank correlation coefficient $\rho$ between the vector of human judgments and the scores of the system (see Table 4 for an example).

### 3.2    RT: Synonyms and Hypernyms

**Description of the Dataset.** This dataset follows the structure of the BLESS dataset [25]. Each target word has the same number of related and unrelated source words. The dataset contains 9 548 relations for 1 008 nouns (see Table 2). Half of these relations are synonyms and hypernyms from the RuThes-lite thesaurus [37] and half of them are unrelated words. To generate negative pairs we used the automatic procedure described in Panchenko et al. [32]. We filtered out false negative relations for 1 008 source words with the help of human annotators. Each negative relation in this subset was annotated by at least two annotators: Masters' students of an NLP course, native speakers of Russian.

As the result, we provide a dataset featuring 9 548 relations of 1 008 source words, where each source word has the same number of negative random relations and positive (synonymous or hypernymous) relations. In addition, we provide a larger dataset of 114 066 relations for 6 832 source words, where negative relations have not been verified manually.

---

[4] Annotation guidelines for the HJ dataset: http://russe.nlpub.ru/task/annotate.txt.

**Using the Dataset.** To evaluate a similarity measure using this dataset one should (1) calculate relatedness scores for each pair in the dataset; (2) first sort pairs by the score; and then (3) calculate the average precision metric:

$$AveP = \frac{\sum_r P(r)}{R},$$

where $r$ is the rank of each non-random pair, $R$ is the total number of non-random pairs, and $P(r)$ is the precision of the top-$r$ pairs. See Table 4 and [32] for examples. Besides, the dataset can be used to train classification models for predicting hypernyms and synonyms using the binary $s_{ij}$ scores.

### 3.3   AE: Cognitive Associations

**Description of the Dataset.** The structure of this dataset is the same as the structure of the RT dataset: each source word has the same number of related and unrelated target words. The difference is that, related word pairs of this dataset were sampled from a Russian web-based associative experiment.[5] In the experiment, users were asked to provide a reaction to an input stimulus source word, e.g.: man $\rightarrow$ woman, time $\rightarrow$ money, and so on. The strength of association in this experiment is quantified by the number of respondents providing the same stimulus-reaction pair. Associative thesauri typically contain a mix of synonyms, hyponyms, meronyms and other types, making relations asymmetric. To build this dataset, we selected target words with the highest association with the stimulus in Sociation.org data. Like with the other datasets, we used only single-word nouns. Similarly to the RT dataset, we automatically generated negative word pairs and filtered out false negatives with help of annotators.

As the result, we provide a dataset featuring 3 002 relations of 340 source words (see Table 2), where each source word has the same number of negative random relations and positive associative relations. In addition, we provide the larger dataset of 86 772 relations for 5 796 source words, where negative relations were not verified manually.

**Using the Dataset.** Evaluation procedure using this dataset is the same as for the RT dataset: one should calculate the average precision $AveP$. Besides, the dataset can be used to train classification models for predicting associative relations using the binary $s_{ij}$ scores.

## 4   Machine Judgements About Semantic Relatedness

### 4.1   MJ: Machine Judgements of Word Pairs

**Description of the Dataset.** This dataset contains 12 886 word pairs of 1 519 source words coming from HJ, RT, and AE datasets. Only 398 word pairs from

---

[5] The associations were sampled from the sociation.org database in July 2014.

the HJ dataset have continuous scores, while the other pairs which come from the RT and the AE datasets have binary relatedness scores. However, for training and evaluation purposes it is desirable to have continuous relatedness scores as they distinguish between the shades of relatedness. Yet, manual annotation of a big number of pairs is problematic: the largest dataset of this kind available to date, the MEN, contains 3 000 word pairs. Thus, unique feature of the MJ dataset is that it is at the same time large-scale, like BLESS, and has accurate continuous scores, like WordSim-353.

To estimate continuous relatedness scores with high confidence without any human judgements, we used 105 submissions of the shared task on Russian semantic similarity (RUSSE). We assumed that the top-scored systems can be used to bootstrap relatedness scores. Each run of the shared task consisted of 12 886 word pairs along with their relatedness scores. We used the following procedure to average these scores and construct the dataset:

1. Select one best submission for each of the 19 participating teams for HJ, RT and AE datasets (total number of submissions is 105).
2. Rank the $n = 19$ best submissions according to their results in HJ, RT and AE: $r_k = n + 1 - k$, where $k$ is the place in the respective track. The best system obtains the rank $r_1 = 19$; the worst one has the rank $r_{19} = 1$.
3. Combine scores of these 19 best submissions as follows: $s'_{ij} = \frac{1}{n} \sum_{k=1}^{n} \alpha_k s_{ij}^k$, where $s_{ij}^k$ is the similarity between words $(w_i, w_j)$ of the $k$-th submission; $\alpha_k$ is the weight of the $k$-th submission. We considered three combination strategies each discounting differently teams with low ranks in the final evaluation. Thus the best teams impact more the combined score. In the first strategy, the $\alpha_k$ weight is the rank $r_k$. In the second strategy, $\alpha_k$ equals to the exponent of this rank: $\exp(r_k)$. Finally, in the third strategy, the weight equals to the square root of rank: $\sqrt{r_k}$. We tried to use $AveP$ and $\rho$ as weights, but this did not lead to better fit.
4. Union pairs $(w_i, w_j, s'_{ij})$ of HJ, RT and AE datasets into the MJ dataset. Table 1 presents example of the relatedness scores obtained using this procedure.

**Evaluation of the Dataset.** Combination of the submissions using any of the three methods yields relatedness scores that outperforms all single submissions of the shared task (see Table 3). Note that ranks of the systems were obtained using the HJ, RT and AE datasets. Thus we can only claim that MJ provides continuous relatedness scores that fit well to the binary scores. Among the three weightings, using inverse ranks provides the top scores on the HJ and the AE datasets and the second best scores on the RT dataset. Thus, we selected this strategy to generate the released dataset.

**Using the Dataset.** To evaluate a relatedness measure using the MJ dataset, one should (1) calculate relatedness scores for each pair in the dataset; (2) calculate Spearman's rank correlation $\rho$ between the vector of machine judgments

**Table 3.** Performance of three combinations of submissions of the RUSSE shared task compared to the best scores for the HJ/RT/AE datasets across all submissions

|  | HJ, $\rho$ | RT, $AveP$ | AE, $AveP$ |
|---|---|---|---|
| The best RUSSE submissions for resp. datasets | 0.762 | 0.959 | 0.985 |
| **MJ: $\alpha_k$ is the rank $r_k$** | **0.790** | 0.990 | **0.992** |
| MJ:$\alpha_k$ is the exponent of rank $\exp(r_k)$ | 0.772 | **0.996** | 0.991 |
| MJ: $\alpha_k$ is the sqrt of rank $\sqrt{r_k}$ | 0.778 | 0.983 | 0.989 |

and the scores of the evaluated system. Besides, the dataset can be used to train regression models for predicting semantic relatedness using the continuous $s_{ij}$ scores.

## 4.2 RDT: Russian Distributional Thesaurus

While four resources presented above are accurate and represent different types of semantic relations, their coverage (222–1 519 source words) makes them best suited for evaluation and training purposes. In this section, we present a large-scale resource in the same $(w_i, w_j, s_{ij})$ format, the first open Russian distributional thesaurus. This resource, thanks to its coverage of 932 896 target words can be directly used in NLP systems.

**Description of the Dataset.** In order to build the distributional thesaurus, we used the Skip-gram model [38] trained on a 12.9 billion word collection of Russian texts extracted from the digital library lib.rus.ec. According to the results of the shared task on Russian semantic relatedness [32,39], this approach scored in the top 5 among 105 submissions, obtaining different ranks depending on the evaluation dataset. At the same time, this method is completely unsupervised and language independent as we do not use any preprocessing except tokenization, in contrast to other top-ranked methods e.g. [40] who used extra linguistic resources, such as dictionaries.

Following our prior experiments [39], we selected the following parameters of the model: minimal word frequency – 5, number of dimensions in a word vector – 500, three or five iterations of the learning algorithm over the input corpus, context window size of 1, 2, 3, 5, 7 and 10 words. We calculated 250 nearest neighbours using the cosine similarity between word vectors for the 1.1 million of the most frequent tokens. Next we filtered all tokens with non-Cyrillic symbols which provided us a resource featuring 932 896 source words. In addition to the raw tokens we provide a lemmatized version based on the PyMorphy2 morphological analyzer [41]. We performed no part of speech filtering as it can be trivially performed if needed.

Figure 1 visualizes top 20 nearest neighbours of the word "физика" (physics) from the RDT. One can observe three groups of related words: morphological

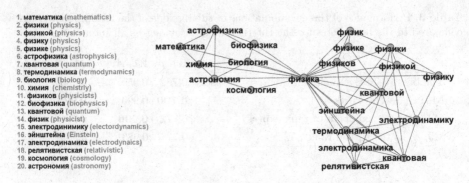

**Fig. 1.** Visualization of the 20 most semantically related words to the word "физика" (physics) in the Russian Distributional Thesaurus in the form of a list (on the left) and an ego-network (on the right)

variants (e.g. "физике", "физику"), physical terms, e.g. "квантовая" (quantum) and "термодинамика" (thermodynamics) and names of other scientific disciplines, e.g. "математика" (mathematics), "химия" (chemistry). Note that the thesaurus contains both raw tokens as displayed in Fig. 1 and lemmatized neighbours.

An important added value of our work is engineering. While our approach is straightforward, training a large-scale Skip-gram model on a 12.9 billion tokens corpus with three iterations over a corpus takes up to five days on a `r3.8xlarge` Amazon EC2 instance featuring 32 CPU cores and 244 GB of RAM. Furthermore, computation of the neighbours takes up to a week for only one model using the large 500 dimensional vectors, not to mention the time needed to test different configurations of the model. Besides, to use the word embeddings directly, one needs to load more than seven millions of the 500 dimensional vectors, which is only possible on a similar instance to `r3.8xlarge`. On the other hand, the resulting RDT resource is a CSV file that can be easily indexed in an RDBMS system or an succinct in-memory data structure and subsequently efficiently used in most NLP systems. However, we also provide the original word vectors for non-standard use-cases.

**Evaluation.** We evaluated the quality of the distributional thesaurus using the HJ, RT and AE datasets presented above. Furthermore, we estimated precision of extracted relations for 100 words randomly sampled from the vocabulary of the HJ dataset. For each word we extracted the top 20 similar words according to each model under evaluation resulting in 4 127 unique word pairs. Each pair was annotated by three distinct annotators with a binary choice as opposed to a graded judgement, i.e. an annotator was supposed to indicate if a given word pair is plausibly related or not.[6] In this experiment, we used an open source

---

[6] Annotation guidelines are available at http://crowd.russe.nlpub.ru.

**Table 4.** Evaluation of different configurations of the Russian Distributional Thesaurus (RDT). The upper part of the table reports performance based on correlations with human judgements (HJ), semantic relations from a thesaurus (RT), cognitive associations (AE) and manual annotation of top 20 similar words assessed with precision at $k$ ($P@k$). The lower part of the table reports result of the top 4 alternative approaches from the RUSSE shared task

| Model | #tok. | HJ, $\rho$ | RT, $AvgP$ | AE, $AveP$ | P@1 | P@5 | P@10 | P@20 |
|---|---|---|---|---|---|---|---|---|
| win10-iter3 | 12.9B | **0.700** | **0.918** | **0.975** | 0.971 | **0.971** | 0.944 | **0.912** |
| win10-iter5 | 2.5B | 0.675 | 0.885 | 0.970 | **1.000** | **0.971** | **0.947** | 0.910 |
| win5-iter3 | 2.5B | 0.678 | 0.886 | 0.966 | **1.000** | 0.953 | 0.935 | 0.881 |
| win3-iter3 | 2.5B | 0.680 | 0.887 | 0.959 | 0.971 | 0.953 | 0.935 | 0.884 |
| 5-rt-3 [40] | – | **0.763** | **0.923** | **0.975** | – | – | – | – |
| 9-ae-9 [32] | – | 0.719 | 0.884 | 0.952 | – | – | – | – |
| 9-ae-6 [32] | – | 0.704 | 0.863 | 0.965 | – | – | – | – |
| 17-rt-1 [32] | – | 0.703 | 0.815 | 0.950 | – | – | – | – |

crowdsourcing engine [42].[7] Judgements were aggregated using a majority vote. In total, 395 Russian-speaking volunteers participated in our crowdsourcing experiment with the substantial inter-rater agreement of 0.47 in terms of Krippendorff's alpha. The dataset obtained as a result of this crowdsourcing is publicly available (see download link below).

**Discussion of the Results.** Evaluation of different configurations of the distributional thesaurus are presented in Table 4 and Fig. 2. The model trained on the full 12.9 billion tokens corpus with context window size 10 outperforms other models according to HJ, RT, AE and precision at 20 metrics. We used this model to generate the thesaurus presented in Table 2. However, the model trained on the 2.5 billion tokens sample of the full lib.rus.ec corpus (20% of the full corpus) yields very similar results in terms of precision. Yet, this model show slightly lower results according to other benchmarks. Models based on other context window sizes yield lower results as compared to these trained using the context window size 10 (see Fig. 2).

## 5   Conclusion

In this paper, we presented five new language resources for the Russian language, which can be used for training and evaluating semantic relatedness measures, and to create NLP applications requiring semantic relatedness. These resources were used to perform a large-scale evaluation of 105 submissions in a shared task on Russian semantic relatedness. One of the best systems identified in this

---

[7] http://mtsar.nlpub.org.

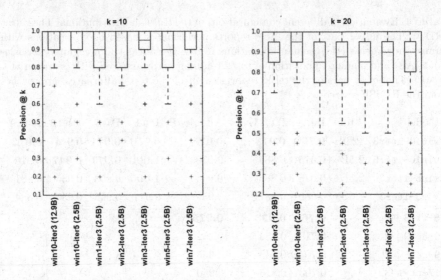

**Fig. 2.** Precision at $k \in \{10, 20\}$ top similar words of the RDT based on the Skip-gram model with 500 dimensions evaluated using crowdsourcing. The plot shows dependence of the performance of size of the context window (window size $1-10$) and size of the training corpus (2.5 and 12.9 billions of tokens) and number of iterations during training (3 or 5)

evaluation campaign was used to generate the first open Russian distributional thesaurus. Manual evaluation of this thesaurus, based on a large-scale crowd-sourcing with native speakers, showed a precision of 0.94 on the top 10 similar words. All introduced resources are freely available for download.[8] Finally, the methodology for bootstrapping datasets for semantic relatedness presented in this paper can help to construct similar resources in other languages.

**Acknowledgements.** We would like to acknowledge several funding organisations that partially supported this research. Dmitry Ustalov was supported by the Russian Foundation for Basic Research (RFBR) according to the research project no. 16-37-00354. Denis Paperno was supported by the European Research Council (ERC) 2011 Starting Independent Research Grant no. 283554 (COMPOSES). Natalia Loukachevitch was supported by Russian Foundation for Humanities (RFH), grant no. 15-04-12017. Alexander Panchenko was supported by the Deutsche Forschungs-gemeinschaft (DFG) under the project "Joining Ontologies and Semantics Induced from Text (JOIN-T)".

# References

1. Budanitsky, A., Hirst, G.: Evaluating WordNet-based measures of lexical semantic relatedness. Comput. Linguist. **32**(1), 13–47 (2006)

---

[8] http://russe.nlpub.ru/downloads.

2. Pedersen, T., Pakhomov, S.V., Patwardhan, S., Chute, C.G.: Measures of semantic similarity and relatedness in the biomedical domain. J. Biomed. Inform. **40**(3), 288–299 (2007)

3. Gabrilovich, E., Markovitch, S.: Computing semantic relatedness using Wikipedia-based explicit semantic analysis. In: Proceedings of the 20th International Joint Conference on Artifical Intelligence, IJCAI 2007, pp. 1606–1611. Morgan Kaufmann Publishers Inc. (2007)

4. Batet, M., Sánchez, D., Valls, A.: An ontology-based measure to compute semantic similarity in biomedicine. J. Biomed. Inform. **44**(1), 118–125 (2011)

5. Bär, D., Biemann, C., Gurevych, I., Zesch, T.: UKP: computing semantic textual similarity by combining multiple content similarity measures. In: Proceedings of the First Joint Conference on Lexical and Computational Semantics, vol. 1: Proceedings of the Main Conference and the Shared Task, vol. 2: Proceedings of the Sixth International Workshop on Semantic Evaluation, SemEval 2012, pp. 435–440. Association for Computational Linguistics (2012)

6. Tsatsaronis, G., Varlamis, I., Vazirgiannis, M.: Text relatedness based on a word thesaurus. J. Artif. Intell. Res. **37**(1), 1–40 (2010)

7. Patwardhan, S., Banerjee, S., Pedersen, T.: Using measures of semantic relatedness for word sense disambiguation. In: Gelbukh, A. (ed.) CICLing 2003. LNCS, vol. 2588, pp. 241–257. Springer, Heidelberg (2003). doi:10.1007/3-540-36456-0_24

8. Hsu, M.-H., Tsai, M.-F., Chen, H.-H.: Query expansion with ConceptNet and WordNet: an intrinsic comparison. In: Ng, H.T., Leong, M.-K., Kan, M.-Y., Ji, D. (eds.) AIRS 2006. LNCS, vol. 4182, pp. 1–13. Springer, Heidelberg (2006). doi:10.1007/11880592_1

9. Panchenko, A.: Similarity measures for semantic relation extraction. Ph.D. thesis, UCLouvain (2013)

10. Miller, G.A.: WordNet: a lexical database for English. Commun. ACM **38**(11), 39–41 (1995)

11. Rubenstein, H., Goodenough, J.B.: Contextual correlates of synonymy. Commun. ACM **8**(10), 627–633 (1965)

12. Miller, G.A., Charles, W.G.: Contextual correlates of semantic similarity. Lang. Cogn. Processes **6**(1), 1–28 (1991)

13. Finkelstein, L., Gabrilovich, E., Matias, Y., Rivlin, E., Solan, Z., Wolfman, G., Ruppin, E.: Placing search in context: the concept revisited. In: Proceedings of the 10th International Conference on World Wide Web, WWW 2001, pp. 406–414. ACM (2001)

14. Agirre, E., Alfonseca, E., Hall, K., Kravalova, J., Paşca, M., Soroa, A.: A study on similarity and relatedness using distributional and WordNet-based approaches. In: Proceedings of Human Language Technologies: The 2009 Annual Conference of the North American Chapter of the Association for Computational Linguistics, NAACL 2009, pp. 19–27. Association for Computational Linguistics (2009)

15. Gurevych, I.: Using the structure of a conceptual network in computing semantic relatedness. In: Dale, R., Wong, K.-F., Su, J., Kwong, O.Y. (eds.) IJCNLP 2005. LNCS (LNAI), vol. 3651, pp. 767–778. Springer, Heidelberg (2005). doi:10.1007/11562214_67

16. Hassan, S., Mihalcea, R.: Cross-lingual semantic relatedness using encyclopedic knowledge. In: Proceedings of the 2009 Conference on Empirical Methods in Natural Language Processing, EMNLP 2009, vol. 3, pp. 1192–1201. Association for Computational Linguistics (2009)

17. Postma, M., Vossen, P.: What implementation and translation teach us: the case of semantic similarity measures in WordNets. In: Proceedings of the Seventh Global Wordnet Conference, pp. 133–141 (2014)

18. Jin, P., Wu, Y.: Semeval-2012 task 4: evaluating Chinese word similarity. In: Proceedings of the First Joint Conference on Lexical and Computational Semantics, vol. 1: Proceedings of the Main Conference and the Shared Task, vol. 2: Proceedings of the Sixth International Workshop on Semantic Evaluation, SemEval 2012, pp. 374–377. Association for Computational Linguistics (2012)

19. Yang, D., Powers, D.M.W.: Verb similarity on the taxonomy of WordNet. In: Proceedings of the Third International WordNet Conference – GWC 2006, Masaryk University, pp. 121–128 (2006)

20. Meyer, C.M., Gurevych, I.: To exhibit is not to loiter: a multilingual, sense-disambiguated wiktionary for measuring verb similarity. In: Proceedings of COLING 2012: Technical Papers, The COLING 2012 Organizing Committee, pp. 1763–1780 (2012)

21. Hill, F., Reichart, R., Korhonen, A.: SimLex-999: evaluating semantic models with (genuine) similarity estimation. Comput. Linguist. **41**(4), 665–695 (2015)

22. Bruni, E., Tran, N.K., Baroni, M.: Multimodal distributional semantics. J. Artif. Intell. Res. **49**(1), 1–47 (2014)

23. Ferraresi, A., Zanchetta, E., Bernardini, S., Baroni, M.: Introducing and evaluating ukWaC, a very large web-derived corpus of English. In: Proceedings of the 4th Web as Corpus Workshop (WAC-4): Can we beat Google? pp. 47–54 (2008)

24. Faruqui, M., Dyer, C.: Community evaluation and exchange of word vectors at wordvectors.org. In: Proceedings of the 52nd Annual Meeting of the Association for Computational Linguistics: System Demonstrations, pp. 19–24. Association for Computational Linguistics (2014)

25. Baroni, M., Lenci, A.: How we BLESSed distributional semantic evaluation. In: Proceedings of the GEMS 2011 Workshop on GEometrical Models of Natural Language Semantics, GEMS 2011, pp. 1–10. Association for Computational Linguistics (2011)

26. Van de Cruys, T.: Mining for meaning: the extraction of lexicosemantic knowledge from text. Ph.D. thesis, University of Groningen (2010)

27. Biemann, C., Riedl, M.: Text: now in 2D! A framework for lexical expansion with contextual similarity. J. Lang. Model. **1**(1), 55–95 (2013)

28. Sahlgren, M.: The word-space model: using distributional analysis to represent syntagmatic and paradigmatic relations between words in high-dimensional vector spaces. Ph.D. thesis, Stockholm University (2006)

29. Griffiths, T.L., Steyvers, M.: Prediction and semantic association. In: Advances in Neural Information Processing Systems, vol. 15, pp. 11–18. MIT Press (2003)

30. Rapp, R., Zock, M.: The CogALex-IV shared task on the lexical access problem. In: Proceedings of the 4th Workshop on Cognitive Aspects of the Lexicon (CogALex), pp. 1–14. Association for Computational Linguistics and Dublin City University (2014)

31. Kiss, G.R., Armstrong, C., Milroy, R., Piper, J.: An associative thesaurus of English and its computer analysis. In: The Computer and Literary Studies, pp. 153–165. Edinburgh University Press (1973)

32. Panchenko, A., Loukachevitch, N.V., Ustalov, D., Paperno, D., Meyer, C.M., Konstantinova, N.: RUSSE: the first workshop on Russian semantic similarity. In: Computational Linguistics and Intellectual Technologies: Papers from the Annual conference "Dialogue", vol. 2, pp. 89–105. RGGU (2015)

33. Resnik, P.: Using information content to evaluate semantic similarity in a taxonomy. In: Proceedings of the 14th International Joint Conference on Artificial Intelligence, IJCAI 1995, vol. 1, pp. 448–453. Morgan Kaufmann Publishers Inc. (1995)

34. Lin, D.: An information-theoretic definition of similarity. In: Proceedings of the Fifteenth International Conference on Machine Learning, ICML 1998, pp. 296–304. Morgan Kaufmann Publishers Inc. (1998)

35. Patwardhan, S., Pedersen, T.: Using WordNet-based context vectors to estimate the semantic relatedness of concepts. In: Proceedings of the Workshop on Making Sense of Sense: Bringing Psycholinguistics and Computational Linguistics Together, pp. 1–8. Association for Computational Linguistics (2006)

36. Zesch, T., Müller, C., Gurevych, I.: Using Wiktionary for computing semantic relatedness. In: Proceedings of the 23rd National Conference on Artificial Intelligence, AAAI 2008, vol. 2, pp. 861–866. AAAI Press (2008)

37. Loukachevitch, N.V., Dobrov, B.V., Chetviorkin, I.I.: RuThes-Lite, a publicly available version of Thesaurus of Russian language RuThes. In: Computational Linguistics and Intellectual Technologies: papers from the Annual conference "Dialogue", pp. 340–349. RGGU (2014)

38. Mikolov, T., Sutskever, I., Chen, K., Corrado, G.S., Dean, J.: Distributed representations of words and phrases and their compositionality. In: Advances in Neural Information Processing Systems, vol. 26, pp. 3111–3119. Curran Associates, Inc. (2013)

39. Arefyev, N., Panchenko, A., Lukanin, A., Lesota, O., Romanov, P.: Evaluating three corpus-based semantic similarity systems for Russian. In: Computational Linguistics and Intellectual Technologies: Papers from the Annual conference "Dialogue", vol. 2, pp. 106–118. RGGU (2015)

40. Lopukhin, K.A., Lopukhina, A.A., Nosyrev, G.V.: The impact of different vector space models and supplementary techniques on Russian semantic similarity task. In: Computational Linguistics and Intellectual Technologies: Papers from the Annual conference "Dialogue", vol. 2, pp. 115–127. RGGU (2015)

41. Korobov, M.: Morphological analyzer and generator for Russian and Ukrainian languages. In: Khachay, M.Y., Konstantinova, N., Panchenko, A., Ignatov, D.I., Labunets, V.G. (eds.) AIST 2015. CCIS, vol. 542, pp. 320–332. Springer, Heidelberg (2015). doi:10.1007/978-3-319-26123-2_31

42. Ustalov, D.: A crowdsourcing engine for mechanized labor. In: Proceedings of the Institute for System Programming, vol. 27, no. 3, pp. 351–364 (2015)

# Evaluating Distributional Semantic Models with Russian Noun-Adjective Compositions

Polina Panicheva[✉], Ekaterina Protopopova, Grigoriy Bukia,
and Olga Mitrofanova

St. Petersburg State University, St. Petersburg, Russia
ppolin86@gmail.com, {protoev,gregorybookia}@yandex.ru,
o.mitrofanova@spbu.ru

**Abstract.** In the paper vector-space semantic models based on Word2Vec word embeddings algorithm and a count-based association-oriented algorithm are evaluated and compared by measuring association strength between Russian nouns and adjectives. A dataset of nouns and associated adjectives is used as the test set for pseudodisambiguation task. Models are trained with corpora of Russian fiction. A measure of lexical association anomaly is applied evaluating similarity between the initial noun and the resulting attributive phrase. Results of association strength are reported for models characterized by different parameter values; the best parameter value combinations are proposed. The test exemplars producing the error rate are manually annotated, and the model errors are categorized in terms of their linguistic nature and compositionality features.

**Keywords:** Distributional semantics · Vector-space semantic models · Vector-space representation evaluation · Association measures · Selectional restrictions

## 1    Introduction

Current research in the field of distributional semantics reveals a strong shift towards studying meaning of complex linguistic units (constructions, clauses, sentences) with the help of vector space models and their modifications [7,12,14,15,17], etc. Rapid increase of the interest towards this task was noticed by SemEval 2014 competition[1] which included evaluation of compositional distributional semantic models of full sentences for English.

One of the successful projects based on distributional and model-theoretic semantics is COMPOSES (M. Baroni and colleagues, Trento University, Italy[2]) which is aimed at modelling linguistic units in semantic space with the help of compositional operations. Experiments were carried out for tagged English corpora, the toolkit comprising morphosyntactic parser. The model is based on

---

[1] http://alt.qcri.org/semeval2014/task1/.
[2] http://clic.cimec.unitn.it/composes/.

© Springer International Publishing AG 2017
D.I. Ignatov et al. (Eds.): AIST 2016, CCIS 661, pp. 236–247, 2017.
DOI: 10.1007/978-3-319-52920-2_22

corpus statistics: nouns are represented as vectors, while adjectives correspond to functions mapping input items to compositional structures [1].

An outstanding discussion of various mathematical operations applied to determine compositional meaning in English is presented in [5]. The authors focus their attention on tensor-based models where relational words (verbs, adjectives) are regarded as tensors. The given research involves quantum mechanical algorithms in modelling compositional meaning of clauses.

A powerful algorithm based on distributional semantics for synonym detection and word sense disambiguation was developed and implemented by Language Technology Group (Ch. Biemann and colleagues, Technical University of Darmstadt, Germany). JoBimText[3] is a web application which processes corpora in German, English and Russian. The toolkit provides automatic selection of contextual synonyms and defines the number and scope of meanings for a target word. The model comprises graph-based clustering algorithm [2]. As well as COMPOSES software, JoBimText tool provides morphosyntactic parsing of the English and German input texts, but this option is absent for Russian corpora, complicating the usage of the toolkit in real-life fine-grained applications.

Russian computational linguistics teams have been involved in the development of independent resources using distributional semantic models.

Serelex semantic model (A. Panchenko and colleagues, Bauman Moscow State Technical University, Université Catholique de Louvain, Belgium[4]) is incorporated with the information retrieval system which gets a target word as an input and gives a list of its associates as an output [11]. Serelex operates over English, French and Russian web-documents and provides ranked contextual correlates for a target word with regard to an original similarity measure based on lexical-syntactic patterns.

RusVectores[5] is a toolkit providing data on context vectors generated by the models trained on the Russian National Corpus, News and Web corpora. RusVectores extracts contextual associates of target words, measures semantic distances in pairs of words, performs operations on vectors, provides visualization of semantic relations in clusters, etc. The core machine learning algorithms used by RusVectores are Skip-Gram and CBOW within the Word2Vec tool [8,9].

In the paper we apply the word-embeddings semantic models based on Russian texts to a relatively new task of measuring association strength and compare the results to a distributional approach specifically oriented at representing this kind of phenomena. The paper is organized as follows: related work and the distinctive features of our approach are described in Sect. 2; Sect. 3 contains the description of the dataset and the procedure with which it was automatically obtained; two methods of association measurement are presented in Sect. 4; Sect. 5 contains the results, including detailed analysis of model errors; the overall conclusions about the models, their errors and the dataset are presented in Sect. 6.

---

[3] http://maggie.lt.informatik.tu-darmstadt.de/jobimtext/.

[4] http://serelex.it-claim.ru/.

[5] http://ling.go.mail.ru/dsm/ru/.

## 2    Related Work

Large-scale research dealing with the development and elaboration of distributional semantic models for Russian corpora, implementation and evaluation of various semantic relatedness and association measures gave rise to the RUSSE competition[6] [10]. Comparison of functioning toolkits provides keen insight into language-specific features responsible for semantic relations within a text as well as of compositional models of Russian constructions.

However, semantic relatedness evaluation only involves paradigmatic relations between lexical units. Although paradigmatic relations and association strength between lexical items is strongly related to compositionality issues in semantic vector spaces [1,16], to our knowledge there has been no evaluation of vector-space models applied to syntagmatic relations in Russian. In this paper we set out to fill this gap by evaluating the performance of Word2Vec models with various parameter values in comparison with a count-based association measure on a task of association strength measurement between Russian nouns and adjectives.

We apply the testing dataset described in [3]. The authors perform association strength measurement using a very small corpus of 350K sentences and only accounting for information of adjective-noun collocations, yielding high performance in association strength prediction task by using very limited resources and time. We compare an approach performing a different task: not limited to a single type of syntactic relations, it builds a semantic model of the Russian language based on a larger corpus of 11M sentences. It uses co-occurrence information only limited by a pre-set window size and, unlike the count-based approach which is restricted to attributive phrases, represents the multi-dimensional information contained in the corpus. Word2Vec model thus requires larger data and more time for training. Association strength testbed is used for evaluating the syntagmatic or compositional output of the Word2Vec and count-based models.

Association strength measurement is closely related to identification of abnormal lexical compositions [16] and automatic lexical error detection [6]. The authors of the latter works present a number of measures for evaluating semantic anomaly of constructions in semantic vector space. We adopt one of the measures and apply it to a semantic space with reduced dimensionality produced by Word2Vec.

## 3    Dataset

We evaluate distributional semantic models by measuring association strength between nouns and adjectives with the dataset described in [3]. The training corpus consists of fiction texts from Moshkov's library[7] uploaded in 2014. The volume of the corpus is 11 million 600 thousand sentences, or 140 million words excluding punctuation; it is referred to as **Corpus A** below. The testing dataset

---

[6] http://russe.nlpub.ru/.
[7] http://www.lib.ru/.

is based on a smaller corpus containing fiction texts (350K sentences) from Moshkov's library uploaded in 2009 (**corpus B**). The test set consists of 1000 word triplets, 500 for target nouns and 500 for target adjectives. The triplets are combined following the pseudo-disambiguation procedure:

1. 500 nouns are randomly chosen from corpus B.
2. For every noun a random adjective associated with it at least 5 times in the corpus is selected.
3. For the resulting noun-adjective pair another adjective is selected so that:
   – it does not collocate with the target noun;
   – it has the nearest frequency to the initial adjective.
4. Every sentence containing the target noun and adjective regardless of their order and distance between them is deleted from both training corpora A and B.

Thus we obtain a list of word triplets *(noun, adjective1, adjective2)*, with the pair *(noun, adjective1)* representing an acceptable noun-attribute association, and the pair *(noun, adjective2)* an unacceptable one. We also obtain training subcorpora by deleting all the sentences from the training corpus which contain any acceptable pair of words irrespective of their order and distance between them.

**Table 1.** Examples of association strength dataset

| Target noun triplets | | | |
|---|---|---|---|
| N | Target noun | Acceptable adjective | Unacceptable adjective |
| 1 | америка 'America' | колониальный 'colonial' | обильный 'abundant' |
| 2 | бдение 'vigil' | ночной 'nocturnal' | личный 'personal' |
| 500 | ярость 'fury' | немой 'mute' | передний 'anterior' |
| **Target adjective triplets** | | | |
| N | Target adjective | Acceptable noun | Unacceptable noun |
| 1 | алюминиевый 'aluminic' | пудра 'powder' | опера 'opera' |
| 2 | брюшной 'abdominal' | пресс 'muscle' | внешность 'appearance' |
| 500 | шерстяной 'woolen' | носочек 'sock' | кража 'robbery' |

We follow the same procedure for obtaining adjective-based triplets, with 500 random adjectives, an acceptable and an unacceptable noun associates for every target adjective. A separate subcorpus with artificially deleted associations also applies in this case. The examples of the test data are presented in Table 1.

# 4  Vector Space and Count-Based Approaches to Association Measurement

## 4.1  Vector Space Approach to Association Strength Measurement

We adopt the compositional approach investigated in [6,16]. The semantic vector space is constructed with Word2Vec algorithm available in Python Gensim library [13]. The assumption is that the vector representing a noun-adjective composition is meaningful if it is closely related to the head of the composition, i.e. the initial noun. The similarity measure between the composition and the head noun is expected to positively reflect the acceptability of the noun-adjective association. The acceptable compositions are expected to be rated more similar to the initial head nouns than the unacceptable ones. However, with normalized vectors, as in Word2Vec approach, the monotonicity of the functions *Similarity1* and *Similarity2* is the same, although *Similarity2* measures the simple cosine similarity between noun and adjective (see Eqs. (1), (2)).

$$Similarity1(noun, adj) = cos(noun + adj, noun); \tag{1}$$

$$Similarity2(noun, adj) = cos(noun, adj). \tag{2}$$

Quantifying similarity between the noun and the adjective in the same vector space yields here the same result as when quantifying similarity between the initial noun and the attributive noun phrase. The latter formula is linguistically motivated and naturally interpreted, which is not so obvious about the former. Further research is necessary to identify whether it's a weakness introduced by normalized vector space, the same vector space containing both nouns and adjectives, or other reasons.

### 4.2   Count-Based Association Measure

In [3] the authors propose an association measure based on distributional word properties concerning only a fixed construction, namely, the noun-adjective association (referred to as **D** below). The basic assumption is that if two words relevant for a construction slot collocate in texts with similar words (contexts inside the attributive phrase, i.e. attributive adjectives for nouns, and governing nouns for adjectives) relevant for another slot, the probability of the first target word to be combined with the contexts of the second target word and vice versa is high, even when some pairs are not observed in texts. This idea is formally expressed by the notion of confusion probability, which is computed as follows. Given the contexts of the first word $c(x_1)$ and the second one $c(x_2)$, their confusion probability is equal to P:

$$\mathbb{P}\{x_1 \sim x_2\} = \frac{|c(x_1) \bigcap c(x_2)|^2}{|c(x_1)||c(x_2)|}.$$

The association strength between two words in a collocation occurring in a corpus is computed using the Mutual Information score, reported to perform higher than alternatives in [3]. The count-based association measure between a noun and an adjective $D(a, n)$ in a collocation is then defined as an average of such counts over all confusable words weighted by the confusion probability.

## 5   Association Strength Results

### 5.1   Evaluation Procedure

Firstly, we train the skip-gram semantic space based on the pseudo-disambiguation subcorpus of corpus A, i.e. excluding all the actual co-occurrences of the nouns with adjectives in question. Preliminary experiments have shown that training the Word2Vec models with the smaller corpus B decreases the performance dramatically. For every testing triplet we quantify similarity between the target and the alternative association words according to Eq. (1) above. For target noun triplets we compare the resulting *Similarity (noun, adj1)* and *Similarity (noun, adj2)*. For target adjective triplets respectively *Similarity (noun1, adj)* and *Similarity (noun2, adj)* are compared. Thus we obtain the measure referred to as **W2V** below. We also measure the count-based association measure **D**(noun, adj) for the respective word pairs in the triplet.

Correct result is registered for the triplet if the acceptable word pair (noun, adj) is scored with higher similarity/association than the unacceptable one, based on skip-gram and count-based distributional measures. We then quantify the accuracy as the portion of the correct results for 500 noun and adjective triplets.

## 5.2   Word Embeddings Association Strength Results

We train Word2Vec word-embeddings with the following parameters:

1. Skip-Gram model is used. CBOW is reported to perform worse for sparse data, which is confirmed by our preliminary experiments.
2. The co-occurrence window size ranges from 1 to 10.
3. We test different dimensionality value in the range {25, 50, 100, 150, 300}.

We evaluate 50 Skip-Gram models both for the target nouns association strength experiment and 50 models for the target adjectives experiment.

The pseudo-disambiguation results for noun-adjective association strength for target nouns are presented in Table 2; Table 3 contains the results for target adjectives. The best results for every dimensionality value are highlighted in bold. The best results for every window size are shown in italics.

**Table 2.** Word embeddings pseudo-disambiguation accuracy of target nouns

| Window | Dimensionality | | | | |
|---|---|---|---|---|---|
| | 25 | 50 | 100 | 150 | 300 |
| 1 | 0,756 | 0,758 | 0,778 | 0,770 | *0,790* |
| 2 | 0,762 | 0,790 | 0,802 | *0,811* | 0,802 |
| 3 | 0,784 | *0,809* | 0,807 | 0,796 | *0,809* |
| 4 | 0,784 | 0,819 | *0,821* | 0,815 | 0,809 |
| 5 | **0,790** | 0,809 | *0,823* | **0,821** | 0,817 |
| 6 | 0,772 | ***0,827*** | 0,817 | 0,802 | 0,811 |
| 7 | **0,790** | 0,819 | *0,825* | 0,819 | **0,819** |
| 8 | 0,786 | 0,813 | *0,819* | *0,819* | 0,813 |
| 9 | 0,788 | 0,817 | ***0,825*** | 0,807 | 0,807 |
| 10 | 0,786 | 0,815 | *0,823* | 0,809 | 0,798 |

**Table 3.** Word embeddings pseudo-disambiguation accuracy of target adjectives

| Window | Dimensionality | | | | |
|--------|------|------|------|------|------|
|        | 25   | 50   | 100  | 150  | 300  |
| 1      | 0,765 | *0,794* | 0,783 | 0,790 | 0,771 |
| 2      | 0,767 | 0,800 | 0,800 | *0,814* | 0,777 |
| 3      | 0,769 | **0,810** | *0,810* | 0,806 | 0,794 |
| 4      | 0,771 | 0,798 | 0,806 | ***0,822*** | **0,796** |
| 5      | 0,771 | *0,808* | *0,808* | 0,806 | **0,796** |
| 6      | 0,769 | **0,810** | 0,804 | 0,799 | 0,794 |
| 7      | 0,761 | 0,804 | *0,810* | 0,798 | 0,794 |
| 8      | **0,777** | 0,798 | ***0,816*** | 0,810 | 0,785 |
| 9      | 0,763 | 0,785 | 0,792 | 0,783 | *0,794* |
| 10     | 0,761 | 0,781 | *0,804* | 0,800 | 0,790 |

The models for target nouns and adjectives yield consistently similar results. The best models perform with around 82% accuracy, with the highest result for nouns, **82,7%**, achieved with **window size = 6** and **dimensionality = 50**; and for **adjectives, 82,2%**, with **window size = 4** and **dimensionality = 150**.

The highest performing window size for target nouns ranges from 5 to 9, with the leading window size of 7 containing the best results for 3 out of 5 dimensionality sizes. For target adjectives the highest performing window size ranges from 3 to 8, with the leading window size of 4 and 8 each yielding the best results for 2 out of 5 dimensionality sizes.

The best performing dimensionality follows the same pattern: medial dimensionality sizes tend to perform better. 25 dimensions do not result in the highest performance for any window size. For dimensionality ranging in {50, 100, 150, 300} the best results for different window size values are achieved respectively {2, 6, 2, 2} times for nouns and {4, 5, 2, 1} times for adjectives. The models with dimensionality size 100 consistently achieve the highest accuracy for most of the window sizes both for target nouns and adjectives.

For comparison, we also apply a high-performing skip-gram model in RUSSE task available online [8] which is trained with RNC data.[8] The dimensionality is set to 300, window size = 5. The resulting accuracy with our dataset reached 0,7869 for target nouns and 0,7986 for target adjectives. The performance of the RNC-trained model is comparable to our results; however, they are lower than the best-performing models trained with fiction texts and in the case of target nouns lower than the fiction-based models with the same parameter values. A possible reason is that both training and test set in our experiment belong to the same genre. This could be due to the fact that moderate dimensionality sizes of 100 and 150 are better applicable to this type of task.

---

[8] http://ling.go.mail.ru/misc/dialogue_2015.html#rnc.

## 5.3   Count-Based Association Strength Results

The same experiment was conducted using the count-based approach. Its accuracy when training on a small corpus B was discussed in detail in [3]. Training on a much larger corpus A results in 2% improvement. Table 4 reports the results of the best performing count-based association model based on Mutual Information score.

Table 4. Count-based pseudo-disambiguation accuracy

|                  | Corpus A | Corpus B |
| ---------------- | -------- | -------- |
| Target noun      | 0, 762   | 0, 748   |
| Target adjective | 0, 792   | 0, 774   |

## 5.4   Error Analysis

To provide interpretation and comparison for the results of both methods, the erroneous triplets, where the unacceptable noun-adjective combination was rated with higher similarity/association value than the acceptable one, were extracted. The errors of the count-based and best performing Word2Vec similarity models were compared.

First of all, the erroneous triplets for target nouns were annotated with one of the four cases:

1. **Both correct** case: both adjectives form an acceptable association with the current noun.
2. **None correct** case: none of the adjectives in the triplet forms an acceptable association with the current noun.
3. **Correct data** case: the adjective attested as the correct associate in the pseudo-disambiguation corpus is the only acceptable option in the triplet, according to the manual annotation. The test triplet is thus manually evaluated as correct.
4. **Incorrect data** case: the adjective chosen as the incorrect associate in the pseudo-disambiguation corpus is the only acceptable option. The test triplet is thus manually evaluated as incorrect, as it was previously automatically annotated in an opposite way to the human judgements.

Similar annotation was performed for erroneous triplets with target adjectives. Statistics of the types of mistakes for each experiment are presented in Table 5.

Obviously, only the third kind of errors is relevant when considering the disadvantages of each method. We proceed to analyze only these mistakes of both approaches. First of all, it should be noticed that there are several common mistakes made by both systems. The detailed statistics is provided in Table 6.

**Table 5.** Types of errors by association measurement models

|  | Target noun | | Target adjective | |
| --- | --- | --- | --- | --- |
|  | W2V | D | W2V | D |
| Both correct | 39 | 51 | 38 | 33 |
| None correct | 12 | 13 | 12 | 20 |
| Correct data | 31 | 41 | 29 | 42 |
| Incorrect data | 10 | 11 | 9 | 9 |
| Total | 92 | 116 | 88 | 104 |

**Table 6.** Common and specific errors of the models

|  | Target noun | Target adjective |
| --- | --- | --- |
| Common mistakes | 11 | 11 |
| W2V-specific mistakes | 20 | 18 |
| D-specific mistakes | 30 | 31 |

## 5.5  Error Classification

A number of common mistakes were found among the relevant ones. They can be divided into two groups concerning their reason: acceptable combinations representing rare or occasional metaphorical expressions, see Table 7.

**Table 7.** Common errors of the models

| Target | Reason | N | Example | Translation |
| --- | --- | --- | --- | --- |
| Noun | Occasional metaphor | 45% | круглая сирота | total orphan |
|  | General meaning | 55% | европейский квартал | european quarter |
| Adjective | Occasional metaphor | 45% | информационная чума | informational boom |
|  | General meaning | 55% | ограниченная сфера | restricted sphere |

The rest of the errors, i.e. specific ones, were ordered by the acceptable combination frequency. The classification is presented in Table 8. In both experiments a group of very low frequency (acceptable) combinations can be found: either rare themselves or containing a rare word, where even a native speaker is scarcely able to construct a sentence where these combinations are justified (see low association frequency combinations, Table 8). The high frequency combinations are constructions in the sense of Construction Grammar [4]: their meaning is not additive and can not be simply derived from the meaning of the constituents (high frequency, Table 8).

The medium frequency combinations contain real errors which are due to the algorithm structure or assumptions: the word embeddings similarity measure fails to extract such combinations, because the constituents appear to have

**Table 8.** Specific errors of the models

| Target | Association frequency | N | Example | Translation |
|--------|----------------------|-----|---------|-------------|
| **Word-embeddings specific errors** | | | | |
| Noun | High | 20% | последнее издыхание | last breath |
| | Medium | 25% | большая мышца | big muscle |
| | Low | 55% | немой клон | mute clone |
| Adjective | High | 29% | трезвая голова | reasonable person |
| | Medium | 23% | копировальный центр | copy center |
| | Low | 48% | безработный фанатик | unemployed fanatic |
| **Count-based specific errors** | | | | |
| Noun | High | 26% | родовая схватка | birth spasm |
| | Medium | 20% | тёмное предчувствие | dark presentiment |
| | Low | 54% | сухая конвульсия | dry convulsion |
| Adjective | High | 13% | жевательная резинка | chewing gum |
| | Medium | 50% | сумасшедшая история | crazy story |
| | Low | 37% | суеверный закон | superstitious law |

too few intersecting contexts. The errors of the count-based model may also be explained by the underlying assumption that similar words occur in similar noun-adjective contexts. In this case the words similar to the target one do not usually collocate with its acceptable combination. Consider, for example, the expression 'стоматологический + кресло' (dental chair) with the target noun. Virtually neither of the words similar to 'кресло' (chair — table, armchair etc.) can be observed in texts collocating with 'стоматологический' (dental), so the whole combination gets a low score.

## 6   Conclusions

We have applied the task of measuring association strength between Russian nouns and adjectives as a testbed for comparing semantic models based on word embeddings with a count-based association measure. The training and test sets were derived from fiction corpora. The word embeddings semantic models perform with high accuracy on the noun-adjective association strength task. The resulting accuracy ranges from 75,6 to 82,7% for different values of window size and dimensionality of the models. The models with window and dimensionality size around the medial values tend to perform higher for both adjective and noun keyword tasks. The best results achieved by the models were above 82% and had window size of 6 and 4 and dimensionality of 50 and 150, for nouns and adjectives respectively. The count-based model is slightly outperformed by word embeddings, with the accuracy of the former reaching 76–79%.

We have manually analysed cases where the models produced errors. Although the number of common mistakes of both models is small, the reasons for the mistakes by different models are similar. The first group is represented by rare or metaphorical expressions, including constructions with lack of compositionality in their meaning (about 40%). The other group consists of associations where distributional properties of words are likely to produce an error, as the distributional properties of associates have no intersection. While the latter group of errors can hardly be addressed by purely distributional methods, the errors of the former group could be corrected by additional corpus enrichment and by accounting for metaphorical usage.

The error analysis also provides valuable insights on the nature of the automatically derived test dataset. Although obviously representing association strength in a definite volume of fiction texts, it should be carefully interpreted in tasks involving general semantic regularities, with metaphorical and other non-compositional expressions accounted for in a fine-grained manner.

**Acknowledgments.** The reported study is supported by RFBR grant № 16-06-00529 "Development of a linguistic toolkit for semantic analysis of Russian text corpora by statistical techniques".

# References

1. Baroni, M., Bernardi, R., Zamparelli, R.: Frege in space: a program of compositional distributional semantics. Linguist. Issues Lang. Technol. **9** (2014)
2. Biemann, C.: Unsupervised and knowledge-free natural language processing in the structure discovery paradigm. Ph.D. thesis, Universität Leipzig (2007)
3. Bukia, G., Protopopova, E., Mitrofanova, Θ.: A corpus-driven estimation of association strength in lexical constructions. In: Sergey Balandin, T.T., Trifonova, U. (eds.) Proceedings of the AINL-ISMW FRUCT, pp. 147–152. FRUCT Oy, Finland (2015). http://fruct.org/publications/ainl-abstract/files/Buk.pdf
4. Goldberg, A.: Constructions: a construction grammar approach to argument structure (1994)
5. Kartsaklis, D., Sadrzadeh, M., et al.: Prior disambiguation of word tensors for constructing sentence vectors. In: Proceedings of EMNLP, pp. 1590–1601 (2013)
6. Kochmar, E., Briscoe, T.: Capturing anomalies in the choice of content words in compositional distributional semantic space. In: Proceedings of RANLP, pp. 365–372 (2013)
7. Kolb, P.: Disco: a multilingual database of distributionally similar words. In: Proceedings of KONVENS-2008, Berlin (2008)
8. Kutuzov, A., Andreev, I.: Texts in, meaning out: neural language models in semantic similarity task for russian. arXiv preprint arXiv:1504.08183 (2015)
9. Mikolov, T., Sutskever, I., Chen, K., Corrado, G.S., Dean, J.: Distributed representations of words and phrases and their compositionality. In: Advances in Neural Information Processing Systems, pp. 3111–3119 (2013)
10. Panchenko, A., Loukachevitch, N., Ustalov, D., Paperno, D., Meyer, C., Konstantinova, N.: Russe: the first workshop on Russian semantic similarity. In: Proceeding of the Dialogue 2015 Conference (2015)

11. Panchenko, A., Romanov, P., Morozova, O., Naets, H., Philippovich, A., Romanov, A., Fairon, C.: Serelex: search and visualization of semantically related words. In: Serdyukov, P., Braslavski, P., Kuznetsov, S.O., Kamps, J., Rüger, S., Agichtein, E., Segalovich, I., Yilmaz, E. (eds.) ECIR 2013. LNCS, vol. 7814, pp. 837–840. Springer, Heidelberg (2013). doi:10.1007/978-3-642-36973-5_97

12. Pekar, V., Staab, S.: Word classification based on combined measures of distributional and semantic similarity. In: Proceedings of the Tenth Conference on European Chapter of the Association for Computational Linguistics, vol. 2, pp. 147–150. Association for Computational Linguistics (2003)

13. Rehurek, R., Sojka, P.: Software framework for topic modelling with large corpora (2010)

14. Sahlgren, M.: The word-space model: using distributional analysis to represent syntagmatic and paradigmatic relations between words in high-dimensional vector spaces (2006)

15. Schütze, H.: Dimensions of meaning. In: Proceedings of Supercomputing 1992, pp. 787–796. IEEE (1992)

16. Vecchi, E.M., Baroni, M., Zamparelli, R.: (Linear) maps of the impossible: capturing semantic anomalies in distributional space. In: Proceedings of the Workshop on Distributional Semantics and Compositionality, pp. 1–9. Association for Computational Linguistics (2011)

17. Widdows, D., Cohen, T.: The semantic vectors package: new algorithms and public tools for distributional semantics. In: 2010 IEEE Fourth International Conference on Semantic Computing (ICSC), pp. 9–15. IEEE (2010)

# Applying Word Embeddings to Leverage Knowledge Available in One Language in Order to Solve a Practical Text Classification Problem in Another Language

Andrew Smirnov[1,3] and Valentin Mendelev[2,3(✉)]

[1] STC-Innovations, Saint Petersburg, Russia
smirnov-a@speechpro.com
[2] Speech Technology Center, Saint Petersburg, Russia
mendelev@speechpro.com
[3] ITMO-University, Saint Petersburg, Russia

**Abstract.** A text classification problem in Kazakh language is examined. The amount of training data for the task in Kazakh is very limited, but plenty of labeled data in Russian are available. Language vector space transform is built and used to transfer knowledge from Russian into Kazakh language. The obtained classification quality is comparable to that of an approach that employed sophisticated automatic translation system.

**Keywords:** Language vector space · Word embedding · Text classification · Low resource

## 1 Introduction

Text classification tasks are ubiquitous and essential for modern technologies. One may need to categorize documents, detect sentiment, intention or desired action etc. This paper is devoted to users' requests classification for customer support. In modern contact centers all initial appeals are fed to a classifier that determines a request topic and performs an action which may be to forward a message to a responsible staff member or to generate a unique answer automatically.

We investigate a problem of building a classifier when training data in a target language are scarce but significant amount of labeled requests in another language is available. Kazakh language was chosen to be the target language while training data were in Russian (source language).

Distributed vector representations (or word embeddings) have become a very useful tool in various Natural Language Processing (NLP) tasks, including language modeling, sentiment analysis, word-sense disambiguation, word similarity and synonym detection (e.g. [1,3,5,11,12]). Overcoming data sparsity problem, word embeddings represent words as low-dimensional dense vectors. Methods of

© Springer International Publishing AG 2017
D.I. Ignatov et al. (Eds.): AIST 2016, CCIS 661, pp. 248–254, 2017.
DOI: 10.1007/978-3-319-52920-2_23

their construction (e.g. [8,10], see also [12] for survey of classical vector space models) require only large corpora of unlabeled text data.

Word embeddings are effective in capturing linguistic regularities that generalize across different languages and thus can serve as a way to transfer linguistic knowledge from one language to another. In particular, distributed representations can be used to transfer limited labeled information from a high-resource to a low-resource language.

Several methods have been proposed to train and align bi- and multilingual word embeddings. In [9] monolingual models of languages are built separately and then a linear projection between the vector spaces of languages is learned on a small bilingual dictionary. More sophisticated approaches [2,4,6] optimize monolingual and cross-lingual objectives simultaneously (e.g. by minimizing the sum of monolingual loss functions and the cross-lingual one). To train such a model one need a parallel corpus aligned at a sentence level.

The quality of bilingual vector representations can be evaluated on a cross-lingual document classification task. A common setup was introduced by Klementiev et al. [6]. They use a subset of English and German sections of the Reuters RCV1/RCV2 corpora [7]. There are four topics in the corpus. The classifier is trained on documents belonging to one language and tested on documents in the other language. Coulmance et al. [2] reported a classification accuracy for several bilingual word embeddings models.

This paper deals with a problem similar to the one investigated in [6] but with significant peculiarities that bring a new level of complexity to the task. We classify users' requests. The requests are short phrases, a typical request consists only of several words; the number of target classes can be fairly large (up to 40); the target and source languages belong to different language families. We are also not in possession of any parallel Russian-Kazakh text corpus.

In the rest part of the paper several ways to transfer knowledge are introduced and investigated. It is shown that even simple transferring techniques allow to increase the classification accuracy significantly.

The problem under investigation is described in Sect. 2. In Sect. 3 a brief description of word embeddings approach is provided. Section 4 is devoted to knowledge transfer technique which is the main point this work. Experiment results are presented in Sect. 5 and followed by a discussion and conclusions.

## 2 Problem Description

All experiments were conducted on users' requests collected by a customer support services of major Russian telecommunication companies. Every client calling on a support line is asked to describe her problem in free form, after that the utterance is converted to a text by an automatic speech recognition (ASR) system and classification algorithm is applied to assign one of predefined possible message topics to the utterance. The call is then redirected to a suitable branch of a voice menu or a staff member dedicated to the topic.

We had 6000 such requests in Russian with manually assigned class labels. A total number of classes was 40. Though a lot of labeled data are helpful to build an accurate classifier it is not always possible to obtain them. An ideal classifier building process should not require labor at all. One may imagine 3–4 possible formulations for each of the target topics and that should be enough. After all, humans effectively discriminate messages referring to different topics without special training. Of course humans do use huge amount of language knowledge which is not accessible to algorithms in many languages. On one hand, we don't want to prepare a lot of labeled samples for each new customer support automation project, and on the other, we cannot rely on wordnets and big textual corpora which may simply not exist for a language of interest.

The aim of this work is to investigate means to transfer domain knowledge available in one language to another while building call steering classifier for an abstract Kazakh telecommunication company.

## 3    Word Embeddings for Text Classification

Distributed vector representation of words can be understood as a mapping from each word in the dictionary to a vector in a low-dimensional embedded space. The main advantage of the approach is that words with similar distributional properties are mapped to close vectors. Thus distributed representations provide information about lexico-semantic relatedness which make their application to NLP extremely successful.

Skip-gram and continuous bag of words (CBOW) [8] are two algorithms to learn monolingual word embeddings from raw text. The training objective of CBOW is to predict the word by its context in the sentence while the training objective of the skip-gram model is to predict surrounding words by the word itself. The training of CBOW is known to be faster but skip-gram learns better representation when the training corpus is small. According to our experiments monolingual word embeddings learned by the CBOW model give better results in terms of classification accuracy on both Russian and Kazakh request datasets. All results reported hereafter were obtained with the CBOW model.

For inducing monolingual word embeddings we use the word2vec tool [13]. The training data for the Russian model consist of transcribed spontaneous conversational speech, fiction and news articles (~200 M tokens). The Kazakh model was trained on the latest dump of Kazakh Wikipedia and Kazakh news articles (~30 M tokens). Both models did manage to capture semantic information for the target domain of knowledge which is telecommunication (see Table 1 for some

**Table 1.** Example words from telecommunication domain along with their 3 nearest neighbors, using the cosine distance between embeddings

| Russian model | | | | Kazakh model | | | |
|---|---|---|---|---|---|---|---|
| base word | стс | adsl | мегафон | base word | тнт | adsl | интернет |
| closest | тнт | адсл | билайн | closest | телебағдарламасында | dsl | мобильді |
| words | нтв | dsl | мтс | words | нтв | dslam | онлайн |
| | рифей | etth | мегафона | | comedy | fddi | интернетке |

examples). The dimension of vector representation is 200 for the Russian model and 100 for Kazakh.

We shift the weighted average of word vectors to zero and apply the whitening transform. We found that this preprocessing procedure improves the classification results.

Vectors for sentences are obtained by averaging words vectors.

## 4   Knowledge Transfer

Two questions regarding transferring knowledge may be raised in light of call steering problem: what transfer destination to choose and how exactly to conduct the transfer. One may choose to use existing classifier in the source language and transfer requests from the target language as they arrive. In this case transferring mechanism should be very fast. Another possibility is to transfer training samples from the source language into the target language, and then train the classifier and use it in a production system. In this work the latter approach is adopted.

The following transferring mechanisms have been considered: automatic translation with a sophisticated translation engine and semantic vector space transformation.

For a given target and source languages there may not exist parallel text corpora sufficient to train bilingual word embeddings jointly but the translation of the most common words can usually be obtained without significant effort. At the same time the algorithm of aligning monolingual word embeddings by learning the mapping from small bilingual dictionary performs well on translation tasks [9]. We follow that approach to transfer information about class labels from Russian to Kazakh language.

We train the transformation matrix $A$ that maps word vectors from Russian to Kazakh vector space by minimizing $min_A \sum_{i=0}^{N} ||Av_i^r - v_i^k||^2$, where $v_i^r$ is the vector of Russian word, the $v_i^k$ is the vector of its translation to Kazakh and $N$ is the size of the dictionary.

To construct the Russian-Kazakh dictionary we chose the 5000 most frequent words from the available training corpus of users' requests in Russian and translated them with the Google translation service. About 1000 translations that were not present in the dictionary of our Kazakh word embeddings model have been discarded.

To examine the quality of the transformation matrix we splitted the rest 4000 words into 3500 words train and 500 words test sets. The accuracy of translation depends on the dimensionality of vector spaces, in [9] it was observed that for the best performance the word vectors trained on the source language should be several times (around 2x–4x) larger than the word vectors trained on the target language. We experimented with the dimensions form 50 to 500 for the Kazakh language and from 100 to 1000 for Russian. The best performance on the test set was observed with 100-dimensional Kazakh word vectors and 200-dimensional Russian vectors. The accuracy of translation was 38% on the train set and 13% on the test set. Note that this accuracy is underestimated: we count only exact match as a successful translation, while synonym translations are counted as mistakes.

## 5    Experiments

A small part of our Russian user request data was manually translated into Kazakh to construct a development set of 120 examples (40 classes) and 3 test sets containing 120 examples each assigned to 40, 20 and 10 classes respectively. Test sets were not disjoint. The rest of the available requests in Russian were translated into Kazakh in different ways (by manual translation, Google translate or by transformation matrix). We include into the training set only requests with classes that are presented in the corresponding test set (about 1000 examples for 10 classes, 2000 for 20 classes, and 5900 for 40 classes). The classification of Kazakh requests was performed by the K-nearest neighbors classifier. The parameters of the classifier as well as the setting of word embeddings training process were tuned on the development set.

We compare the accuracy of classification in several setups. The setups are denoted by $Test\_<type\ of\ translation>$ or $Test\_<type\ of\ translation> + Train\_<type\ of\ translation>$ according to the data used for training and testing the classifier. The option $Test\_<type\ of\ translation>$ means that the leave-one-out cross-validation was performed on the test set translated to Kazakh with the method specified by $<type\ of\ translation>$. For the option $Test\_<type\ of\ translation> + Train\_<type\ of\ translation>$ the procedure is very similar except that for each example in the test set the classifier is trained not only on the rest of the examples from the test set but also on the examples from the corresponding train set translated to Kazakh with the method specified by $<type\ of\ translation>$ in $Train\_<type\ of\ translation>$.

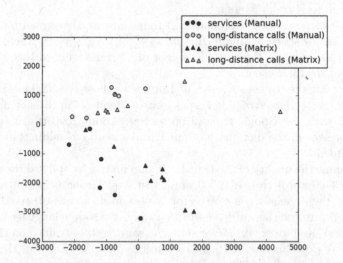

**Fig. 1.** Distributed representations in the Kazakh vector space of users' requests for two classes (services and long-distance calls) and two types of translation (manual and by the translation matrix). The vectors were projected down to two dimensions by PCA.

**Table 2.** Accuracy of classification for various setups.

| Number of classes | Test_Manual | Test_Matrix | Test_Rus |
|---|---|---|---|
| 40 | 23.3 | 11.2 | 34.5 |
| 20 | 45 | 29.2 | 56.7 |
| 10 | 63 | 40.3 | 79.2 |
| Number of classes | Test_Manual + Train_Matrix | Test_Manual + Train_GT | Test_Rus + Train_Rus |
| 40 | 35.3 | 44.8 | 60.3 |
| 20 | 50 | 52.5 | 76,7 |
| 10 | 63 | 72.3 | 88.3 |

The possible <*types of translation*> are: manual (*Manual*), by transformation matrix (*Matrix*), by Google translate (*GT*) and no translation at all (*Rus*). The *Rus* option means that the classifier is trained and tested on the original Russian requests. Figure 1 visualize relative locations of *Test_Manual* and *Train_Matrix* vectors for 2 classes. The results are presented in Table 2.

One can see that under considered circumstances the target language classifier performs worse than the source language classifier (*Test_Manual* and *Test_Rus* columns in Table 2). Matrix-based knowledge transferring significantly improves results in the target language (*Test_Manual* and *Test_Manual+ Train_Matrix* columns in Table 2) while automatic translation based transferring understandably produces even better results (*Test_Manual+ Train_Matrix* and *Test_Manual+ Train_GT* columns in Table 2). Last column in Table 2 shows the maximum accuracy that can be achieved on the studied data with the employed classification algorithm. One may notice that additional training examples are more significant when the number of classes is bigger.

## 6   Discussion

As follows from Table 2 knowledge transferring effect is more pronounced when the amount of classes is bigger (the amount of samples per class is smaller). This statement is even more true for matrix based transferring. Such behavior seems natural because nearest neighbors algorithm is not able to use new information after a certain amount of samples has been provided to it. This saturation may come later for more complex classification algorithms.

It is also obvious that the observed accuracy values are too low to deploy the classifier in real life applications. This problem may and should be addressed from different sides which include at least employing more sophisticated classification algorithm, building better vector spaces and optimizing language transfer mapping.

# 7   Conclusions

In this paper a problem of building a text classifier with no or very limited amount of training data in a target language was studied. It has been shown that transferring knowledge from another language to the target one is an efficient way to increase the classification accuracy. Such transferring may be accomplished with word embeddings in language vector spaces and does not necessarily require expensive language resources or human labor. Experiments with call steering data and Russian-Kazakh language pair have demonstrated substantial classification accuracy gains when the proposed scheme is applied.

**Acknowledgments.** This work was financially supported by the Ministry of Education and Science of the Russian Federation, Contract 14.579.21.0008, ID RFMEFI57914X0008.

# References

1. Bengio, Y., Ducharme, R., Vincent, P., Janvin, C.: A neural probabilistic language model. J. Mach. Learn. Res. **3**, 1137–1155 (2003)
2. Coulmance, J., Marty, J.-M., Wenzek, G., Benhalloum, A.: Trans-gram, fast cross-lingual word-embeddings. In: Proceedings of the Empirical Methods in Natural Language Processing (2015)
3. Erk, K., Pad, S.: A structured vector space model for word meaning in context. In: Proceedings of EMNLP (2008)
4. Gouws, S., Bengio, Y., Corrado, G.: Bilbowa: fast bilingual distributed representations without word alignments. In: Proceedings of the 25th International Conference on Machine Learning, vol. 15, pp. 748–756 (2015)
5. Huang, E.H., Socher, R., Manning, C.D., Ng, A.Y.: Improving word representations via global context and multiple word prototypes. In: Proceedings of ACL, pp. 873–882. ACL (2012)
6. Klementiev, A., Titov, A., Bhattarai, B.: Inducing crosslingual distributed representations of words. In: International Conference on Computational Linguistics (COLING), Bombay, India (2012)
7. Lewis, D.D., Yang, Y., Rose, T., Li, F.: Rcv1: a new benchmark collection for text categorization research. J. Mach. Learn. Res. **5**, 361–397 (2004)
8. Mikolov, T., Chen, K., Corrado, G., Dean, J.: Efficient estimation of word representations in vector space. In: Proceedings of Workshop at ICLR (2013)
9. Mikolov, T., Le, Q.V., Sutskever, I.: Exploiting similarities among languages for machine translation (2013). http://arXiv.org/abs/1309.4168
10. Pennington, J., Socher, R., Manning, C.D.: Glove: global vectors for word representation. In: Proceedings of the Empirical Methods in Natural Language Processing (2014)
11. Socher, R., Pennington, J., Huang, E.H., Ng, A.Y., Manning, C.D.: Semi-supervised recursive autoencoders for predicting sentiment distributions. In: Proceedings of EMNLP, pp. 151–161. ACL (2011)
12. Turney, P.D., Pantel, P., et al.: From frequency to meaning: vector space models of semantics. J. Artif. Intell. Res. **37**(1), 141–188 (2010)
13. https://code.google.com/archive/p/word2vec/

# Analysis of Images and Video

# Image Processing Algorithms with Structure Transferring Properties on the Basis of Gamma-Normal Model

Inessa Gracheva[✉] and Andrey Kopylov

Tula State University, Tula, Russian Federation
gia1509@mail.ru, And.Kopylov@gmail.com

**Abstract.** Within the framework of the Bayesian approach, the general problem of image processing can be represented as a problem of estimation of the hidden component of the two-component random field on the basis of realization of its observable component, that is an analyzed image. Nonstationary gamma-normal model of the two-component random field showed good results in processing quality and computation time by solving the problem of image denoising. This paper proposes to extend the initial formulation for solving problems requiring transferring structure of the intermediate image on the processing result. Haze removal problem, HDR image compression and edges refinement of an image are considered as practical examples of such problems.

**Keywords:** Gamma-normal model · Structure transferring features · Linear time filtering

## 1 Introduction

Edge-preserving smoothing is one of the key technique of image preprocessing for a variety of practical problems, from simple image denoising and restoration to HDR imaging and haze removal in enhanced vision systems. The basic difficulty with smoothing methods is that they tend to blur local image structure during processing. The conflict between noise elimination and image degradation gives rise to variety of papers on the subject presenting different kind of compromises in this collision.

It is clear, that any additional information about image structure, which can be exploited for the smoothing algorithm, makes it possible to improve the quality of the processing result. In many practical tasks the information about image structure can be extracted from some additional "guided" image. For example, in the haze removal problem a rough transmission map, obtained by the dark channel method [11], acts as source image while the initial hazy image can be used for getting structural information. In the HDR compression problem a low-contrast image is a subject of processing and the HDR layer is a source of information about local image structure. In edges refinement for segmentation task, a rough classifier output should be refined on the basis of initial image as

© Springer International Publishing AG 2017
D.I. Ignatov et al. (Eds.): AIST 2016, CCIS 661, pp. 257–268, 2017.
DOI: 10.1007/978-3-319-52920-2_24

a "guide". Note, that all these tasks need to transfer structure from one image to another.

Thereafter a new corresponding class of filters, called structure-transferring filters [1, 2] arises recently in the literature on the data analysis, signal and image processing. A work on anisotropic diffusion [3] was one of the first papers, proposed the utilizing of additional structural information. The estimates of the gradient of processed image were used to control the diffusion process in order to preserve local data patterns. Weighted least squares method in [4] can also be considered as such a filter. In that paper the input image itself is used to adjust the weights of the filter, instead of the intermediate result, as in [3]. Finally, joint bilateral filter [5] and the Guided filter [1], which currently occupies a leading position among the filters of this class, explicitly use additional image, that carries information about the structure, as a guide image. The main disadvantage of the Joint bilateral filter and the Guided filter is the presence of artifacts, which visually manifested in the form of halos around the edges of the objects. The availability of such artifacts is characteristic of all filters with finite impulse response that is explained by the Gibbs effect. In the recent paper [2] an attempt to overcome this fact is made by using the weighted global average parameters of the corresponding linear model, but at the same time, the computational complexity is increased.

In this paper we elaborate Bayesian approach to image processing and other types of ordered data with structure transferring properties. The framework of Markov random fields makes it possible to take into account the structure, extracted from the "guide" data array, through setting an appropriate probabilistic relationships between the elements of processed data. The Markov model is necessary to train (i.e. to evaluate it parameters) on the input data array and "guide" data array.

Modern learning methods for the Markov models have one important disadvantage - high computational complexity. This is what limits their use in real systems of computer vision and image processing. For example, EM algorithm requires that the aposterior distributions of the hidden model components were calculated on E-step, what is an NP-hard problem itself. In the case of variation approach coordinate descent methods are often used, and iterative gradient methods [6,7] are utilized for calculating of the minimum energy value at fixed parameters. The approximate methods serve as means to struggle with this limitation. They allow to find the approximate estimate of the minimum of the energy, such as in work [8] by estimation parameters of active Markov fields, which ideologically close to the method used in this paper.

## 2    Related Work

In this work we use a generalization of the Bayesian approach to the image analysis described in the paper [9]. The aim of processing can be represented as a transformation of the original image $Y = (y_t, t \in T)$, defined on a subset of the two-dimensional discrete space $T = \{t = (t_1, t_2) : t_1 = 1, ..., N_1, t_2 = 1, ..., N_2\}$, into a secondary array $X = (x_t, t \in T)$, which would be defined on the same argument set and take values from a set depending on the problem. We will consider an analyzed image $Y = (y_t, t \in T)$ and result of processing $X = (x_t, t \in T)$ as, respectively, the observed and hidden component of the two-component random field $(X, Y)$.

Probabilistic properties of a two-component random field $(X, Y)$ are completely determined by the joint conditional probability density $\Phi(Y|X, \delta)$ of original functions $Y = (y_t, t \in T)$ with respect to the secondary data $X = (x_t, t \in T)$, and the a prior joint distribution $\Psi(X|\Lambda, \delta)$ of hidden component $X = (x_t, t \in T)$.

Following [9], let the joint conditional probability density $\Phi(Y|X, \delta)$ be in the form of Guassian distribution:

$$\Phi(Y|X, \delta) = \frac{1}{\delta^{(N_1 N_2)/2}(2\pi)^{(N_1 N_2)/2}} \exp(-\frac{1}{2\delta} \sum_{t \in T} (y_t - x_t)^2), \tag{1}$$

where $\delta$ is the variance of the observation noise, which is unknown.

The a priori joint distribution of the hidden component $X = (x_t, t \in T)$ is also assumed to be Gaussian. But the variances $r_t$ of hidden variables is assumed to be different at different points $t \in T$ of the hidden field $X$. The unknown variances $r_t, t \in T$, are considered as proportional to the variance of the observation noise $\delta$, unknown as well, with the proportionality coefficients $r_t = \lambda_t \delta$ acting as factors of the unknown local variability of the sought for processing result. Under this assumption, we come to the improper a priori density:

$$\Psi(X|\Lambda, \delta) \propto \frac{1}{\left(\prod_{t \in T} \delta\lambda_t\right)^{1/2} (2\pi)^{(N_1 N_2)/2}}$$

$$\times \exp\left(-\frac{1}{2} \sum_{t', t'' \in V} \frac{1}{\delta\lambda_t}(x_{t'} - x_{t''})^2\right), \tag{2}$$

where $V$ is the neighborhood graphs of image elements having the form of a lattice.

Finally, we assume the inverse factors $1/\lambda_t$ to be a priori independent and identically gamma-distributed on the positive half-axis $\lambda_t \geq 0$.

$$\gamma(1/\lambda_t|\delta, \lambda, \mu) \propto (1/\lambda_t)^{\frac{2\mu+1}{2\delta\mu}} \exp\left(-\frac{\lambda}{2\delta\mu}(1/\lambda_t)\right),$$

where $\lambda$ and $\mu$ is, respectively, the basic average factors and the ability of instantaneous factors to change along the image plane.

The mathematical expectation and variance of gamma-distribution $E(1/\lambda_t) = \frac{(1+\delta)\mu+1}{\lambda}, Var(1/\lambda_t) = 2\delta\mu\frac{(1+\delta)\mu+1}{\lambda^2}$.The a priori distribution density of the entire field of the factors:

$$G(\Lambda|\delta, \lambda, \mu) = \exp\left[-\frac{1}{2\delta\mu} \sum_{t \in T} \left(\lambda\frac{1}{\lambda_t} + \frac{1}{\lambda}\ln\lambda_t\right)\right], \tag{3}$$

The independent prior distribution of each instantaneous inverse factors $1/\lambda_t$ is almost completely concentrated around the mathematical expectation $1/\lambda$ if $\mu \to 0$. On the contrary, with $\mu \to \infty$ coefficient $1/\lambda$ have tends to the almost uniform distribution.

So, we have completely defined the joint prior normal gamma-distribution of both hidden fields $X = (x_t, t \in T)$ and $\Lambda = (\lambda_t, t \in T)$:

$$H(X, \Lambda | \delta, \lambda, \mu) = \Psi(X | \Lambda, \delta) G(\Lambda | \delta, \lambda, \mu).$$

Coupled with the conditional density of the observable field (1), it makes basis for Bayesian estimation of the field $X = (x_t, t \in T)$.

The joint a posteriori distribution of hidden elements $P(X, \Lambda | Y, \delta, \alpha, \vartheta)$, namely, those of field $X = (x_t, t \in T)$ and its instantaneous factors $\Lambda = (\lambda_t, t \in T)$, is completely defined by (1), (2) and (3). The Bayesian estimate is independent of the observation noise variance $\delta$ and can be obtained by solving the following optimization task:

$$\begin{cases} (\hat{X}, \hat{\Lambda} | \lambda, \mu) = \arg\min_{X, \Lambda} J(X, \Lambda | Y, \lambda, \mu), \\ J(X, \Lambda | Y, \lambda, \mu) = \sum_{t \in T} (y_t - x_t)^2 \\ + \sum_{t', t'' \in V} \left\{ \frac{1}{\lambda_{t'}} \left[ (x_{t'} - x_{t''})^2 + \lambda/\mu \right] + (1 + 1/\mu) \ln \lambda_{t'} \right\}. \end{cases} \quad (4)$$

As it has been shown in [9], the growing value of parameter $\mu$ endows this criterion with a pronounced tendency to keep the majority of estimated factors $\hat{\lambda}_t$ close to the basic low value $\lambda$ and to allow single large outliers, keeping thereby the local structure of processed data.

An iterative Gauss-Seudel procedure is used in [9] to solve the optimization problem (4).

## 3  Structure Transferring Filtering

We expand here formulation of the problem given in the previous section. As mentioned above, the field of factors $\Lambda = (\lambda_t, t \in T)$ serves as a measure of local variability of a hidden field $X = (x_t, t \in T)$. As it can be seen from the criterion (4), $\lambda_{t'}, t' \in T$ actually plays the role of a penalty on the difference between values of two corresponding neighboring variables $x_{t'}^g$ and $x_{t''}^g$, $(t', t'') \in V$. Thus $\Lambda = (\lambda_t, t \in T)$ being estimated with the help of an additional guided image, can be used to transfer the structure of local relations between elements of the guided image to the result of processing.

Then $X = (x_t, t \in T)$ is fixed, criterion (4) gives the following equation for optimal $\Lambda$ with fixed structural parameters $\mu$ and $\lambda$:

$$\hat{\lambda}_{t'}(X^g, \lambda, \mu) = \lambda \frac{(1/\lambda)(x_{t'}^g - x_{t''}^g)^2 + 1/\mu}{1 + 1/\mu}, (t', t'') \in V. \quad (5)$$

The additional guided image can serve here as $X^g$ to objectify its structure by $\hat\Lambda = (\hat\lambda_t, t \in T)$. The estimates $\hat\Lambda = (\hat\lambda_t, t \in T)$, in turn, gives the optimal estimates of the field $\hat X = (\hat x_t, t \in T)$:

$$\hat X = (\hat x_t, t \in T) = \arg\min_X J(X, \Lambda | Y, \lambda, \mu)$$

$$= \arg\min_X \left\{ \sum_{t \in T} (y_t - x_t)^2 + \sum_{t', t'' \in V} \frac{1}{\lambda_t} (x_{t'} - x_{t''})^2 \right\}. \quad (6)$$

Figure 1 illustrates this process on the example of edges refinement task, which can be treated as a reference task for this type of problems. The rough initial mask, obtained by some segmentation algorithm or represents the output of some classifier, plays the role of analyzed image $Y = (y_t, t \in T)$, input image is taken as the "guide" image $X^g = (x_t^g, t \in T)$. Figure 2 shows the behavior of our algorithm. We observe that our method reasonably well recovers the hair, even though the analyzed image $Y = (y_t, t \in T)$ is binary and very rough.

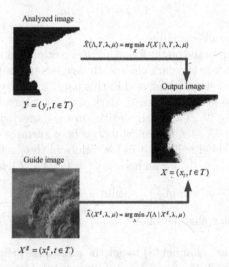

**Fig. 1.** Illustrations of the structure transferring feathering.

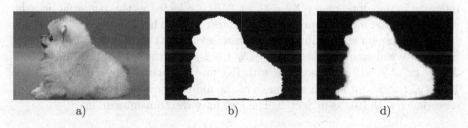

**Fig. 2.** The structure-transferring filtering. (a) Original image. (b) Binary mask. (c) Result of the structure-transferring filtering ($\lambda = 10^{-9}$, $\frac{1}{\mu} = 2 * 10^{-5}$).

The structure transferring property of the proposed algorithm lets us to expand the traditional for the used mathematical framework set of applications on new problems such as dehazing (Sect. 3), HDR image compression (Sect. 4) and segmentation (Sect. 5).

# 4   Image Haze Removal Algorithm

Image haze removal is a difficult problem that requires an inference method or prior knowledge of the scene. The amount of fog observed on an image depends on the distance from the object to the camera, wavelength of the light and the size of the scattering particles in the atmosphere [10]. In the haze removal problem the original haze image plays the role of guide image $X^g = (x_t^g, t \in T)$, $X^g \in R^3$, and dark channel image is taken as the analyzed image $Y = (y_t, t \in T)$, $Y \in R$. The transmission map $X = (x_t, t \in T)$, $X \in R$ (is the medium transmission describing the portion of the light that is not scattered and reaches the camera) linked with the "dehaze" version $\tilde{X} = (\tilde{x}_t, t \in T)$, $\tilde{X} \in R$ through the atmospheric model [12],

$$x_t^g = \tilde{x}_t x_t + (1 - x_t)a, \tag{7}$$

where $a \in R^3$ is atmospheric light.

The dark channel is based on the following observation about shading of an original image [11]. To obtain the dark channel, image has to be splitted into patches of some size ($5 \times 5$ pixels patch is used in this paper). Most of these blocks have very low intensity of some pixels (called "dark pixels") at least in one color channel (RGB), as show in Fig. 3b. The low intensities in the dark channel are mainly due to three factors: (a) shadows; (b) colorful objects or surfaces (any object lacking color in any color channel will result in low values in the dark channel); (c) dark objects or surfaces. Formally, for an image $X^g = (x_t^g, t \in T)$, we dene

$$y_t = \min_{c \in \{r,g,b\}} \min_{j \in \Omega(t)} ((X^g)_j^c), \tag{8}$$

where $(X^g)^c$ is a color channel of $X^g = (x_t^g, t \in T)$ and $\Omega(t)$ is a local patch centered at $t$.

We will use the dark channel (8) to get the atmospheric light estimation as in paper [11]. First we pick the top 0.1% brightest pixels in the dark channel $Y = (y_t, t \in T)$. These pixels are best define the haze. Among these pixels with highest intensity in the guide image $X^g = (x_t^g, t \in T)$ is selected as the atmospheric light $a$. Note that these pixels may not be brightest in the whole image.

We will use for estimating of the transmission map the algorithm described in Sect. 2. We will estimate factors $\hat{\Lambda} = (\hat{\lambda}_t, t \in T)$ with the help of the guide image $X^g = (x_t^g, t \in T)$ in accordance with (5), and smooth the dark channel image $Y = (y_t, t \in T)$ for obtain estimates of the hidden sequence $\hat{\Lambda} = (\hat{\lambda}_t, t \in T)$ in accordance with (6), as in Sect. 2. Then we have the transmission map

$$x_t = 1 - \omega(\hat{x}_t), \tag{9}$$

where $\omega$ is application-based. We x it to 0.95 for all results reported in this paper.

Now we have all of the desired parameters by (9), then the dehazing image is obtained with

$$\tilde{x}_t = \frac{x_t^g - a}{x_t} + a$$

The estimation of the factors is important for preserving scene depth discontinuities in the veiling (hence transmission) estimate. In Fig. 3 we demonstrate the importance of the factor estimates for image dehazing. Notice, that after image dehazing the edges of the objects are preserved, as shown in Fig. 3(c). The blurred depth discontinuities in the transmission estimate causes the image dehazing have well-defined edges of objects (see Fig. 3(d)).

**Fig. 3.** Image Dehazing. (a) Guide image. (b) Observe image. (c)Transmission map. (d) Dehaze image ($\lambda = 10^{-12}, \frac{1}{\mu} = 10^{-2}$). (Color figure online)

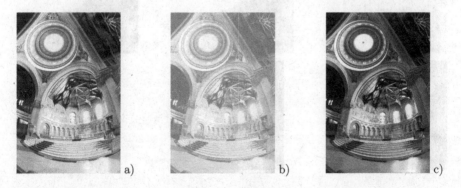

**Fig. 4.** HDR image compression. (a) Original HDR image. (b) Low-contrast image; (c) Result of HDR image processing with use our algorithm ($\lambda = 10^{-12}, \frac{1}{\mu} = 5 * 10^{-3}, \alpha = 0.3$).

# 5  HDR Image Compression

When drawing or painting, many artists capture visual appearance with a "coarse-to-fine" sequence of boundaries and shading. Many begin with a sketch of large, important scene features and then gradually add finer, more subtle details. Initial sketches hold sharply dened boundaries around large, smoothly shaded regions for the largest, highest contrast, and most important scene features. The artist then adds more shadings and boundaries to build up ne details and "fill in" the visually empty regions and capture rich detail everywhere. This method works particularly well for high contrast scenes because it permits separate contrast adjustments at every stage of increasing detail and renement. The artist may also emphasize or mute scene components, to control their prominence and direct the viewer's attention.

The algorithm is based on use of the nonstationary gamma-normal model in the framework of the Bayesian approach to image processing as described in Sect. 1. For HDR image compression we will estimate the sequence of factors $\hat{\Lambda} = (\hat{\lambda}_t, t \in T)$ (5) by the "guide" image $X^g = (x_t^g, t \in T)$ and the hidden component $\hat{X} = (\hat{x}_t, t \in T)$ (6) by the analyzed image $Y = (y_t, t \in T)$. Thus, we get the analyzed image $Y = (y_t, t \in T)$, which in the HDR compression problem is a low-contrast image (see Fig. 4(b)) and "guide" image $X^g = (x_t^g, t \in T)$ is the analyzed image (see Fig. 4(a)). And then calculate new values output image

$$\tilde{x}_t = y_t - \alpha(\hat{x}_t + mean(\hat{x})), \tag{10}$$

where $\tilde{X} = (\tilde{x}_t, t \in T)$ is result of HDR image processing, $\alpha$ is contrast sensitivity and will compress contrasts for values $< 1$.

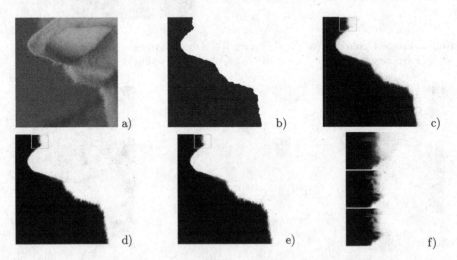

**Fig. 5.** Comparisons on structure-transferring filtering. (a) Original image; (b) Binary mask; (c) Our algorithm ($\lambda = 10^{-9}$, $\frac{1}{\mu} = 5 * 10^{-5}$); (d) Guided filter ($r = 60, \varepsilon = 10^{-6}$) [1]; (e) Fast guided filter ($r = 60, \varepsilon = 10^{-6}, s = 4$) [15]; (f) In the zoom-in patches, our algorithm compare with the Guided filter [1] and the fast Guided filter [15].

The developed procedure simulates the process flow diagram of the artist, namely to selectively stresses parts from the scene, which were not visible on the analyzed image, as shown in Fig. 4. The level of detail images is regulated depending on the structural parameters $\mu$ and $\lambda$ and values of the scaling factor $\alpha$.

## 6   Experimental Results

In this section, we conduct a series of experiments for different applications including structure transferring feathering (Fig. 5), image dehazing (Fig. 6) and HDR image compression (Fig. 7) to verify the effectiveness of the proposed filtering method. And also compare developed algorithm with Guided filter [1] and fast Guided filter [15] by quality processing and computation time.

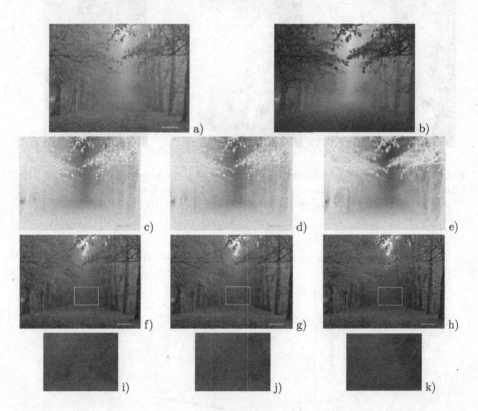

**Fig. 6.** Haze removal. (a) Original image; (b) Dark channel image; (c), (d), (e)The raw transmission map is refined by the Guided filter ($r = 20, \varepsilon = 10^{-3}$) [1], the fast Guided filter ($r = 20, \varepsilon = 10^{-3}, s = 4$) [1], our algorithm ($\lambda = 5 * 10^{-9}, \frac{1}{\mu} = 9 * 10^{-5}$) respectively; (f), (g), (h) Dehaze using the Guided filter [1], the fast Guided filter [15], our algorithm respectively; (i), (j), (k) In the zoom-in patches, we compare with the Guided filter [1] and the fast Guided filter [15].

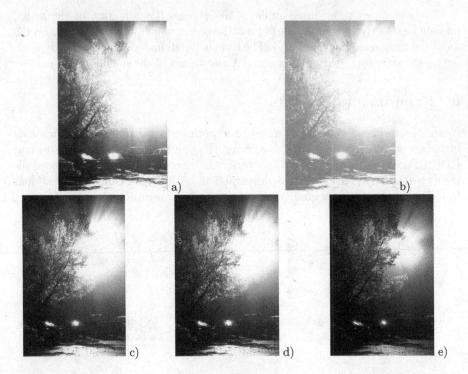

**Fig. 7.** HDR compression. (a) The original image; (b) The low-contrast image; (c) The Guided filter output ($r = 16, \varepsilon = 0.02^2$) [1]; (d) The fast Guided filter ($r = 16, \varepsilon = 0.02^2, s = 4$) [15]; (e) Our algorithm result ($\lambda = 10^{-5}, \frac{1}{\mu} = 10^{-3}, \alpha = 0.2$).

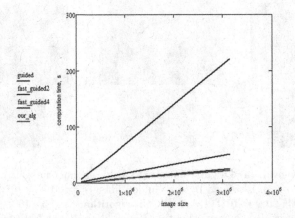

**Fig. 8.** Computation time of Guided filter [1], fast Guided filter (with $s = 2$) [15], fast Guided filter (with $s = 4$) [15] and our algorithm for processing of images of different sizes.

Experimental comparison of algorithms were performed on a Core i3 workstation with 1.6-GHz in MATLAB. Increase in Fig. 5 shows a smoother structure than our method with Guided filter [1]. And conversely, an increase in Fig. 6 demonstrates the lack of artifacts and a clearer result image. Comparison of algorithms by computation time (Fig. 8) shows that presented in this paper the method a speedup > 10x than the joint guided filter, a speedup > 2x than the fast guided filter when a subsampling ratio $s = 2$, and have roughly the same time for $s = 4$.

## 7    Conclusion

In this paper, we propose a new image filtering method with structure transferring properties based on the nonstationary gamma-normal model. The advantage of our approach is, first, a rigorous probabilistic formulation of the problem by using the normal-gamma model and the exact substantive interpretation of parameters. In addition, for fixed parameters criterion is quadratic pair-wise separable function, which allows to use highly computationally efficient optimization procedures on the basis of Kalman filter - interpolator or dynamic programming [14]. Experimental results on different applications verify that the proposed method comparable and in some cases outperforms the Guided filter by quality processing and computation time.

**Acknowledgements.** This research is funded by RFBR grants, 16-07-01039 and 16-57-52042.

## References

1. He, K., Sun, J., Tang, X.: Guided image filtering. IEEE Trans. Pattern Anal. Mach. Intel. **35**(6), 1397–1409 (2013)
2. Zhang, J., Cao, Y., Wang, Z.: A new image filtering method: nonlocal image guided averaging. In: 2014 IEEE International Conference on Acoustics, Speech and Signal Processing, vol. 2012, pp. 2479–2483 (2014)
3. Perona, P., Malik, J.: Scale-space and edge detection using anisotropic diffusion. IEEE Trans. Pattern Anal. Mach. Intell. **12**(7), 629–639 (1990)
4. Farbman, Z., et al.: Edge-preserving decompositions for multi-scale tone and detail manipulation. ACM Trans. Graph. **27**(3), 1145–1155 (2008)
5. Petschnigg, G., et al.: Digital photography with flash and no-flash image pairs. ACM Trans. Graph. **23**(3), 664–673 (2004)
6. Tappen, M.F.: Utilizing variational optimization to learn Markov random fields. In: Proceedings of IEEE Conference on Computer Vision and Pattern Recognition, pp. 1–8 (2007)
7. Geman, D., Reynolds, G.: Constrained restoration and the recovery of discontinuities. IEEE Trans. Pattern Anal. Mach. Intell. **14**(3), 367–383 (1992)
8. Barbu, A.: Training an active random field for real-time image denoising. IEEE Trans. Image Process. **18**(11), 2451–2462 (2009)

9. Gracheva, I., Kopylov, A., Krasotkina, O.: Fast global image denoising algorithm on the basis of nonstationary gamma-normal statistical model. In: Khachay, M.Y., Konstantinova, N., Panchenko, A., Ignatov, D.I., Labunets, V.G. (eds.) AIST 2015. CCIS, vol. 542, pp. 71–82. Springer, Heidelberg (2015). doi:10.1007/978-3-319-26123-2_7

10. Cheng, Y.J., et al.: Visibility enhancement of single hazy images using hybrid dark channel prior. In: 2013 IEEE International Conference on Systems, Man, and Cybernetics (SMC), pp. 3627–3632. IEEE (2013)

11. Kaiming, H., Jian, S., Tang, X.: Single image haze removal using dark channel prior. In: IEEE International Conference on Computer Vision and Pattern Recognition (CVPR), pp. 1956–1963 (2009)

12. Kristofor, B., Gibson, B., Truong, Q.N.: Fast single image fog removal using the adaptive Wiener filter. In: International Conference on Information Processing (ICIP) (2013)

13. Larin, A., Seredin, O., Kopylov, A., Kuo, S.-Y., Huang, S.-C., Chen, B.-H.: Parametric representation of objects in color space using one-class classifiers. In: Perner, P. (ed.) MLDM 2014. LNCS (LNAI), vol. 8556, pp. 300–314. Springer, Heidelberg (2014). doi:10.1007/978-3-319-08979-9_23

14. Mottl, V.V., et al.: Hidden tree-like quasi-Markov model and generalized technique for a class of image processing problems. In: Proceedings of the 13th International Conference on Pattern Recognition, 1996, vol. 2, pp. 715–719. IEEE (1996)

15. He, K., Sun, J.: Fast Guided Filter. arXiv: 1505.00996v1 [cs.CV], 5 May 2015

# Two Implementations of Probability Anomaly Detector Based on Different Vector Quantization Algorithms

Anna Denisova[✉]

Laboratory of Earth Remote Sensing Technologies, Samara National Research
University, 34, Moskovskoe Shosse, Samara 443086, Russian Federation
denisova_ay@geosamara.com
http://www.ssau.ru

**Abstract.** This article continues studies of probability anomaly detector method which was presented in author's previous works. Here two implementations of this method are introduced. The implementations are based on different vector quantization algorithms. Description of both algorithms and results of experimental research of their parameters are provided. Both implementations are compared with well known RX anomaly detector on synthetic hyperspectral images.

**Keywords:** Probability anomaly detector · Vector quantization · Aggregation functions · Hyperspectral images

## 1 Introduction

Hyperspectral images are multichannel images and are produced by Earth remote sensing systems. Each channel of hyperspectral image is obtained for a narrow spectral band. Pixel is represented by n-dimensional vector, that corresponds to the reflected light spectrum.

In hyperspectral image analysis an anomaly is usually regarded as a small region which spectral description is significantly different from spectral description of a background. The most important issues for the anomaly detection algorithms are background model and measure of differences between anomaly and background. There are several common background models for anomaly detection [1]:

(1) Local normal background model (Reed-Xaoli Detector (RXD) [2]). It assumes that background pixels have normal distribution.
(2) Gaussian mixture model (Gaussian mixture model-Generalized likelihood ratio test Anomaly detector (GMM-GLRT) [3], Cluster based Anomaly detector [4,5]). It supposes that background consists of several classes each being characterized by normal distribution. Anomaly detection algorithms, based on this model, use segmentation before calculating parameters of Gaussian mixture.

© Springer International Publishing AG 2017
D.I. Ignatov et al. (Eds.): AIST 2016, CCIS 661, pp. 269–280, 2017.
DOI: 10.1007/978-3-319-52920-2_25

(3) Linear spectral mixture model (Orthogonal Subspace Projection Anomaly Detector (OSP Detector) [6], Signal Subspace Processing Anomaly Detector (SSP) [7]). In this model pixels are considered as linear combination of fixed spectra (endmembers) which can be associated with some materials existing on the scene.

(4) Local normal model in feature space (Kernel-RX Anomaly detector [8]) corresponds to the methods that use transformations of initial image to feature space before anomaly detection, for example principal component analysis. It is assumed that in feature space background can be described by means of normal distribution.

(5) Nonparametric background models (SVDD Anomaly Detector [9,10]) are based on background estimation from a training sample.

Anomaly detection algorithms can be divided into global and local algorithms that use the whole image or its small part for background parameters evaluation.

Each of listed above models of background has its own limitations, therefore an algorithm, that does not use any suggestions about background or training sample for background estimation, should be developed.

Proposed in previous work [11], probability anomaly detection (PAD) method is a global algorithm. It treats pixels with the lowest probability as anomalies. So the anomaly in this case is a small object that has an unusual spectral signature in the scene. Probability criterion does not require any information about background. So the aim is to provide an implementation of the probability detection method and give their performance evaluation.

The paper is organized as follows. First of all the description of common method is given, then two its implementations are considered. In following section an overview of the algorithms parameters selection and experimental comparison with RXD are provided.

## 2    Probability Anomaly Detector

Hyperspectral image pixel is a high dimensional vector, therefore it is hard to organize search for pixels with lowest probability without special transformations of initial data. Besides the probability of presence of two identical pixels on hyperspectral images is too low because of high dimensionality of data and noise. Thus search for unique pixels with low probability must work with quantized values of pixels. Vector quantization can result in a decrease in dynamic range of pixel values and, at the same time, it saves proximity of pixel values to their quantized representations.

Let us denote a pixel of hyperspectral image as $x\,(n_1, n_2) \in R^n$ and quantized value of this pixel as $\bar{q} = (q_0, q_1, ..., q_{n-1})^T$. Calculation of probabilities for each quantized value in quantization table is equivalent to calculation of a histogram. In our case histogram is a table of probabilities for every realized quantization value. The problem of using and building such histogram is that in case of high dimensions the amount of all possible quantized values is enormous and it is difficult to save the histogram in computer memory and to operate with it. But the number of nonzero values of the histogram is no more than the total number of pixels in image. Thus it is enough to store a list of quantized values really derived from pixels of image. The length of this list will be shorter than or equal to the total amount of pixels in the image.

To construct this short histogram representation a transformation of quantized pixel into integer number should be defined. This transformation is necessary for indexing of unique quantized values of image. In this article integer hash functions $f(\bar{q}) : R^n \mapsto Z$ are proposed as a variant of required transformation. Hashing of quantized vector is used for quick search of appropriate histogram value and rapid calculation of probabilities.

Final anomaly measure for each pixel is computed for by aggregation of probabilities of adjacent pixels. The aim of aggregation is to delineate local areas of pixels with anomaly signatures. Let us denote a set of pixels inside sliding window in current position as $I(n_1, n_2)$ and a set of their quantized values as $\{\bar{q}_i\}_{i \in I(n_1, n_2)}$. So the result for the current window position can be computed as Eq. (1):

$$P(n_1, n_2) = 1 - min_{i \in I(n_1, n_2)} P(\bar{q}_i).$$  (1)

High values of $P$ correspond to anomalous values and low ones correspond to background. Except the minimum function in (1) we can use other functions, for example, sum, maximum, median etc. The question of choice of aggregation function is omitted in this article due to lack of space.

## 3   PAD Modification with Uniform Quantizing (PAD UQ)

PAD UQ uses uniform quantizing with $K$ levels for each image channel as a vector quantization algorithm. In this case one histogram interval corresponds to n-dimensional parallelepiped bounded by neighboring quantization levels. Quantization levels are defined separately for each image component. The minimum $x_{l_{min}}$ and $x_{l_{max}}$ maximum values per channel are computed, and quantization level is calculated as follows:

$$q_l(n_1, n_2) = \left[ K \frac{x_l(n_1, n_2) - x_{l_{min}}}{x_{l_{max}} - x_{l_{min}}} \right],$$  (2)

where $[y]$ is integer part of a number $y$. The value $x_{l_{max}}$ corresponds to the last quantization level $K$.

Uniform quantizing of each image component leads to transformation of the initial image into a set of integer vectors with unknown amount of unique values. To construct histogram of quantized values we use a hash table as a data structure. Hash tables can grow dynamically as new unique quantized values are discovered. If the hash values of two quantized vectors are the same, these vectors are considered to be identical. In real terms, collisions of hash values are possible. However, the amount of actual hash values is much lower than the number of all possible hash values, that is why amount of possible collisions can be regarded as insufficient.

Each component of a quantized vector is a number from 0 to $K - 1$ and the whole vector can be represented as an integer number in numerical system with base $K$:

$$f(\bar{q}) = \sum_{i=0}^{n-1} q_i K_i. \tag{3}$$

But input data usually have a hundreds of spectral channels and the number representation for quantized vector is so huge, that it could not be saved with standard long integer types. The following hash functions can be used to make it shorter:

(1) modulo hashing

$$f(\bar{q}) = \sum_{i=0}^{n-1} q_i K_i \bmod M, \tag{4}$$

where $M$ is a prime number. In this case hash values are varied from 0 to $M - 1$. M must be greater than amount of pixels on image as much as possible. Maximum prime number used for hashing is $M = 2147483647$. It was used in this study for the experiments. It is sufficient to compute division by modulo $M$ after each multiplication by $K$ for each summand, i.e., it doesnt requires special methods for multiplication of big numbers.

(2) multiplicative modulo hashing:

$$f(\bar{q}) = \sum_{i=0}^{n-1} \alpha q_i K_i \bmod M, \tag{5}$$

where $\alpha$ is real constant. It is often defined as $\alpha = 0.618033$ according to the golden ratio rule. This method sets middle-order digits of a big number as its leading orders. In the case of simple modulo hashing high-order digits are leading.

(3) hash functions for strings. It is known, that strings are represented as integer arrays in computer memory, therefore any algorithm for string hashing can be used, for example, Horner algorithm. In present study the following method is applied:

1. Initialization: $h = 0$ and $a = 127$.
2. For all $n = 0, ..., N - 1 : h = (a * h + q_n) \bmod M$.

Hashing and computing histogram can run at the same time. After histogram of quantized pixels is computed, each output pixel is assigned to anomaly measure defined by Eq. (1).

## 4  PAD Modification with Vector Quantization Based on Agglomerative Clusterization (PAD AC)

Second modification of PAD algorithm uses vector quantization algorithm with fixed codebook size. In case of fixed codebook hashing is not needed, because indexes of codes in codebook can be used as hash.

There are several vector quantization algorithms with fixed quantization table, for example, LGB-algorithm [12] and k-means [13]. High computational complexity is considered to be the common disadvantage of these algorithms, which examine all image pixels many times while codebook is building. LBG-algorithm decomposes all pixel by means of grid. Each cell in grid has its centroid and an error of pixels representation by means of centroids is minimum possible error for given size of codebook. LBG algorithm is locally convergent and the result depends on quality of initial codebook. K-means algorithm clusterizes an image and codebook is built from centroids of clusters. Further quantization decomposes pixels on clusters by minimum distance between pixel and cluster centroid.

To avoid limitations of LGB and k-means algorithms a new algorithm of vector quantization with agglomerative clusterization is proposed:

(1) Denote maximum size of codebook as $M$ and initial value of quantization error as $\varepsilon$. Initially codebook $Q = \{x_{c_1}\}$, $x_{c_1} = x(0,0)$ includes only one first image pixel.

(2) For each other pixel $x(n_1, n_2)$ a cluster $C_m$, to which this pixel belongs, is defined:

$$x(n_1, n_2) \in C_m, \text{ if } d(x(n_1, n_2), x_{c_m}) < \varepsilon. \tag{6}$$

Pixel is inserted in the first cluster for which condition (6) holds. At the same time center of the cluster is recomputed and is assigned to an average value of pixels belonging to the current cluster. If there are no clusters for which condition (6) holds for current pixel, then step 3 should be executed, else step 2 should be repeated for next pixel until all pixels of the image are examined.

(3) If there are no clusters for which condition (6) holds for current pixel, and codebook size doesnt exceed maximum size, a new cluster $C_m$ with center in current pixel is created. If the size of codebook exceeds maximum size
3.1 value of threshold $\varepsilon$ should be increased,
3.2 pairs of clusters with distance between centers less than $\varepsilon$ should be merged. Released codebook positions are used for new cluster with center in current pixel.

As a result of image segmentation, pixels are fragmented into irregular cells. Center of each cell is computed as average value of pixels belonging to the cell. Quantization is made by checking condition (6), where $\varepsilon$ is a value or a quantization error from the last step of algorithm.

It should be mentioned that the histogram of quantized values can be calculated during clusterization process since amount of cluster elements is saved every time for quick cluster center computation in case of cluster merging.

The main advantage of this algorithm is that the image is scanned only once during clusterization. Also threshold of representation error is equal for each cluster and it adjusts while algorithm is running.

As a result codebook and probabilities of each code in it define histogram of quantized pixel values. Anomaly value is computed using a selected aggregation function.

## 5    Experimental Research

It is difficult to make ground truth experiments with real hyperspectral images to evaluate numerical values of algorithms performance, because it requires to construct special polygons and acquire expensive image data from satellites. That is why all experiments in this paper are described for synthetic hyperspectral images to get quantitative characteristics of the developed algorithms. The images, used in experiments, have a linear spectral mixture model of background. To construct background with different correlation properties we used particular correlation functions to generate model coefficients. Coefficients of linear spectral mixture model are generated as stationary random field with biexponential autocorrelation function (ACF) and stationary random field with gauss autocorrelation function (ACF). Test images are generated by means of the following steps:

(1) selecting background spectral signatures. In this paper for background we used spectral signatures from IGCP spectral library [14]: ACTINO-LITE_AM3000, ALUNITE_AL705, BUDDINGTONITE_NHB2301, CAL-CITE_CO2004, CHLORITE_CH2402.
(2) generating stationary random fields with given correlation for each signature. These images are interpreted as coefficients in linear spectral mixture model;
(3) generating hyperspectral background image according to linear spectral mixture model [15]. Coefficients of linear spectral mixture should be normalized to fulfill sum-to-one and non-negativity constraints;
(4) embedding anomalies. In this article 5 anomalies are embedded into each test image. They have square form with size $3 \times 3$ pixels. Each anomaly corresponds to its own signature that doesnt belong to background signatures. Anomaly signatures for described experiments are GOETHITE_FE2600, JAROS ITE_JR2501, MUSCOVITE_IL107, NOTRONITE_SMN454, SIDERITE_COS2002.

Teh test images have size $128 \times 128$ pixels and 99 spectral channels. The results of the experiments are described for 20 images with correlation coefficients from 0.4 to 0.99. The probability of anomaly signatures is 0.00055. The examples of test images are presented on Fig. 1.

To evaluate quality of anomaly detection algorithms following values are used:

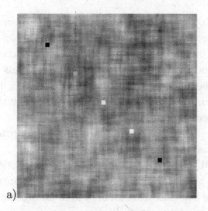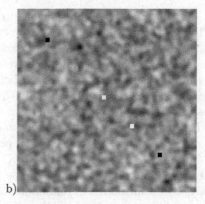

**Fig. 1.** Examples of synthetic hyperspectral images with linear spectral mixture model (a) with biexponential correlation function of coefficients, (b) with gauss correlation function of coefficients. Correlation coefficient is 0,9

- $TP$ probability of detection anomaly pixel as anomaly,
- $FP$ probability of detection nonanomaly pixel as anomaly.

To compute these characteristics binary image is used. It is obtained from output image by thresholding. Anomaly pixels, that have anomaly measure more than the value of the threshold, are assigned to one, and other values are set to zero.

Due to lack of space the study of different aggregation functions is omitted, and all experiments are described in case of aggregation window with a size equal to 1 pixel.

The results of selection of appropriate quantization level number for PAD UQ algorithm are presented in Table 1. The value of the threshold for binary processing is 0,999 and modulo hash function is used. Threshold corresponds to real probability of anomaly signatures on image. Too large threshold doesnt allow to find anomalies. Retrieved quantization values are practically unique because of the high dimensionality of data, if the number of quantization levels $K > 2$ and error of false detection is enormous. Nevertheless for $K = 2$ false detection probability is about 10% for both background models.

**Table 1.** Average probability of true and false detection for the case of PAD UQ algorithm with different numbers of quantization levels

| Number of quantization levels | Biexponential ACF | | Gauss ACF | |
|---|---|---|---|---|
| | TP | FP | TF | FP |
| 2 | 1 | 0.098 | 1 | 0.104 |
| 3 | 1 | 0.378 | 1 | 0.387 |
| 4 | 1 | 0.471 | 1 | 0.504 |

Also for PAD UQ the influence of hash function was tested. The results in Table 2 shows, that false detection error for multiplicative modulo hashing is much lower than for other hash functions, but not all anomaly pixels are detected. The difference between multiplicative modulo hashing and other algorithms is that multiplication on real constant between 0 and 1 results in greater weighting of central spectral components in hash number, and spectral signatures, having most differences in the middle part of spectra, can be distinguished better.

In the experiments with noised images PAD UQ algorithm provided unsatisfactory results with false detection probability greater than 10%.

To improve characteristics of PAD UQ algorithm dimensionality reduction transformations of initial image can be applied, for example principal component analysis transformation or minimum noise fraction transform. These transformations significantly reduces number of image channels and noise.

**Table 2.** Average probability of true and false detection for the case of PAD UQ algorithm for different hash functions

| Hash function | K=2 | |
|---|---|---|
| | TP | FP |
| Modulo hash | 1 | 0.10119 |
| Multiplicative modulo hash | 0.8 | 0.00005 |
| Horner algorithm | 1 | 0.10630 |

Adjustable parameters of PAD AC algorithm are quantization table size M and initial quantization error $\varepsilon$. Experiments shows that the best threshold for binary processing in this case equals to 0.999. All anomaly pixels with such threshold are detected with probability $TP = 1$, and probability of false detection $FP$ increases if the threshold also increases.

Figure 2 demonstrates the dependence of false detection probability $FP$ on initial quantization error in case of different sizes of quantization table and binary threshold equal to 0.999.

From diagrams in Fig. 2 it is clear that it is sufficient to set quantization table size to $M = 10$. Too large quantization table produces more tiny clusters that leads to growth of false detections. Intuitively the size of codebook can be selected a little bit more than assumed number of classes (object types) in scene. It should be stressed that PAD AC method has very low probability of false detections (it is less than 2.5%) when probability of true detection equals to 1.

The results for PAD AC algorithm in case of noisy images are presented in Table 3. In this experiment the size of anomalies varies from $3 \times 3$ pixels to $7 \times 7$. The threshold for binary processing is 0.995. Table 3 shows that PAD AC is very effective until signal to noise ratio is less than 150.

To compare proposed methods with common one RXD [2] algorithm is selected. An anomaly measure in RXD algorithm is Mahalonobis distance

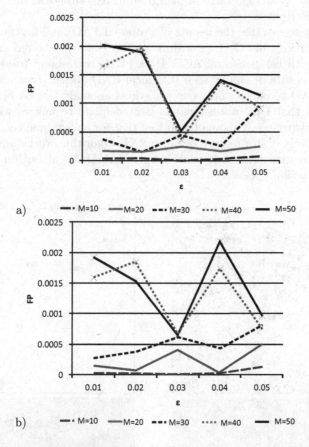

**Fig. 2.** Average probability of false detection for binary threshold 0.999 and PAD AC algorithm with background coefficients autocorrelation function: (a) biexponential ACF, (b) gauss ACF

**Table 3.** Average probability of true and false detection for the case of PAD UQ algorithm in case of noise on images

| Sizes of anomaly areas in pixels | $3 \times 3$ | | $5 \times 5$ | | $7 \times 7$ | |
|---|---|---|---|---|---|---|
| Signal to noise ratio | TP | FP | TP | FP | TP | FP |
| ∞ | 1 | 0.00003 | 1 | 0 | 1 | 0 |
| 250 | 1 | 0.00008 | 0.98 | 0.00004 | 1 | 0 |
| 150 | 1 | 0.00018 | 0.9996 | 0.00008 | 0.99735 | 0.00007 |
| 15 | 0.41 | 0 | 0.4 | 0 | 0.4 | 0 |

between pixel and average value of background. It is supposed that background is normally distributed.

Figure 3 demonstrates the results of proposed PAD modifications and local and global versions of RXD algorithm for the example showed on Fig. 1 for background with biexponential ACF. Dark pixels correspond to higher degree of anomaly measure. It can be seen that local RXD doesnt recognize anomalies and global RXD recognizes only one embedded anomaly (lowest black square). The reason is that local normal model is inadequate to images with complex correlated background. It should be noticed that for background correlation 0.95 and lower especially in presence of noise RXD algorithm works unsatisfactory. It couldnt recognize anomalies, while proposed PAD AC algorithm shows good quality of detection.

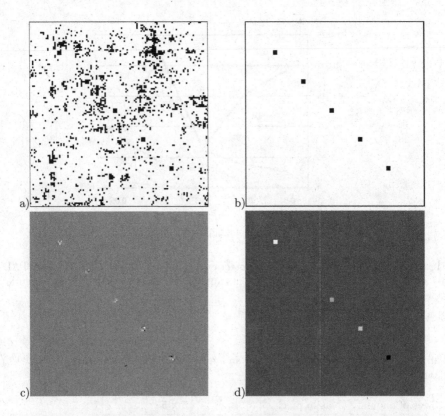

**Fig. 3.** The results of PAD modifications (a) PAD UQ and (b) PAD AC and the results of RXD algorithm: (c) local with window $5 \times 5$, (d) global for input image from Fig. 1a

As for PAD UQ method false detection regions have the same size and form as anomaly regions. For images with higher or lower correlation than 0.9 false detections are also about 10% of all image pixels, but their shape and size are much more different, than in case of correlation 0.9. For background correlation bigger than 0.9 false detection regions are fewer and bigger in size and in case of correlation less than 0.9 there are many false detection regions with tiny size. This means that the results of PAD UQ algorithm are totally dependent on background correlation properties, and it is difficult to analyze visually output of the algorithm if anomaly size can be compared with small texture elements of background.

# 6 Conclusion

In the article two implementations of PAD algorithm are described. Implementations are based on different vector quantization methods.

First PAD UQ algorithm applies vector quantization method based on uniform quantization of each image component and hashing of quantized vectors to compute histogram of image and evaluate probabilities of each pixel. Hashing is used to organize compact histogram storage by means of hash table and to compute probabilities rapidly. But the method is inefficient in case of noise and depends on background correlation properties. To increase productivity of the algorithm PCA and MNF transformations can be applied to initial image. Future development of this method can be dealt with incorporation of dimensionality reduction and denoising algorithms in it.

Second proposed implementation of PAD method uses vector quantization based on agglomerative clusterzation. The result of clusterzation is quantization table of fixed size. The parameters of PAD AC algorithm are initial quantization error and size of codebook. The experiments showed that PAD AC gives high quality of anomaly detection when signal to noise ratio is less than 150.

A comparison of both proposed algorithms with RXD is provided. The experiments shows that RXD algorithm works unsatisfactory for complex correlated background, but PAD AQ gives excellent result. For described in paper example RXD algorithm detects background as anomaly region for most of the pixels.

**Acknowledgements.** This work was financially supported by the Russian Science Foundation (RSF), grant no. 14-31-00014 Establishment of a Laboratory of Advanced Technology for Earth Remote Sensing.

# References

1. Matteoli, S., Diani, M., Corsini, G.: A tutorial overview of anomaly detection in hyperspectral images. IEEE Aerosp. Electron. Syst. Mag. **25**(7), 5–28 (2010)
2. Reed, I.S., Yu, X.: Adaptive multiple-band CFAR detection of AN optical pattern with unknown spectral distribution. IEEE Trans. Acoust. Speech Signal Process **38**(10), 1760–1770 (1990)
3. Stein, D.W., Beaven, S.G., Hoff, L.E., Winter, E.M., Schaum, A.P., Stocker, A.D.: Anomaly detection from hyperspectral imagery. IEEE Signal Process. Mag. **19**(1), 58–59 (2002)

4. Ashton, E.A.: Detection of subpixel anomalies in multispectral infrared imagery using an adaptive bayesian classifier. IEEE Trans. Geosci. Remote Sens. **36**(2), 506–517 (1998)
5. Carlotto, M.J.: A cluster-based approach for detecting man-made objects and changes in imagery. IEEE Trans. Geosci. Remote Sens. **43**(2), 374–384 (2005)
6. Chang, C.I.: Orthogonal subspace projection (OSP) revisited: a comprehensive study and analysis. IEEE Trans. Geosci. Remote Sens. **43**(3), 502–518 (2005)
7. Ranney, K.I., Soumekh, M.: Hyperspectral anomaly detection within the signal subspace. IEEE Geosci. Remote Sens. Lett. **3**(3), 312–316 (2006)
8. Kwon, H., Nasrabadi, N.: Kernel RX-Algorithm: a nonlinear anomaly detector for hyperspectral imagery. IEEE Trans. Geosci. Remote Sens. **43**(2), 388–397 (2005)
9. Banerjee, A., Burlina, P., Diehl, C.: A support vector method for anomaly detection in hyperspectral imagery. IEEE Trans. Geosci. Remote Sens. **43**(8), 2282–2291 (2006)
10. Burnaev, E., Erofeev, P., Smolyakov, D.: Model selection for anomaly detection. In: 8th International Conference on Machine Vision. pp. 987525–987525-6. International Society for Optics and Photonics (2015)
11. Denisova, A.Y., Myasnikov, V.V.: Anomaly detection for hyperspectral imaginary. Comput. Opt. **38**(2), 287–296 (2014)
12. Chang, C.C., Hu, Y.C.: A fast LBG codebook training algorithm for vector quantization. IEEE Trans. Consum. Electron. **44**(4), 1201–1208 (1998)
13. Umnyashkin, S.V.: Theoretical Basics of Digital Signal Processing and Representing. Forum, Moscow (2009)
14. Clark, R.N., Swayze, G.A., Wise, R., Livo, E., Hoefen, T., Kokaly, R., Sutley, S.J.: USGS digital spectral library splib06a. U.S. Geological Survey, Digital Data Series 231
15. Chang, C.I.: Hyperspectral Data Processing: Algorithm Design and Analysis. Wiley, New York (2013)

# Threefold Symmetry Detection in Hexagonal Images Based on Finite Eisenstein Fields

Alexander Karkishchenko and Valeriy Mnukhin$^{(\boxtimes)}$

Southern Federal University, 105/42, Bolshaya Sadovaya,
344006 Rostov-na-Donu, Russia
karkishalex@gmail.com, mnukhin.valeriy@mail.ru

**Abstract.** This paper considers an algebraic method for symmetry analysis of hexagonally sampled images, based on the interpretation of such images as functions on "Eisenstein fields". These are finite fields $\mathbb{GF}(p^2)$ of special characteristics $p = 12k + 5$, where $k > 0$ is an integer. Some properties of such fields are studied; in particular, it is shown that its elements may be considered as "discrete Eisenstein numbers" and are in natural correspondence with hexagons in a $(p \times p)$-diamond-shaped fragment of a regular plane tiling. The concept of logarithm in Eisenstein fields is introduced and used to define a "log-polar"-representation of hexagonal images. Next, an algorithm for threefold symmetry detection in gray-level images is proposed.

**Keywords:** Image analysis · Threefold symmetry · Hexagonal image · Eisenstein numbers · Finite fields · Log-polar coordinates · Polar representation

## 1 Introduction

Symmetry is a central concept in many natural and man-made objects and plays a crucial role in visual perception, design and engineering. That is why symmetry detection and analysis is a fundamental task in such fields of computer science as machine vision, medical imaging, pattern classification and image database retrieval [1,2]. Numerous applications have successfully utilized symmetry for model reduction [3], segmentation [4], shape matching [1], etc.

When discussing image symmetries, the crucial difference between *continuous* and *discrete* cases must be taken into account. Indeed, a continuous object can be characterized by its symmetry group, which is invariant to rotation, scale and translation transformations. At the same time, for digital images we may talk only about some *measure of symmetry*, which depends on rotations and scaling. For example, first three parts of Fig. 1 demonstrate effects of rotation on digital images with square sampling lattice. To reduce the effects, one can use hexagonal lattice-based images, see the right side of Fig. 1. Indeed, it is known [14] that hexagonal lattice has some advantages over the square lattice which can have implications for analysis of images defined on it. These advantages are as follows:

© Springer International Publishing AG 2017
D.I. Ignatov et al. (Eds.): AIST 2016, CCIS 661, pp. 281–292, 2017.
DOI: 10.1007/978-3-319-52920-2_26

**Fig. 1.** Change of symmetry measure under rotations

- **Isoperimetry.** As per the honeycomb conjecture, a hexagon encloses more area than any other closed planar curve of equal perimeter, except a circle.
- **Additional equidistant neighbours.** Every hexagonal pixel has six equidistant neighbours with a shared edge. In contrast, a square pixel has only four equidistant neighbours with a shared edge or a corner. This implies that curves can be represented in a better fashion on the hexagonal lattice.
- **Additional symmetry axes.** Every hexagon in the lattice has 6 symmetry axes in contrast with squares, which have only 4 axes. This implies that there will be less ambiguity in detecting symmetry of images.

In general, the hexagonal structure provides a more flexible and efficient way to perform image translation and rotation without losing image information [28].

A considerable amount of research in hexagonally sampled images processing (HIP) is taking place now [14] despite the fact that there are no hardware resources that currently produce or display hexagonal images. For this, software resampling is in use, when the original data is sampled on a square lattice while the desired image is to be sampled on a hexagonal lattice. Moreover, hybrid systems involving both types of sampling are useful for taking advantage of both in real-life applications. (Note that for the sake of brevity, the terms *square image* and *hexagonal image* will be used throughout to refer to images sampled on a square lattice and hexagonal lattice, respectively.)

When developing algorithms for image analysis, both in square and hexagonal grid, it is quite common to proceed on the assumption of continuity of images. Then powerful tools of continual mathematics, such as complex analysis and integral transforms, can be efficiently used. However, its application to digital images often leads to systematic errors associated with the inability to adequately transfer some concepts of continuous mathematics to the discrete plane. As examples we can point to such concepts as rotation in the plane and the polar coordinate system. Being natural and elementary in the continuous case, they lose these qualities when one tries to define them accurately on a discrete plane. As a result, formal application of continuous methods to digital images could be complicated by systematic errors [5,6]. This raises the issue of the development of methods initially focused on discrete images and based on tools of algebra and number theory.

One of such methods is considered in this paper. It is based on the interpretation of hexagonal images as functions on "Eisenstein fields". These are finite fields $\mathbb{GF}(p^2)$ of special characteristics $p = 12k+5$, where $0 \leq k \in \mathbb{Z}$. We show that elements of such fields may be considered as "discrete Eisenstein numbers" and are in natural correspondence with hexagons in a $(p \times p)$-diamond-shaped fragment of a regular plane tiling. Hence, non-negative functions on Eisenstein fields may be considered as hexagonal images of prime "sizes" $p = 5, 17, \ldots, 113, \ldots, 257$, ets. (Note that this is not a limitation since every image can be trivially extended to an appropriate size.)

The significance of such approach is based on the fact that finite Eisenstein fields inherit some properties of the continuous complex field. In particular, we show that the concept of principal value complex logarithm can be transferred to Eisenstein fields. As an immediate result, we derive discrete "log-polar"-coordinates in hexagonal images. It occurs that the corresponding "log-polar"-representation of hexagonal images can be used for the analysis of their symmetry in the same fashion as for continuous images in [2,5–7] and for square-sampled digital images in [17,20–23].

In this paper we consider the most simple case of threefold symmetry detection, which, nevertheless, has important practical applications. The proposed algorithm is optimal for hexagonal images but can be also used for regular square images after resampling (in fact, that is how examples of the last section have been worked out).

## 2    Background

A lot of research has been done in the areas of symmetry detection (see, for example, [1–5]) and of hexagonal image processing, but only a few, if any, have been deal with symmetry of hexagonal images (in any case, the book [14], which seems to be the most exhaustive HIP survey at the moment, contains no references to the subject). Possibly it is related with the fact of existence different addressing schemes in hexagonal lattices. In particular, I. Her suggested 3-coordinate scheme that exploits the implicit symmetry of the hexagonal lattice and introduced a series of geometric transformations such as scaling, rotation and shearing on the corresponding 3-coordinate system ([14,27, p. 20]), but apparently not studied symmetry of hexagonal images; see also [28].

In this paper we exploit finite fields to produce a new coordinate scheme in fragments of hexagonal lattices. It should be noted that there has been a significant amount of work in applications of algebraic structures and number-theoretic transforms in image processing; such methods have been around for 40 years [26] now and have received extensive theoretical and experimental treatment [24,25]. In particular, in 1993 H.G. Baker [19] proposed the ring of Gaussian integers for square image processing; later "finite complex fields" were in use [16,18,20–23]. Nevertheless, applications of "finite Eisenstein fields" in HIP seem to be new.

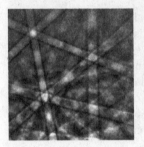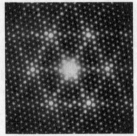

**Fig. 2.** Examples of EM-images with local three-fold symmetry. From left to right: (a–b) electron diffraction pattern of a nickel titanium crystal and of $PbCr_3S_4$ crystals; (c) high-resolution transmission electron microscope image of a $PbCr_3S_4$ crystal (from http://www.mineralsocal.org/micro/ and http://www.globalsino.com/EM/).

The threefold symmetry detection is currently claimed in various areas of crystallography [9,10], virology [11], analysis of electron microscope images [12], ets. In particular, Fig. 2 demonstrates some examples of EM images with local threefold symmetry.

## 3    Finite Fields of Eisenstein Integers

Let $\mathbb{Z}$ and $\mathbb{C}$ be the ring of integers and the complex field respectively, let $\mathbb{Z}_n = \mathbb{Z}/n\mathbb{Z}$ be a residue class ring modulo an integer $n \geq 2$, and let $\mathrm{GF}(p^m)$ be a Galois field with $p^m$ elements, where $p$ is a prime and $m > 0$ is an integer.

In number theory [13, Chap. 1.4] a *Gaussian integer* is a complex number $z = a + bi \in \mathbb{C}$ whose real and imaginary parts are both integers. Further, *Eisenstein integers* are complex numbers of the form $z = a + b\omega$, where $a, b \in \mathbb{Z}$ and $\omega = \exp(2\pi i/3) \in \mathbb{C}$ is a primitive (non-real) cube root of unity, so that $\omega^3 = 1$ and $\omega^2 + \omega + 1 = 0$. Note that within the complex plane the Eisenstein integers may be seen to constitute a triangular lattice, in contrast with the Gaussian integers, which form a square lattice.

Both Gaussian and Eisenstein integers, with ordinary addition and multiplication of complex numbers, form, respectively, subrings $\mathbb{Z}[i]$ and $\mathbb{Z}[\omega]$ in the field $\mathbb{C}$. Unfortunately, lack of division in these rings significantly restricts its applicability to image processing problems [19]. So it is natural to look for finite fields, whose properties would be in some respect similar to properties of $\mathbb{Z}[i]$ and $\mathbb{Z}[\omega]$. In fact, it is known that if $p$ is a prime number such that $p \equiv 3 \bmod 4$, then the factor ring $\mathbb{C}(p) \overset{\text{def}}{=\!\!=} \mathbb{Z}_p[x]/(x^2+1) \simeq \mathrm{GF}(p^2)$ is a "finite complex field". Its applications in analysis and processing of square sampled digital images have been considered in [16,18,20–23].

We will use now the similar approach to construct "finite fields of Eisenstein integers". Indeed, it easy to show that if $p = 12k + 5$ is a prime, then the polynomial $x^2 + x + 1$ is irreducible over $\mathbb{Z}_p$ but $x^2 + 1 = 0$ is not. As an immediate corollary, the next definition follows.

**Definition 1.** *Let $p \geq 5$ be a prime number such that $p \equiv 5$ (mod 12). Then the finite field*

$$\mathbb{E}(p) \stackrel{\text{def}}{=} \mathbb{Z}_p[x]/(x^2 + x + 1) \simeq GF(p^2)$$

*will be called* Eisenstein field. *Elements of* $\mathbb{E}(p)$ *will be called* discrete Eisenstein numbers.

Thus, Eisenstein fields have $p^2$ elements, where

$$p = 5, 17, 29, 41, 53, 89, 101, 113, 137, 149, 173, 197, 233, 257, \ldots.$$

In particular, there are 44 fields $\mathbb{E}(p)$ for $5 \leq p < 1000$. Elements of Eisenstein fields are of the form $z = a + b\omega$, where $a, b \in \mathbb{Z}_p$ and $\omega$ denotes the class of residues of $x$, so that $\omega^2 + \omega + 1 = 0$. The product of $a + b\omega \in \mathbb{E}(p)$ and $c + d\omega \in \mathbb{E}(p)$ is given by

$$(a + b\omega)(c + d\omega) = (ac - bd) + (bc + ad - bd)\omega,$$

the addition is straightforward. The number $z^* = a + b\omega^2 = (a - b) - b\omega \in \mathbb{E}(p)$ is *conjugate* to $z$, and the product $zz^* = a^2 - ab + b^2 \in \mathbb{Z}_p$ is the *norm* $N(z)$ of $z$. (Note that in $\mathbb{E}(p)$ the concept of modulus $|z| = \sqrt{N(z)}$ is not defined.) It is easy to show that $N(z_1 z_2) = N(z_1)N(z_2)$ and $N(z) = 0 \Leftrightarrow z = 0$. Thus, nonzero elements $z \neq 0$ have inverses $z^{-1} = z^* N(z)^{-1}$, and so division is defined in $\mathbb{E}(p)$. The next Fig. 3 demonstrates the Eisenstein field $\mathbb{E}(5)$.

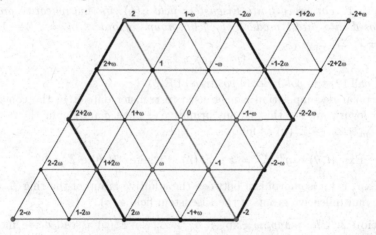

**Fig. 3.** The finite Eisenstein field $\mathbb{E}(5)$.

# 4 Polar Decompositions of Eisenstein Fields

Let us use the analogy between $\mathbb{C}$ and $\mathbb{E}(p)$ to represent elements of $\mathbb{E}(p)$ in an "exponential form". For this, let us recall the algebraic method of introducing log-polar coordinate system onto a continuous complex plane.

Let $\mathbb{C}^*$ be the multiplicative group of complex numbers and $\mathbb{R} = \langle \mathbb{R}, + \rangle$ be the additive group of reals. Note that the correspondence

$$0 \neq z = re^{i\theta} = e^{\ln r + i\theta} \leftrightarrow (l, \theta), \quad \text{where} \quad l = \ln r \in \mathbb{R} \quad \text{and} \quad 0 \leq \theta < 2\pi,$$

between non-zero complex numbers $z$ and its log-polar coordinates $(l, \theta)$ may be considered as an isomorphism

$$\mathbb{C}^* \simeq \mathbb{R} \times (\mathbb{R}/2\pi\mathbb{Z}). \tag{1}$$

In fact, *any* direct product decomposition of $\mathbb{C}^*$ produces a representation of complex numbers.

Let us transfer the previous construction to $\mathbb{E}(p)$. For this, note that since $\mathbb{E}(p)$ is a finite field, its multiplicative group $\mathbb{E}^*(p) = \mathbb{E}(p) \smallsetminus \{0\}$ is cyclic [15, p. 314] and is generated by a primitive element $g$. For example, it is easy to check that $g = 1 + 3\omega$ is primitive in $\mathbb{E}^*(p)$ when $p = 5, 17, 89, 101, 257$, and $g = 1 + 5\omega$ is primitive for $p = 29, 53, 113, 233$, ets.

We need the following elementary verifying result.

**Lemma 1.** *For every* $p = 12k + 5$, *numbers* $m = 2(p - 1) = 8(3k + 1)$ *and* $n = (p + 1)/2 = 3(2k + 1)$ *are relatively prime*, $\gcd(m, n) = 1$.

Note that $mn = p^2 - 1 = |\mathbb{E}^*(p)|$ and so for $\mathbb{E}^*(p)$ there appears [15, p. 163] the following analogue of the decomposition (1).

**Theorem 1.** *For every finite Eisenstein field* $\mathbb{E}(p)$, *its multiplicative group is decomposed into direct product of cyclic groups of orders* $m = 2(p - 1)$ *and* $n = (p + 1)/2$,

$$\mathbb{E}^*(p) = \langle g \rangle \simeq \mathbb{Z}_m \times \mathbb{Z}_n. \tag{2}$$

We will call (2) the *polar decomposition* of $\mathbb{E}(p)$.

The polar decomposition can be used to transfer onto $\mathbb{E}(p)$ the concept of complex logarithm. For this, fix any primitive element $g$ and define the mapping $\text{Exp}_g : \mathbb{Z}_m \times \mathbb{Z}_n \to \mathbb{E}^*(p)$ as follows:

$$\text{Exp}_g(l, \theta) = g^{nl + m\theta} = z \in \mathbb{E}^*(p), \quad \text{where} \quad (l, \theta) \in \mathbb{Z}_m \times \mathbb{Z}_n. \tag{3}$$

Thus, $\text{Exp}_g$ is an isomorphism between the additive group of the ring $\mathbb{Z}_m \times \mathbb{Z}_n$ and the multiplicative group of the Eisenstein field $\mathbb{E}(p)$.

**Definition 2.** *The mapping* $\text{Exp}_g : \mathbb{Z}_m \times \mathbb{Z}_n \to \mathbb{E}^*(p)$ *is called the* modular exponent to base g, *and its inverse mapping* $\text{Ln}_g : \mathbb{E}^*(p) \to \mathbb{Z}_m \times \mathbb{Z}_n$ *is the* modular logarithm to base g. *Its domain* $\mathbb{Z}_m \times \mathbb{Z}_n$ *is called the* polar domain.

The "*basic logarithmic identity*" follows immediately:

$$\text{Ln}_g(z_1 z_2) = \text{Ln}_g(z_1) + \text{Ln}_g(z_2). \tag{4}$$

Note that $\text{Ln}_g(0)$ is not defined, and to evaluate $(l, \theta) = \text{Ln}_g(z)$ for any $z = g^s \in \mathbb{E}^*(p)$ one just needs to solve the Diophantine equation $nx + my = s$ and set

$$(l, \theta) = (x \bmod m, \ y \bmod n) \in \mathbb{Z}_m \times \mathbb{Z}_n. \tag{5}$$

For example, let $g = 1 + 3\omega \in \mathbb{E}^*(5)$ and $z = g^2 = 2 + 2\omega$. Then $s = 2$, $m = 8$, $n = 3$ and the equation $3x + 8y = 2$ has the evident solution $x = -2$, $y = 1$. Hence, $\text{Ln}_g(2 + 2\omega) = (-2 \bmod 8, 1 \bmod 3) = (6, 1) \in \mathbb{Z}_8 \times \mathbb{Z}_3$. Note that depending on $g$, either $\text{Ln}_g(\omega) = (0, n/3)$ or $\text{Ln}_g(\omega) = (0, 2n/3)$.

We will consider the pair $(l, \theta) \in \mathbb{Z}_m \times \mathbb{Z}_n$ as "log-polar coordinates" of the corresponding discrete Eisenstein number $z$. It is useful to consider $\mathbb{Z}_m \times \mathbb{Z}_n$ as a $m \times n$-discrete torus, see Fig. 4.

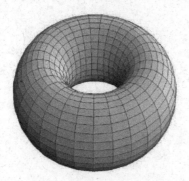 

**Fig. 4.** A discrete torus and a fragment of the picture "*Torus and Sphere*" by Tom Wilcox (2008) as an illustration to the concept of extended polar domain.

# 5   Hexagonal Images as Functions on Eisenstein Fields

Let $\mathbb{E}(p)$ be any Eisenstein field of a characteristic $p = 12k + 5$ and let $f(z) : \mathbb{E}(p) \to \mathbb{R}$ be any real-valued function on $\mathbb{E}(p)$. Due to our geometric interpretation of Eisenstein fields, we call $f(z)$ a *hexagonal gray-level image of size $p \times p$*, or just *hexagonal image* briefly.

Other way to describe hexagonal images is based on polar decompositions of Eisenstein fields. Namely, we may associate with a $p \times p$-hexagonal image $f(z)$ a square-sampled image $\psi$ of size $m \times n$.

For this, fix any primitive element $g \in \mathbb{E}^*(p)$ and define a function $\psi : \mathbb{Z}_m \times \mathbb{Z}_n \to \mathbb{R}$ such that

$$\psi(\text{Ln}_g(z)) = f(z), \qquad 0 \neq z \in \mathbb{E}^*(p). \tag{6}$$

**Definition 3.** *The transform $\mathscr{P}_g[f] = \psi$ will be called* log-polar transform to base $g$ *of the hexagonal image $f$, or just its* polar transform $\mathscr{P}$ *briefly. The square image $\psi$ will be called the* polar form *of $f$.*

Thus, the polar form of $f$ may be considered as some arrangement of all its pixels except $f(0,0)$ in the form of a $m \times n$-matrix, as it is shown in Fig. 5. Thus, the image $f$ is "almost" recovered by its polar form $\psi$. In order to achieve full recoverability, let us formally agree to extend the polar domain by an extra

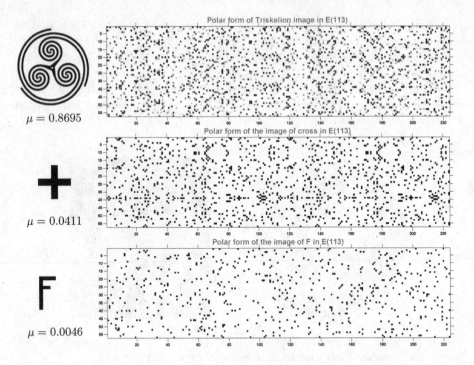

**Fig. 5.** Images and its polar form matrices.

element $\infty$, assuming that $\psi(\infty) = f(0,0)$. Note that in such an *extended polar domain* $\mathbb{Z}_m \times \mathbb{Z}_n \cup \{\infty\}$ the polar transform $\mathscr{P}$ becomes invertible. The linearity of $\mathscr{P}$ is obvious. (Just as discrete torus is an intuitive interpretation of the polar domain, we may interpret the extended polar domain as a torus with a small sphere inside, see the right side of Fig. 4.)

As the following statement shows, the transform $\mathscr{P}$ indeed can be regarded as a discrete analogue of the transition to the log-polar coordinate system.

**Proposition 1.** *If $\mathscr{P}_g[f(z)] = \psi(l, \theta)$, then*

$$\mathscr{P}[f(wz)] = \psi(l - l_0, \theta - \theta_0) \tag{7}$$

*where $0 \neq w \in \mathbb{E}(p)$ and $\mathrm{Ln}(w) = (l_0, \theta_0)$.*

We will apply the previous relation (7) to symmetry analysis in hexagonal images. As is known, a continuous object is said to have *r-fold rotational symmetry* with respect to a point $C$ if a rotation by an angle of $2\pi/r$ around $C$ does not change the object. Unfortunately, this definition does not work for digital images because digital rotation is much more hard to define (see, for example, [8, p. 377] for square images and [27, 28], [14, p. 97] for hexagonal images) As a result, for digital images we may talk only about some *measure of symmetry*, which depends on rotations and scaling.

The geometric interpretation of Eisenstein fields immediately implies the following definition.

**Definition 4.** *A hexagonal image $f$ is said to have* threefold central rotational symmetry *if and only if $f(\omega z) = f(z)$.*

Let $\psi = \mathscr{P}[f]$ be the polar form of an hexagonal $p \times p$-image $f$. We may consider $\psi$ as an $m \times n$-matrix, where $n = 3(2k + 1)$, $m = 8(3k + 1)$ and $k = (p - 5)/12 \in \mathbb{Z}$. Since $n$ is multiple of 3, decompose $\psi$ into three blocks $\psi_0$, $\psi_1$, $\psi_2$ of equal size $m \times n/3$.

**Proposition 2.** *An image $f$ is threefold central symmetric if and only if its polar form $\psi$ can be decomposed into three equal blocks, $\psi_0 = \psi_1 = \psi_2$.*

*Proof.* Since the polar transform is invertible, it follows from Definition 4, that $f$ is threefold symmetric if and only if $\mathscr{P}[f(\omega z)] = \mathscr{P}[f(z)] = \psi(l, \theta)$. But it have been noted in Sect. 4 that $\mathrm{Ln}_g(\omega) = (0, \pm n/3)$ for every primitive $g \in \mathbb{E}^*(p)$. Hence, threefold symmetry of $f$ is equivalent to the condition

$$\psi(l, \theta \pm n/3) = \psi(l, \theta) \quad \text{for all} \quad l \in \mathbb{Z}_m, \ \theta \in \mathbb{Z}_n,$$

or just $\psi_0 = \psi_1 = \psi_2$.

Evidently, for real-world images one can expect only approximate equalities $\psi_0 \simeq \psi_1 \simeq \psi_2$, so that the problem to choose an appropriate measure of symmetry $\mu(f)$ for an image $f$ arise. One of possible ways is the following. For any normalized polar form matrix $\tilde{\psi} = \psi/\max\{\psi\}$ of an image $f$ take

$$\mu(f) = \exp\left(-\alpha x^\beta\right), \quad \text{where} \quad x = \max_{0 \le i < j \le 2} \left\{ \frac{\|\tilde{\psi}_i - \tilde{\psi}_j\|}{\|\tilde{\psi}_i\| + \|\tilde{\psi}_j\|} \right\}. \tag{8}$$

Here $\| \cdot \|$ stands for any matrix norm, and $\alpha$, $\beta$ are nonnegative reals, whose precise values could vary depending on the practical problem to be solved. Then $\mu(f)$ measures the central threefold symmetry of $f$.

*Note 1.* Since the polar form $\psi$ depends on the primitive element $g \in \mathbb{E}^*(p)$, one may ask if $\mu(f)$ also depends on $g$. It is an easy matter to extend the "change of base" formula [20, Prop. 7] to logarithms $\mathrm{Ln}_g(z)$ in Eisenstein fields and to prove invariance of $\mu(f)$ under change of $g$.

*Example 1.* To illustrate the method we consider images $f_1$, $f_2$, $f_3$ of triskelion, cross and letter **F** shown in Fig. 5 on the left. These are square-sampled $113 \times 113$-images, grayscale the first and binary the last two. To embed them into the Eisenstein field $\mathbb{E}(113)$ the standard resampling algorithm [14, Sect. 6.1.1] has been used. Polar form matrices $\psi_i$ of the images, evaluated for $g = 1 + 5\omega \in \mathbb{E}^*(113)$, are in the right side of Fig. 5. The evaluated symmetry measures are the following:

$$\mu(f_1) = 0.8695, \qquad \mu(f_2) = 0.0411, \quad \text{and} \quad \mu(f_3) = 0.0046,$$

where the Frobenius matrix norm was used in (8) and the parameters were taken to be $\alpha = 32$, $\beta = 4$.

Jointly with any of the sliding window methods, the introduced approach can be used to detect centers of local threefold symmetry in images. Note that for fixed $p$, the polar transform is based on precomputations and do not require any arithmetic operations.

**Fig. 6.** Searching centers of local 3-symmetry in the NY Times Square Ball image.

*Example 2.* Figure 6 demonstrates the result of searching centers of local three-fold symmetry in an image of the New York City's Times Square Ball at 2012 New Year's Eve celebration. A sliding $41 \times 41$-window algorithm based on the method discussed above, has been applied to this $488 \times 488$-grayscale image. As the result, $488 \times 488$-matrix $M = (\mu_{ij})$ has been evaluated, where $\mu_{ij}$ is the measure of 3-symmetry with the center at $(i, j)$, for $21 \leq i, j \leq 467$, and $\mu_{ij} = 0$ outside the diapason. Then $M$ has been cutted-off at 0.86 rate and combined with the original image, so that dark spots in the resulting image correspond to points with $\mu \geq 0.86$.

As it is easy to notice, centers of hexagons in the shadowed hemisphere of the Ball have been correctly detected in spite of geometrical distortions. For the illuminated upper hemisphere results are below the cutoff rate due to nonuniformity of the light field in this area. There are also few artifacts; some of them (say, the edge between buildings in the left upper side of the image) are obvious, others are probably related with local features. It indicates necessity of further improvements; however, these are beyond the scope of the paper.

# 6   Conclusion

This paper proposes an algebraic method for the processing of hexagonally sampled images, based on their representation as functions on special finite fields, called "Eisenstein fields". Some properties of such fields are studied; in particular, it is shown that its elements are in natural correspondence with hexagons in a $(p \times p)$-diamond-shaped fragment of a regular plane tiling. The concept of logarithm in Eisenstein fields is introduced and used to define a "log-polar"-representation of hexagonal images. An algorithm for threefold symmetry detection in gray-level images is proposed. However, apart from of this work, there remain issues such as the detailed study of properties of the introduced polar transformation, the study of possibilities of its generalizations, as well as the details of its practical applications including estimation of the symmetry detection quality in the presence of noise. Authors hope to return to the study of these issues in the further papers.

**Acknowledgements.** This research has been partially supported by the Russian Foundation for Basic Research grant no. 16-07-00648. The authors would like to thank the anonymous reviewers for their constructive comments and suggestions.

# References

1. Gool, L., Moons, T., Ungureanu, D., Pauwels, E.: Symmetry from shape and shape from symmetry. Int. J. Robot. Res. **14**(5), 407–424 (1995)
2. Martinet, A., Soler, C., Holzschuch, N., Sillion, F.: Accurate detection of symmetries in 3D shapes. ACM Trans. Graph. **25**(2), 439–464 (2006)
3. Mitra, N.J., Guibas, L.J., Pauly, M.: Partial and approximate symmetry detection for 3D geometry. ACM Trans. Graph. **25**(3), 560–568 (2006)
4. Thrun, S., Wegbreit, B.: Shape from Symmetry. In: Proceedings of International Conference on Computer Vision (ICCV), vol. 2, pp. 1824–1831 (2005)
5. Chertok, M., Keller, Y.: Spectral symmetry analysis. IEEE Trans. Pattern Anal. Mach. Intell. **32**(7), 1227–1238 (2010)
6. Derrode, S., Ghorbel, F.: Shape analysis and symmetry detection in gray-level objects using the analytical Fourier-Mellin representation. Sig. Process. **84**(1), 25–39 (2004)
7. Karkishchenko, A.N., Mnukhin, V.B.: Symmetry recognition in the frequency domain (in Russian). In: 9th International Conference on Intelligent Information Processing, pp. 426–429. TORUS Press, Moscow (2012)

8. Karkishchenko, A.N., Mnukhin, V.B.: Topological filtration for digital images recognition and symmetry analysis (in Russian). J. Mach. Learn. Data Anal. 1(8), 966–987 (2014)
9. Hundt, R., Schön, J.C., Hannemann, A., Jansen, M.: Determination of symmetries and idealized cell parameters for simulated structures. J. Appl. Crystallogr. 32, 413–416 (1999)
10. Spek, A.L.: Structure validation in chemical crystallography. Acta Crystallogr. D65, 148–155 (2009)
11. Zeyun, Y., Bajaj, C.: Automatic ultrastructure segmentation of reconstructed CryoEM maps of icosahedral viruses. IEEE Trans. Image Process. 14(9), 1324–1337 (2005)
12. Kondo, S., Lutwyche, M., Wada, Y.: Observation of threefold symmetry images due to a point defect on a graphite surface using scanning tunneling microscope (STM). Jpn. J. Appl. Phys. 33(9B), 1342–1344 (1994)
13. Ireland, K., Rosen, M.: A Classical Introduction to Modern Number Theory. Springer, Heidelberg (1982)
14. Middleton, L., Sivaswamy, J.: Hexagonal Image Processing: A Practical Approach. Springer, Heidelberg (2005)
15. Dummit, D.S., Foote, R.M.: Abstract Algebra. Wiley, Hoboken (2004)
16. de Souza, R.M.C., de Oliveira, H.M., Silva, D.: The $Z$ Transform over Finite Fields. ArXiv preprint 1502.03371, 11 February 2015
17. de Souza, R.M.C., Farrell, R.G.: Finite field transforms and symmetry groups. Discrete Mathematics 56, 111–116 (1985)
18. Bandeira, J., de Souza, R.M.C: New trigonometric transforms over prime finite fields for image filtering. In: VI International Telecommunications Symposium (ITS2006), pp. 628–633. Fortaleza-Ce, Brazil (2006)
19. Baker, H.G.: Complex Gaussian integers for Gaussian graphics. ACM Sigplan Not. 28(11), 22–27 (1993)
20. Mnukhin, V.B.: Transformations of digital images on complex discrete tori. Pattern Recogn. Image Anal. Adv. Math. Theor. Appl. 24(4), 552–560 (2014)
21. Mnukhin, V.B.: Digital images on a complex discrete torus (in Russian). Mach. Learn. Data Anal. 1(5), 540–548 (2013)
22. Karkishchenko, A.N., Mnukhin, V.B.: Applications of modular logarithms on complex discrete tori in digital image processing (in Russian). Bull. Rostov State Univ. Railway Transp. 3, 147–153 (2013)
23. Mnukhin, V.B.: Fourier-Mellin transform on a complex discrete torus. In: 11th International Conference on Pattern Recognition and Image Analysis: New Information Technologies (PRIA-11-2013), pp. 102–105. Samara, Russia, 23–28 September 2013
24. Labunets, V.G.: Number Theoretic Transforms over Quadratic Fields (in Russian). In: Complex Control Systems, Institute of Cybernetics USSR, Kiev, pp. 30–37 (1982)
25. Varitschenko, L.V., Labunets, V.G., Rakov, M.A.: Abstract Algebraic Systems and Digital Signal Processing (in Russian). Naukova Dumka, Kiev (1986)
26. Creutzburg, R., Labunets, V.G.: The Early Papers on Number-theoretic Transforms. https://www.researchgate.net/publication/229043248
27. Her, I.: Geometric transforms on the hexagonal grid. IEEE Trans. Image Process. 4(9), 1213–1222 (1995)
28. He, X., Jia, W., Hur, N., Qiang, W., Kim, J.: Image translation and rotation on hexagonal structure. In: 6th IEEE International Conference on Computer and Information Technology (CIT 2006), Seoul, p. 141 (2006)

# Reflection Symmetry of Shapes Based on Skeleton Primitive Chains

Olesia Kushnir[1]([✉]), Sofia Fedotova[1], Oleg Seredin[1],
and Alexander Karkishchenko[2]

[1] Tula State University, Tula, Russia
kushnir-olesya@rambler.ru, fedotova.sonya@gmail.com, oseredin@yandex.ru
[2] Southern Federal University, Rostov-on-Don, Russia
karkishalex@gmail.com

**Abstract.** In this paper the novel fast approach to identify the reflection symmetry axis of binary images is proposed. We propose to divide a skeleton of a shape into two parts – the "left" and the "right" sub-skeletons. The left part is traversed counterclockwise and the right one – in clockwise direction. As a result, the "left" and the "right" primitive sub-chains are achieved; they can be compared by the known shape matching procedure based on pair-wise alignment of primitive chains. So, the most similar parts of a skeleton among all possible ones correspond to the most similar parts of a figure which are considered as reflection symmetric parts. The start and the end points of skeleton division into "left" and "right" parts will be the points belonging to a symmetry axis of a figure. Also, the exact brute-force symmetry evaluation algorithm and two its optimizations are suggested for finding ground truth of symmetry axis. All proposed methods were experimentally tested on Flavia leaves dataset.

**Keywords:** Binary image · Reflection symmetry · Shape matching · Skeleton · Chain of primitives · Pair-wise alignment

## 1 Introduction

Analysing binary images it is easy to notice that many, both artificial and natural, objects have intrinsic reflection (bilateral) symmetry property. It is obvious that real-world images rarely can be absolute reflection symmetric. So, it is valuable to detect approximate reflection symmetry and evaluate the measure of approximate reflection symmetry of shape (see Fig. 1). Estimation of shapes symmetry could be used in many applications like analysis of plants growing conditions or tumor detection in medical images processing.

As will be shown in Sect. 2, approximate symmetry detection problem applying to binary images is well known but there are not so many effective methods for solving it. The main ones are based on: (1) Fourier series expansion of parametric contour representation, (2) contour representation by turning function,

© Springer International Publishing AG 2017
D.I. Ignatov et al. (Eds.): AIST 2016, CCIS 661, pp. 293–304, 2017.
DOI: 10.1007/978-3-319-52920-2_27

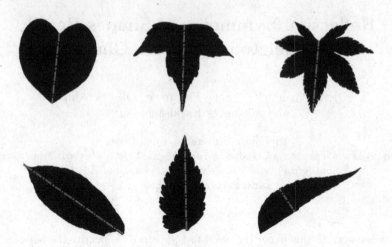

**Fig. 1.** Examples of more (top) or less (bottom) symmetric shapes

(3) contour representation by critical points and computation of similarity measure for two sub-contours via vectors of geodesic distances. All mentioned methods exploit known algorithms of shapes dissimilarity (or similarity) evaluation.

In Sects. 3 and 4 the fast method of reflection symmetry detection is proposed. It uses means of the shapes comparison method based on alignment of skeleton primitive chains. The skeleton of the shape can be divided into two parts – the "left" and the "right" sub-skeletons. They are processed: the left part is traversed counterclockwise and the right one – in clockwise direction. As a result, the "left" and the "right" primitive sub-chains are achieved; they can be compared by the known shape matching procedure based on pair-wise alignment of primitive chains. So, the most similar parts of the skeleton among all possible ones correspond to the most similar parts of the shape which are considered as approximately reflection symmetric parts. The start and the end points of division of the skeleton into "left" and "right" parts can be considered as points belonging to the symmetry axis of the shape.

Evidently, all above-mentioned approaches to the symmetry detection (including the proposed one) are numerical methods. It needs to have the procedure that gives the ground truth for comparison and assessment of results of numerical methods. In Sect. 5 the exact symmetry evaluation algorithm and two its optimizations are proposed. The exact method is based on brute force search of the symmetry axis: lines are drawn through all possible pairs of contour points, each of them are considered as a potential symmetry axis. A line is divided the shape into two parts, each part is represented as a set of pixels. Similarity between two sets is calculated using the Jaccard measure. Two most similar parts are considered as a symmetric ones and the line of division is an axis of shape symmetry. Obviously, the brute force algorithm is very time-consuming, so two optimal modifications were developed – one takes into consideration the semi-perimeters of the shape and another – the center of mass of the shape.

All proposed methods – numerical and exact – are experimentally tested on known Flavia leaves dataset [10], some results are shown in Sect. 6. In Sect. 7 discussions, future plans and conclusion are presented.

## 2 Related Work

The problem of approximate symmetry detection and symmetry measure calculation for gray-scale and color images is well-known [7]. It is usually solved using information about characteristic points of the image. However, there are not so many effective methods for solving the same task for binary images. They based on the certain algorithms of shape matching and (dis)similarity measure calculation. Nevertheless, those algorithms have to be modified because they are applied not to different shapes but to two parts of a shape. A shape is divided into two parts by a line, similarity of two parts is said to be their mirror similarity with respect to that line. The line with respect to which a shape will be divided in two most similar parts is taken as desired symmetry axes.

Proposed in [9] method of reflection symmetry detection is based on parametric contour representation. Several variants of information-preserving parametric contour representation are given. The normalized to $2\pi$ arc length, or perimeter of polygonal contour, is used as a parameter. Fourier series expansion is applied to the representation. Achieved Fourier coefficients of different contours can be compared by calculating their mutual distance. Based on such a method could be calculated pair-wise dissimilarity measure of shapes or approximate asymmetry measure of a shape.

The suggested in [8] approach for reflection symmetry detection based on the method of comparison of polygonal shapes proposed in [1], where contour of the shape is normalized to unity and represented by a turning function. A turning function is a periodic parametric contour representation. A drawback of such representation is its high sensitivity to noise in the boundary. Therefore preliminary smoothing of boundary might be useful when using this representation for comparing shapes. The scaling, translation and rotation invariant dissimilarity metric for two turning functions is introduced. Based on this metric the similarity measure for two shapes is given. This measure could be applied to symmetry measure calculation as maximum of all possible similarity measures between the contour of the shape and the same contour reflected relative to all possible lines. The line which gives the maximum will be the axes of the symmetry.

According to approach stated in [11] contour of a figure is represented as a set of critical points [5]. This set captures all the information required for detecting dissimilarities of contours. A set is computed using Discrete Curve Evolution (DCE) method [5]. Then a set is divided into two parts. The parts that minimize the dissimilarity value, i.e., maximize self-similarity, are used to define a main similarity axis, which for many shapes corresponds to the main axis of reflection symmetry. Proposed dissimilarity measure is motivated by inner distance (or geodesic distance) introduced in [6]. The advantage of geodesic distance is that it is insensitive to the articulation of parts. This property could be very important for computing approximate reflection symmetry measures.

It is worth to notice that all above-mentioned methods of reflection symmetry detection are based on contour representation of the figure.

## 3  Adapting the Shape Comparison Method Based on Alignment of Skeleton Primitive Chains to Calculation of Reflection Symmetry

Unlike the methods mentioned in previous section the proposed approximate method of reflection symmetry detection is skeleton-based one [2]. It uses the procedure of pair-wise skeletons comparison where skeletons of binary images are represented by chains of primitives [4]. To encode a skeleton by chain of primitives the skeleton needs to be traversed counterclockwise. Each primitive represents a traversing edge of the skeleton and consists at least of two normalized numbers: the length of the current edge and the angle between the current and the next edges. Scaling unit for the length is the diameter of minimal circle circumscribed about the skeleton. The angle is normalized to $2\pi$.

To obtain more complex and precise shape representation it is proposed to use a vector of Legendre coefficients $\mathbf{p} = \{p_0, ..., p_n\}$ as third component of a primitive. Legendre coefficients represent the radial skeleton function for each skeleton edge. The radial skeleton function describes width of the shape. So, each primitive is the three-component vector $\omega = \{l, \alpha, \mathbf{p}\}$ [3]. The chain of primitives is obtained while traversing the skeleton edge by edge. The chain representation of the skeleton is invariant under translation, rotation and scaling. Two chains of primitives can be optimally aligned and dissimilarity measure can be evaluated for the corresponded images as described in [4].

For the task of reflection symmetry detection the skeleton of the shape can be divided into two parts – the "left" and the "right" sub-skeletons. The left part is needed to be traversed counterclockwise and the right one – in clockwise direction. As a result, the "left" and the "right" primitive sub-chains are achieved; they can be compared by the shape matching procedure based on pairwise alignment of primitive chains [4]. So, the most similar parts of the skeleton among all possible ones correspond to the most similar parts of the shape which are considered as reflection symmetric parts. The start and the end points of division of the skeleton into "left" and "right" parts can be considered as points belonging to the symmetry axis of the shape.

The important note: to avoid the process of rebuilding primitive sub-chains for each variant of skeleton division while searching symmetric parts it needs to adjust the whole primitive chain of the image.

The algorithm of primitive chain adjustment (see details in Fig. 2):

1. The whole primitive chain is divided into two sub-chains: the bounds of division are corresponded to the start and the end points of division of the skeleton into "left" and "right" sub-skeletons.
2. Primitives in the "right" sub-chain are written in reversed order. This means that the "right" sub-skeleton is traversed clockwise (while the order of

primitives in "left" chain means that the "left" sub-skeleton is traversed counterclockwise).

3. Values of angles in each primitive in the "right" sub-chain have to be adjusted
   to correspond to the clockwise traversing order: the angle value in current
   primitive is substituted by the angle value from the next primitive. The angle
   value in the last primitive is substituted by the angle value from the last
   primitive in the "left" sub-chain.

4. The requirement of proper traversing order is led to changes in vectors of
   Legendre coefficients in each primitive in the "right" sub-chain: it needs to
   write with the opposite sign those coefficients which have odd indices in vector. This means that the radial function coded by changed coefficients is
   symmetric to the initial radial function with respect to Y-axes.

As a result of the adjustment algorithm, the "right" sub-chain is corrected
according to the proper clockwise traversing order of the "right" sub-skeleton.
In Fig. 2 the exemplification of the adjustment algorithm is given.

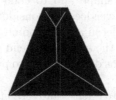

| Sub-skeletons | 1) The whole primitive chain is divided into two sub-chains | 2) Primitives in the "right" sub-chain are written in reversed order | 3) Values of angles in each primitive in the "right" sub-chain are adjusted | 4) Legendre coefficients in each primitive in the "right" sub-chain are changed |
|---|---|---|---|---|
| «Left»: | {0.17 1.0 (1.15 -0.83)} | | | |
| | {0.17 0.41 (1.15 0.83)} | | | |
| | {0.31 0.34 (1.5 0.4)} | | | |
| | {0.51 1.0 (0.59 -0.56)} | | | |
| | {0.51 0.32 (0.59 0.56)} | | | |
| «Right»: | {0.53 1.0 (0.58 -0.53)} | {0.16 0.18 (1.29 0.84)} | {0.16 1.0 (1.29 0.84)} | {0.16 1.0 (1.29 -0.84)} |
| | {0.53 0.34 (0.58 0.53)} | {0.16 1.0 (1.29 -0.84)} | {0.16 0.41 (1.29 -0.84)} | {0.16 0.41 (1.29 0.84)} |
| | {0.31 0.41 (1.5 -0.4)} | {0.31 0.41 (1.5 -0.4)} | {0.31 0.34 (1.5 -0.4)} | {0.31 0.34 (1.5 0.4)} |
| | {0.16 1.0 (1.29 -0.84)} | {0.53 0.34 (0.58 0.53)} | {0.53 1.0 (0.58 0.53)} | {0.53 1.0 (0.58 -0.53)} |
| | {0.16 0.18 (1.29 0.84)} | {0.53 1.0 (0.58 -0.53)} | {0.53 0.32 (0.58 -0.53)} | {0.53 0.32 (0.58 0.53)} |

**Fig. 2.** Binary shape with skeleton, the "left" and the "right" sub-skeletons, the whole
primitive chain and four stages of the "right" primitive chain adjustment. Shaded sub-
chains will be aligned by symmetry detection algorithm

# 4  The Method of Reflection Symmetry Detection Based on Comparison of Skeleton Primitive Chains

The proposed method of reflection approximate symmetry detection is based on the pair-wise alignment of skeleton primitive chains that used in the procedure of binary shapes comparison [4]. For symmetry detection it is utilized the same principle of pair-wise alignment with the help of dynamic programming as it was explained in [4] with one important exception: it needn't to do cyclic shift for one of the primitive chains. In the procedure of shapes comparison the cyclic shift is used to find the best variant of the optimal alignment depending on start primitive in the chain, because comparing shapes can be rotated arbitrary in the image and the start point of skeletons traversal is chosen arbitrary too. As for symmetry detection, the mutual location of two parts of a shape cannot change and it needn't to search for the proper start point of primitive chains alignment.

The algorithm of reflection symmetry detection based on the pair-wise alignment of skeleton primitive chains:

1. Build the skeleton of the shape and to code the skeleton by the chain of primitives [4]. All skeleton vertices are considered as ordered.
2. Choose a vertex of the skeleton as start point.
3. Choose the next vertex of the skeleton as end point. With the start and the end points the skeleton turns divided into two parts: the "left" and the "right" ones.
4. Divide primitive chain into "left" and "right" sub-chains and adjust the "right" sub-chain (see the algorithm suggested above in Sect. 3).
5. Calculate dissimilarity measure of the "left" and the "right" sub-chains using the method of pair-wise alignment of skeleton primitive chains [4] without cyclic shift.
6. Repeat steps 3, 4 and 5 until all skeleton vertices will be passed. Among all possible variants of skeleton division achieved on step 4 choose that one for which is the value of dissimilarity measure calculated on step 5 is the smallest.
7. Repeat steps 2, 3, 4, 5 and 6 until all ordered skeleton vertices will be used as start points on step 2.
8. Among all dissimilarity values achieved on step 6 choose the minimum. It is corresponded with the desired variant of skeleton division. The "left" and the "right" sub-skeletons of this division are represented approximate reflection-symmetric parts of the shape. The start and the end points of division of the skeleton into "left" and "right" parts could be considered as points belonging to the symmetry axis of the shape.

It is obvious that real-world images rarely can be absolute reflection symmetric. The proposed method allows finding symmetry axes not only for absolute symmetric shapes but also for approximate symmetry ones.

# 5    The Exact Algorithm for Detecting Reflection Symmetry of Binary Shapes with Optimization

For the symmetry measure calculation we will use the set-theoretic expression of Jaccard similarity (for binary sets also known as Tanimoto):

$$\mu_T(B) = \frac{|S(B) \cap S(B_r)|}{|S(B) \cup S(B_r)|},  \tag{1}$$

where $B$ is the binary image, and the brightness of the black pixels is coded by 1, and white pixels by 0; $B_r$ is the image obtained by reflection of binary image $B$ relative to straight line, $S(B)$ is the set of pixels of image $B$, the brightness of which is equal to 1. Obviously, $\mu_T(B)$ has the basic "good" properties of measure: $0 \le \mu_T(B) \le 1$, and $\mu_T(B) = 1$ if $B$ is absolutely symmetric, and $\mu_T(B) = 0$ if $B$ and $B_r$ do not intersect.

Suppose that some straight line (axis) is drawn in the image plane, this line divides the pixels of the image into two sets of points. Reflecting the points of the first set (the result doesn't depend on the set choice order) with the brightness of 1 by means of affine transformations relative to the selected axis and comparing the brightness of each corresponding point of the second and reflected sets, it is possible to calculate their intersection and unification: if the black points are combined, then intersection and union are incremented by one; if the combined points have different values (i.e. 0 and 1), then the union is incremented by one. The measure will be the ratio of intersection to the union of the sets that is the Jaccard similarity.

**Definition 1.** *Axis, providing the maximum measure of Jaccard similarity* $\mu_T(B)$, *will be called as **the axis of approximate symmetry**.*

**Definition 2.** *Axis, providing the absolute maximum of Jaccard similarity* $\mu_T(B) = 1$, *will be called as **the axis of absolute symmetry**.*

To evaluate and compare the results of algorithms for determining the reflection symmetry of the binary images, it is necessary to have a precise procedure that will allow computing the axis of approximate symmetry, and the symmetry measure of the image in a reasonable time. Therefore, an exact exhaustive search algorithm was developed and suggested two optimization variants.

It is obvious that the axis should cross the figure, since otherwise the numerator in the formula 1 will be equal to zero and the whole measure of the symmetry will be zero. In addition, the axis of symmetry will cross the contour of an object at least twice, so it makes sense to consider only contour points of the figure. Points of the contour can easily be presented as a sequence. Through the pairs of contour points we will draw lines as candidates in the desired axis of symmetry.

*The exact algorithm based on brute-force search of all lines – candidates to the desired axis of approximate symmetry:*

1. Take a point on the boundary of the shape – an element of the sequence of contour points.

2. Iterate over the remaining points; each of them, together with the point selected in step 1 determines some line. Calculating and storing of measure of symmetry relative to each resulting line. Thus all possible lines passing through the first point will be considered, this point can be excluded from further iterating.
3. The algorithm is repeated for a decremented sequence of contour points.

As a result all the possible lines crossing the figure will be considered. The axis of approximate symmetry of the figure will be one of the lines relative to which the maximum symmetry measure is got.

Exhaustive search algorithm always finds the best axis of symmetry, but the computational complexity of its operation depends on the number of points in the contour $N$ as $O\left(N(N-1)/2\right)$. The average time to process a single image with a resolution of 800 by 600 pixels is about 3 h on a standard laptop. Therefore, two versions of optimization were designed.

*Version 1 – optimization considering semi-perimeters of a figure*
It is obvious that the axis of symmetry splits the absolute reflection-symmetric shape in half, and approximate reflection-symmetric shape – almost in half. Therefore, the length of the contours of each half will be approximately equal. Therefore, it is necessary to iterate over all the lines that divide the contour into two almost equal parts i.e. passing through a neighborhood of the points dividing contour on semi-perimeters.

Thus, the brute-force search with optimization considering semi-perimeters of the figure will be rewritten in the following form:

1. Choose the first point of the contour – $p_1$.
2. Find the second point $p_2$ which together with $p_1$ divides the contour exactly in a half. Find two points that lie at predetermined distance $\varepsilon$ from the second point: $a = p_2 - \varepsilon$, $b = p_2 + \varepsilon$. These points ($a$ and $b$) bound a finite set of contour points.
3. Iterate over all the points located on the segment $[a, b]$, through each of them and the first point $p_1$ draw a line. Calculate and store the measure of symmetry relative to each line.
4. Select as $p_1$ the next point of the sequence of contour points, repeat the algorithm until first half of the sequence of contour points will be passed through as $p_1$.

As a result of the algorithm all the possible lines that divide the contour into two almost equal parts will be examined. The parts that maximize the symmetry measure define the axis of approximate symmetry of the figure.

The complexity of the proposed algorithm with optimization considering semi-perimeters is equal to $O\left(N \cdot \varepsilon\right)$.

*Version 2 – optimization considering the center of mass of a figure*
It is known that the axis of absolute symmetry must pass through the center of mass of the figure. We will use this fact to optimize the exact brute-force search algorithm. We can assume that the axis of approximate symmetry does

not exactly pass through the center of mass, but passes through its certain neighborhood. Thus, it will be sufficient to iterate over all the lines – candidates to a desired axis of symmetry, which locate at a predetermined distance $\varepsilon$ from the center of mass, i.e. which cross a circle with a predetermined radius whose center is the center of mass as shown in Fig. 3(a). In Fig. 3(b) the candidate line does not pass over the neighbor region of center mass and should be excluded from computations. For fast check of "good" candidates it is enough to use only distances among three points – two contour points and the center of mass.

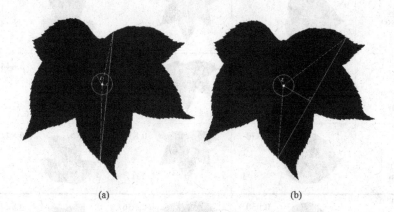

(a)                                        (b)

**Fig. 3.** The sketch for idea of optimization considering the center of mass of a figure

Thus, the brute-force search with optimization considering center of mass of the figure will be rewritten in the following form:

1. Calculate the center of mass of a figure.
2. Fix a point of the contour.
3. Iterate over the remaining points, throw each of them and the point selected in step 2 draw a line. For each line calculate distance from it to the center of mass. If the distance from the line to the center of mass is less than or equal to the radius $\varepsilon$ of the circle, calculate and store the measure of symmetry relative to the line. Thus, all possible lines which are pass through the point selected in step 2 will be considered, this point can be excluded from further iteration.
4. The algorithm is repeated for a decremented sequence of contour points.

As a result all the possible lines which are at a predetermined distance from the center of mass will be examined. Each line divides a figure into two parts. The parts that maximize the symmetry measure define the axis of approximate symmetry of the figure.

The complexity of the above algorithm is also equal to $O\left(N \cdot \varepsilon\right)$, although it depends on the location of the center of mass, which in turn depends on the shape of a figure.

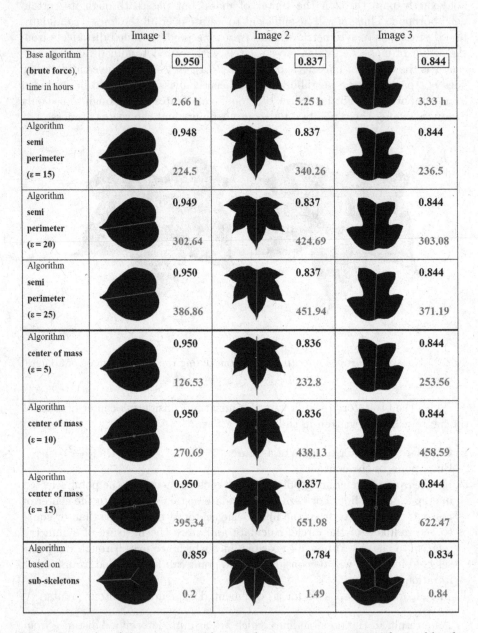

| | Image 1 | | Image 2 | | Image 3 | |
|---|---|---|---|---|---|---|
| Base algorithm (brute force), time in hours | | **0.950** 2.66 h | | **0.837** 5.25 h | | **0.844** 3.33 h |
| Algorithm semi perimeter ($\varepsilon = 15$) | | 0.948 224.5 | | 0.837 340.26 | | 0.844 236.5 |
| Algorithm semi perimeter ($\varepsilon = 20$) | | 0.949 302.64 | | 0.837 424.69 | | 0.844 303.08 |
| Algorithm semi perimeter ($\varepsilon = 25$) | | 0.950 386.86 | | 0.837 451.94 | | 0.844 371.19 |
| Algorithm center of mass ($\varepsilon = 5$) | | 0.950 126.53 | | 0.836 232.8 | | 0.844 253.56 |
| Algorithm center of mass ($\varepsilon = 10$) | | 0.950 270.69 | | 0.836 438.13 | | 0.844 458.59 |
| Algorithm center of mass ($\varepsilon = 15$) | | 0.950 395.34 | | 0.837 651.98 | | 0.844 622.47 |
| Algorithm based on sub-skeletons | | 0.859 0.2 | | 0.784 1.49 | | 0.834 0.84 |

**Fig. 4.** Examples of three images with axes of *approximate symmetry* obtained by the suggested algorithms. First number in cell is symmetry measure and the second one is calculating time in seconds (in the first row - in hours)

# 6  Experimental Results

In Fig. 4 there are examples of three images, axes of *approximate symmetry* obtained by the suggested algorithms, symmetry measures and computing time. Values of parameter $\varepsilon$ are in pixels. Ground-truth symmetry values, obtained by the brute force algorithm, are marked by frames.

The analysis of Fig. 4 demonstrates that fast versions of exact algorithm for searching the symmetry axis significantly reduce computing time. The algorithm based on pair-wise alignment of primitive sub-chains allows to reduce processing time by two-three orders of magnitude while the accuracy level is quite acceptable (see discussions below).

# 7  Discussions and Conclusion

The study of numerous experimental results convinced us that Jaccard symmetry measure is quite acceptable estimation with relation to human visual perception. However in some situations it would be better to suggest alternative symmetry axis based on the experience and the task context. In Fig. 5 there is an example of image for which the best axis according to Jaccard similarity (solid line) does not coincide with the axis drawn according to expert opinion (dash line). Although, difference between them is not great: the measure that corresponds to Jaccard axis $(\mu_1)$ is equal to 0.8828, the measure that corresponds to expert opinion $(\mu_2)$ is equal to 0.8608. We suppose that human eye takes into account some peculiarities of a shape in contrast to Jaccard measure, which uses just integral characteristics of pixel quantities. Consequently, the symmetry axis we achieve with alignment of sub-skeletons could be used as an adequate alternative to ground truth symmetry axis with maximum of Jaccard similarity measure.

The question of finding the upper bounds of $\varepsilon$-parameter values for the fast versions (based on semi-perimeters and center of mass) of exact algorithm needs for additional study. We are planning to evaluate these parameters via some relative characteristic of a figure like radius of circumcircle, radius of maximum inscribed circle, distance between two outermost points, etc.

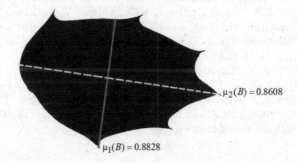

$\mu_2(B) = 0.8608$

$\mu_1(B) = 0.8828$

**Fig. 5.** Discussions about Jaccard symmetry measure

We plan to combine algorithm based on pair-wise alignment of skeleton primitive sub-chains and ideas of fast versions of brute-force algorithm. It gives us a new procedure of symmetry measure and axis evaluation which has acceptable accuracy and calculation speed characteristics, sufficient for using it in real-world computer vision tasks.

**Acknowledgements.** This work is partially supported by Russian Fund for Basic Research, Grants 14-07-00527 and 16-57-52042.

# References

1. Arkin, E.M., Chew, L.P., Huttenlocher, D.P., Kedem, K., Mitchell, J.S.B.: An efficiently computable metric for comparing polygonal shapes. IEEE Trans. Pattern Anal. Mach. Intell. **13**, 209–216 (1991)
2. Blum, H.: A transformation for extracting new descriptors of shape. Models Percept. Speech Vis. Form **19**(5), 362–380 (1967)
3. Kushnir, O., Seredin, O.: Parametric description of skeleton radial function by legendre polynomials for binary images comparison. In: Elmoataz, A., Lezoray, O., Nouboud, F., Mammass, D. (eds.) ICISP 2014. LNCS, vol. 8509, pp. 520–530. Springer, Heidelberg (2014). doi:10.1007/978-3-319-07998-1_60
4. Kushnir, O., Seredin, O.: Shape matching based on skeletonization and alignment of primitive chains. In: Khachay, M.Y., Konstantinova, N., Panchenko, A., Ignatov, D.I., Labunets, V.G. (eds.) AIST 2015. CCIS, vol. 542, pp. 123–136. Springer, Heidelberg (2015). doi:10.1007/978-3-319-26123-2_12
5. Latecki, L.J., Lakämper, R.: Convexity rule for shape decomposition based on discrete contour evolution. Comput. Vis. Image Underst. **73**(3), 441–454 (1999)
6. Ling, H., Jacobs, D.W.: Shape classification using inner-distance. IEEE Trans. PAMI **29**, 286–299 (2007)
7. Liu, J., et al.: Symmetry detection from realworld images competition: summary and results. In: 2013 IEEE Conference on Computer Vision and Pattern Recognition Workshops (CVPRW), pp. 200–205 (2013)
8. Sheynin, S., Tuzikov, A., Volgin, D.: Computation of symmetry measures for polygonal shapes. In: Solina, F., Leonardis, A. (eds.) CAIP 1999. LNCS, vol. 1689, pp. 183–190. Springer, Heidelberg (1999). doi:10.1007/3-540-48375-6_23
9. van Otterloo, P.J.: A contour-oriented approach to digital shape analysis. Ph.D. thesis, Delft University of Technology, Delft, The Netherlands (1988)
10. Wu, S.G., Bao, F.S., Xu, E.Y., Wang, Y.-X., Chang, Y.-F., Xiang, Q.-L.: A leaf recognition algorithm for plant classification using probabilistic neural network. In: 2007 IEEE International Symposium on Signal Processing and Information Technology, pp. 11–16 (2007)
11. Yang, X., Adluru, N., Latecki, L.J., Bai, X., Pizlo, Z.: Symmetry of shapes via self-similarity. In: Bebis, G., Boyle, R., Parvin, B., Koracin, D., Remagnino, P., Porikli, F., Peters, J., Klosowski, J., Arns, L., Chun, Y.K., Rhyne, T.-M., Monroe, L. (eds.) ISVC 2008. LNCS, vol. 5359, pp. 561–570. Springer, Heidelberg (2008). doi:10.1007/978-3-540-89646-3_55

# Using Efficient Linear Local Features in the Copy-Move Forgery Detection Task

Andrey Kuznetsov[1,2]([✉]) and Vladislav Myasnikov[1,2]

[1] Samara University, Samara, Russia
{kuznetsov,vmyas}@geosamara.ru
[2] Image Processing Systems Institute of the Russian Academy
of Sciences (IPSI RAS), Samara, Russia
http://nil97.ssau.ru/

**Abstract.** Digital images are often used to prove some facts or events, but nobody can guarantee their originality. More often we can see in TV news, that some satellite imagery evidences were received to show what has happened. However, we cannot be sure, that these data were not changed by some hackers. In this paper we propose a new algorithm for detection of the most frequently used attack plain copy-move. The algorithm is based on a hash value calculation in a sliding window mode. The hash function is constructed using efficient linear local features that were developed by coauthor V. Myasnikov in 2010. Finally, we present results of conducted experiments and comparison with existing solutions, as well as recommendations for the use of the proposed approach. The main advantage of the proposed solution is 99.95% precision of copy-move blocks detection comparing with existing approaches. Another impact is that it can be easily used for large satellite image analysis as well as ordinary digital images processing because of low computational complexity.

**Keywords:** Copy-move · Forgery · Duplicate · Hash function · Hash table · Galois field · Linear local features

## 1 Introduction

Nowadays digital images play an important role in different spheres of mans life. Digital photos are used as evidence for events, satellite images show territory state, etc. Depending on the importance of images information, they may be changed by intruders for data hiding and providing forgeries for end users of the data.

The most frequently used forgery method is copy and paste the local area of the same image. Such attacks are called copy-move, and copied image regions are called duplicates. There is a number of works [1–4], where copy-move detection algorithms are proposed: the common step of these methods is features development that are invariant to various changes of duplicates (brightness, geometry)

D.I. Ignatov et al. (Eds.): AIST 2016, CCIS 661, pp. 305–313, 2017.
DOI: 10.1007/978-3-319-52920-2_28

and are calculated in an overlapping window mode to reduce computational complexity. In [5,6] another approach to plain copy-move detection is proposed – it is based on hash values calculation in a sliding window mode. In this paper, we propose a copy-move detection algorithm and a new hash function, used for detection. It is quite important to achieve the best results in plain copy-move detection before developing algorithms for transformed copy-move detection, because the easier attack is, the more frequent it is used.

This paper consists of five main sections. Section 2 contains a brief definition of a duplicate and hash functions as well as the structural pattern that is used for image fragment representation. Section 3 is devoted to the description of the proposed copy-move detection algorithm using structural pattern. Section 4 is devoted to the proposed new hash function based on the efficient linear local features (LLF) framework. Section 5 presents results of conducted experiments and comparison of our approach with existing solutions [9]. In Sect. 6 we present conclusions and recommendations for using of the developed algorithm.

## 2    Duplicate and Hash Function Definition

Similar to [6], let us define a digital image with size $M \times N$ as follows

$$f : \mathbb{N}_M \times \mathbb{N}_N \to \mathbb{B}^k,$$

where $\mathbb{B} \equiv \{0,1\}$, $\mathbb{N}_M$ is a set of integers $\{0, 1, ..., M-1\}$, k is a number of bits according to the image pixels data type.

Let us define a structural pattern as a finite set of coordinates $\{(0,0), ..., (m,n)\}$ with the following properties:

– it contains coordinate $(0,0)$;
– it does not contain coordinates with negative values;
– it is quadruple connected.

We introduce a special type of a structural pattern $\aleph(\Lambda, a, b)$, which is defined in the following way:

$$\aleph(\Lambda, a, b) \equiv \bigcup_{(m,n)\in\Lambda} \Pi(a, b, m, n), \tag{1}$$

where the set of coordinates $\Pi(a, b, m, n)$ is defined as follows:

$$\Pi(a, b, m, n) \equiv \left\{ \begin{array}{l} (m,n), (m, n+1), ..., (m, n+b-1), \\ ... \\ (m+a-1, n), ..., (m+a-1, n+b-1) \end{array} \right\}$$

and $\lambda$ is an arbitrary finite set of non-negative coordinates with a guarantee of coordinate pattern $\aleph(\Lambda, a, b)$ properties.

We say that an image contains duplicates by a pattern $\aleph(\Lambda, a, b)$, if there are at least two pairs of coordinates $(m', n')$ and $(m'', n'')$, which satisfy the following set of equalities:

$$f(m' + m, n' + n) = f(m'' + m, n'' + n), \forall(m, n) \in \aleph(\Lambda, a, b).$$

A task of duplicate detection using a pattern $\aleph(\Lambda, a, b)$ corresponds to the matching of all image samples $(m, n)$, which define the upper left corner of an image fragment with a form defined by a pattern $\aleph(\Lambda, a, b)$, with unique numbers $t(m, n) \in \mathbb{N}$, which characterize an image fragment in the following way:

$$t(m, n) \equiv \begin{cases} 0, & \text{no copy-move,} \\ > 0, & \text{copy-move type id.} \end{cases}$$

Under copy-move type number we mean a unique integer value, which is the same for equal image fragments.

*Comment.* It should be noted, that every structural pattern can be represented in several ways. For example, the pattern $\aleph(\{(0, 0)\}, 10, 10)$ defines the same coordinates set as the patterns $\aleph(\{(0, 0), (1, 0)\}, 10, 9)$ and $\aleph(\{(0, 0), (0, 5), (5, 0), (5, 5)\}, 5, 5)$. This representation ambiguity is used further, when we describe the proposed copy-move detection algorithm.

A hash function for an image fragment with a form defined by a pattern $\aleph(\Lambda, a, b)$ is calculated according to the following equation:

$$T : \mathbb{B}^{BPP|\aleph(\Lambda, a, b)|} \rightarrow \mathbb{N}_L,$$

where $\mathbb{N}_L \leq 2^{BPP|\aleph(\Lambda, a, b)|}$ defines a set of possible hash values and $BPP$ corresponds to a bit per pixel value. In other words, the hash function converts pixel intensity values of an image fragment to some positive integer value from the range $[0, L - 1]$. If $L = BPP$, then $T$ is a one-to-one transformation. In [5,6] we proposed two hash functions for rectangular structural patterns $\Pi(a, b, 0, 0)$. Achieved experimental results showed that the best hash function was the function based on 2D Rabin-Karp rolling hash [5]. This hash function is be used in Sect. 5 for comparison with the proposed hash function based on linear local features. The copy-move detection algorithm, which is described in the next section, can be used with all types of hash functions.

## 3   Copy-Move Detection Algorithm

As it was proposed in the previous section, we have a task of *duplicate detection using a pattern* $\aleph(\Lambda, a, b)$. The proposed algorithm analyzes all positions of image fragments $\Pi(a, b, m, n)$ using sliding window mode. This means, that for every point $(m, n)$ we calculate a hash value using intensity values $f(m', n')|_{(m', n') \in \Pi(a, b, m, n)}$. Then the hash value is used for:

1. Updating a hash table $H(T)$, which contains absolute frequencies of hash values $T$ appearance. In other words, this hash table corresponds to a histogram of hash values;

2. Updating an image of hash values $P(m, n)$.

It is obvious, that an image fragment in position $(m', n')$ with a form defined by a pattern $\aleph(\Lambda, a, b)$ is a duplicate only if all the $|\Lambda|$ image fragments of the form

$$f(m + m', n + n')|_{(m',n') \in \Pi(a,b,m,n)}, (m, n) \in \Lambda$$

be duplicates. Thus, the decision rule for image fragment referring to a particular copy-move type is given by:

$$t(m', n') \equiv \begin{cases} 0, & \exists (m, n) \in \Lambda, H(P(m + m', n + n')) \leq 1; \\ P(m', n') + 1, & \text{else.} \end{cases}$$

The proposed algorithm in pseudo code looks in the following way:

```
for i=1:M
   for j=1:N
      bool isDuplicate = true;
      t(i,j) = 0;
      foreach (m,n) in Lambda
         if (H(P(i+m,j+n)) <= 1)
            isDuplicate = true;
            break;
         end
      end

      if (isDuplicate)
         t(i,j) = P(i,j) + 1;
      end
   end
end
```

It can be noted, that due to representation ambiguity of the structural pattern (1) (see *Comment* in Sect. 2) the proposed algorithm is ambiguous. In our previous research [6] we carried out experiments to select the best representation scheme for a specific structural pattern. Analyzing obtained results, the structural pattern size is taken $11 \times 11$, the power of set $|\Lambda|$ is taken 6.

In this paper, we place primary emphasis upon conducting experiments on the quality of the algorithm and the hash function, comparing with existing approaches and previously developed 2D Rabin-Karp rolling hash [5].

## 4    Hash Function Based on Linear Local Features

As it can be seen above, a hash value is a scalar product of pixel intensity values of an image and values of a filter with a finite impulse response (FIR). For these operations, there is a well-developed effective FIR filters constructing framework, which allows obtaining the best way of hash values calculation for a specific task *linear local features* (LLF) of the original image. Description of the efficient LLF framework, as well as the results of its experimental research are described in details in the papers of V. Myasnikov [7,8].

Using efficient LLF framework the following system of linear equations (SLE) is constructed:

$$h(0) = 1,$$

$$h(m) - \sum_{k=1}^{K} a_k h(m-k) = 0, m \in [1, M-1]/\Theta,$$

$$\sum_{k=1}^{K} a_k h(m-k) = 0, m \in [M, M+K-1]/\Theta,$$

$$h(m) - \sum_{k=1}^{K} a_k h(m-k) - \widetilde{\phi}(m) = 0, m \in [1, M-1] \cap \Theta, \qquad (2)$$

$$\sum_{k=1}^{K} a_k h(m-k) + \widetilde{\phi}(m) = 0, m \in [M, M+K-2] \cap \Theta,$$

$$a_K h(M-1) + \widetilde{\phi}(M+K-1) = 0,$$

where $\{h(m)\}_{m=0}^{M-1}$ is a kernel, defined by a finite-length number sequence, $\{a_k\}_{k=1}^{K}$ is a vector of a linear recurrent relation (LRR) coefficients, $\Theta = \{n \in \mathbb{Z}_+ : \phi(n) \neq 0\}$ is a set of irregularities, $M$ is the length of the kernel, $K$ is the number of irregularities. Solutions of the SLE (2) are solutions of efficient LLF building task in the sense of minimizing the selected target function value.

Let the Galois field $\boldsymbol{GF}(p)$ is defined, where $p$ is a prime number. We solve (2) over the field $\boldsymbol{GF}(p)$ and use this solution further to detect plain copy-move. The main characteristic of the field is a prime number $p$ and all the arithmetic operations are carried out over modulo $p$.

We define LRR coefficients $\{a_k\}_{k=1}^{K}$ using one of the following schemes:

1. Fibonacci sequence elements:
   (a) $K = 2, a_1 - 1, a_2 = 2$;
   (b) $K = 3, a_1 - 1, a_2 = 2, a_3 = 3$;
   (c) $K = 4, a_1 - 1, a_2 = 2, a_3 = 3, a_4 = 5$;
2. Polynomial coefficients:

$$a_k = (-1)^k C_K^k.$$

We then select coefficients for predefined $K$ value, define the LLF kernel length $M$ and build (2) for every available tuple $(K, M, \bar{a})$.

The number of different variants of building (2) is $C_{M+K-2}^{K-1}$[7]. According to the proved LLF properties [7] one solution is guaranteed to provide a minimum value of target function, which is the most efficient. The solution of the SLE (2) is a one dimensional kernel $\{h(m)\}_{m=0}^{M-1}$. A two-dimensional kernel $M \times M$ can be obtained by the Cartesian product of one-dimensional kernel $\boldsymbol{h}$ and its transposed value $\boldsymbol{h}^T$.

The main advantage of the proposed algorithm is that it does not miss duplicates. This statement can be proved by the following facts:

1. The algorithm uses sliding window mode;
2. Hash values are calculated for every available position of the sliding window;
3. Copy-move detection decision is made according to the absolute frequency of hash values in the hash table $H(T)$.

## 5   Experiments

### 5.1   Investigation of the Proposed Hash Function Quality

The key parameter of the proposed hash function is the number of LRR $K$. As an indicator of quality for all ongoing research in this section, we use the number of false detected duplicates or collisions (pairs of image fragments that were not obtained by copy-move operation). The achieved results are compared with the solution on the basis of 2D Rabin-Karp rolling hash [6]. Further in tables we use capitalized 'A' for the LRR coefficients calculation scheme based on Fibbonacci sequence and 'B' - for polynomial.

The initial data for all experiments was a dataset of 100 satellite images, received from SPOT-4 ($4600 \times 4600$) with no duplicates. For the selected value of FIR filter length $M = 11$ we first fixed the value $K = 2, \bar{3}, 4$ and the scheme of coefficients $\{a_k\}_{k=1}^{K}$ calculation. Then for every value $K$ we built and solved $C_{M+K-2}^{K-1}$ number of SLE (2). Considering that the calculations are carried over the field $\boldsymbol{GF}(p)$, the prime number $p = 536870923$ was chosen as its characteristic. This value corresponds to the earlier results for the algorithm based on 2D Rabin-Karp rolling hash, where the prime number $p$ was selected close to $2^{29}$ [6]. Every obtained kernel is applied to test images to determine the best values of the vector $\{a_k\}_{k=1}^{K}$, which provides the minimal number of collisions. Efficiency of filters was compared with the results obtained for 2D Rabin-Karp rolling hash using sliding window size $11 \times 11$ the results are shown in Table 1. For $K = \{3, 4\}$ the hash function based on efficient LLF showed better results than 2D Rabin-Karp rolling hash.

Table 1. Collisions number comparison for hash functions with fixed $K$.

| K | The proposed solution | | Irregularities position | | 2D Rabin-Karp rolling hash |
|---|---|---|---|---|---|
| | A | B | A | B | |
| 2 | 7757353 | 12076155 | 11 | 1 | 1550186 |
| 3 | 1500085 | 1500108 | 1,7 | 3,12 | 1550186 |
| 4 | 1348417 | 1347133 | 4,11,13 | 1,4,7 | 1550186 |

## 5.2 Investigation of the Proposed Algorithm Quality

During the next experiment, we analyzed the quality of the copy-move detection algorithm, described in Sect. 3. There was selected a structural pattern $\aleph(\{(0,0),(0,1),(,0),(1,1)\},11,11)$, and the coefficients vector of the SLE (2) corresponded to the best kernel obtained in the previous experiment. Conducted experiment showed that the copy-move detection algorithm based on efficient LLF has less collisions for $K = \{3,4\}$ using both schemes of the SLE (2) coefficients $\{a_k\}_{k=1}^{K}$ calculation comparing with the algorithm based on 2D Rabin-Karp rolling hash (Table 2).

**Table 2.** Collisions number comparison for copy-move detection algorithm.

| K | The proposed solution | | 2D Rabin-Karp rolling hash |
|---|---|---|---|
| | A | B | |
| 2 | 240854 | 1338933 | 1235 |
| 3 | 242 | 230 | 1235 |
| 4 | 232 | 225 | 1235 |

All conducted experiments show that the hash function based on efficient LLF allows obtaining better result (in terms of copy-move detection quality) than 2D Rabin-Karp rolling hash. At the same time, the proposed hash function has greater computational complexity quality improvements were achieved at the expense of 20–25% loss in computation time. This fact is caused by searching for the best SLE (2) solution. When the best kernel is constructed, it can be used for digital images of the type it was generated on.

We compared our results with the results from a survey on popular copy-move detection approaches [9] (Table 3). There were taken 96 images with plain copy-move regions from a benchmark dataset, provided by the authors. Even for these ideal conditions, most of the algorithms (with the exception of those shown in Table 3) have precision less than 90%. Our algorithm provides nearly

**Table 3.** Precision and recall comparison of the proposed approach with existing copy-move detection algorithms.

| Method | Precision, % | Recall, % |
|---|---|---|
| Circle | 92.31 | 100 |
| FMT | 90.57 | 100 |
| Lin | 94.12 | 100 |
| SURF | 91.49 | 100 |
| Zernike | 92.31 | 100 |
| **Our approach** | 99.95 | 100 |

100% precision. This is why there is no need for a postprocessing step to reduce false detection.

We tested our algorithm on several high resolution (0.5 m, 5000 × 5000) satellite images (Fig. 1). The obtained results confirm high quality detection of the algorithm.

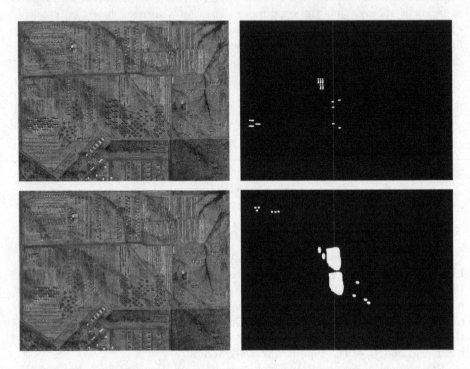

**Fig. 1.** Copy-move forgery (left column) and its detection result (right column).

The computational complexity is quite low. It takes 12 s on an ordinary workstation (Intel Core i5-3470, 8 Gb RAM) to process every single high resolution

**Table 4.** Computational complexity comparison of the proposed approach with existing copy-move detection algorithms.

| Method | Total time, s |
| --- | --- |
| Circle | 5103.43 |
| FMT | 6948.03 |
| Lin | 4785.71 |
| SURF | 1052.12 |
| Zernike | 7065.18 |
| **Our approach** | 420.12 |

satellite image. We compared the computational complexity with the results from a survey on popular copy-move detection approaches [9] (Table 4). There are showed total analysis time values (in seconds) for all 96 images from the dataset in (Table 4).

The results show that the proposed algorithm can be easily used in automatic forgery detection systems in real-time.

# 6    Conclusion

In this paper, we presented a new algorithm for copy-move forgery detection in digital images. A new method of calculating the hash function based on efficient LLF is described. The proposed solution can be used for analysis of all types of digital images without any restrictions on their size and pixel data type. The best structural pattern parameters were taken according to previous experiments [6]. We obtained better results using the number of irregularities $K = \{3, 4\}$ and polynomial scheme of LRR coefficients.

**Acknowledgements.** The proposed copy-move detection algorithm and the hash function based on efficient linear local features (Sects. 2, 3 and 4) were developed with support from the Russian Science Foundation grant №14-31-00014 "Establishment of a Laboratory of Advanced Technology for Earth Remote Sensing". The experimental results (Sect. 5) were obtained with support from the Russian Foundation for Basic Research grant №16-37-00056.

# References

1. Farid, H.: Image forgery detection. IEEE Sig. Process. Mag. **26**, 16–25 (2009)
2. Gong, J., Guo, J.: Exposing region duplication through local geometrical color invariant features. J. Electron. Imaging **24**(3), 033010 (2015)
3. Bayram, S., Sankur, B., Memon, N.: Image manipulation detection. J. Electron. Imaging **15**(4), 041102 (2006)
4. Cao, Y., Gao, T., Fan, L., Yang, Q.: A robust detection algorithm for copy-move forgery in digital images. Forensic Sci. Int. **214**, 33–43 (2012)
5. Vladimirovich, K.A., Valerievich, M.V.: A fast plain copy-move detection algorithm based on structural pattern and 2D Rabin-Karp rolling hash. In: Campilho, A., Kamel, M. (eds.) ICIAR 2014. LNCS, vol. 8814, pp. 461–468. Springer, Heidelberg (2014). doi:10.1007/978-3-319-11758-4_50
6. Glumov, N., Kuznetsov, A., Myasnikov, V.: The algorithm for copy-move detection on digital images. Comput. Opt. **37**(3), 361–368 (2013)
7. Myasnikov, V.: Efficient mutually-calculated features for linear local description of signals and images. In: IASTED International Conference on Automation, Control, and Information Technology - Information and Communication Technology (ACIT-2010), pp. 29–34 (2010)
8. Myasnikov, V.: Constructing efficient linear local features in image processing and analysis problems. Autom. Remote Contr. **72**, 514–527 (2010)
9. Christlein, V., Riess, C., Jordan, J., Riess, C., Angelopoulou, E.: An evaluation of popular copy-move forgery detection approaches. IEEE Trans. Inf. Forensics Secur. **7**(6), 1841–1854 (2012)

# The Color Excitable Schrodinger Metamedium

Ekaterina Ostheimer[1](✉), Valery Labunets[2](✉), and Ivan Artemov[2](✉)

[1] Capricat LLC, Florida, USA
katya@capricat.com
[2] Ural Federal University, Yekaterinburg, Russia
vlabunets05@yahoo.com, etherial.man@gmail.com

**Abstract.** In this work, we apply quantum cellular automata (QCA) to study pattern formation and image processing in quantum-diffusion Schrodinger systems (QDSS) with triplet-valued (color-valued) diffusion coefficients. Triplet numbers have the real part and two imaginary parts (with two imaginary units). They form 3-D triplet algebra. Discretization of the Schrodinger equation gives "lattice based metamaterial models" with various triplet–valued physical parameters. The process of excitation in these media is described by the Schrodinger equations with the wave functions that have values in triplet algebras. If a traditional computer is thought of as a "programmable object", QDSS in the form of QCA is a computer of new kind and is better visualized as a "programmable material". The purpose of this work is to introduce new metamedium in the form of cellular automata. The cells are placed in a 2-D array and they are capable of performing basic arithmetic operating in the triplet algebra and exchanging massages about their state. Cellular automata like architectures have been successfully used for computer vision problems and color image processing. Such metamedia possess large opportunities in processing of color images in comparison with the ordinary diffusion media with the real-valued diffusion coefficients. The latter media are used for creation of the eye-prosthesis (so called the "silicon eye"). The color metamedium suggested can serve as the prosthesis prototype for perception of the color images.

**Keywords:** Image processing · Schrodinger equation · Diffusion · Cellular automata · Triplet algebra · Quantum approach · Non-Euclidean geometry

## 1 Introduction

Quantum cellular automata was already used for image processing purposes by various scientists. For instance, in [1] QCA was considered as a method of implementation of a Directional Marginal Median Filter (DMMF, constructed to reduce impulsive noise) in order to accelerate the computations by achieving the combinational digital logic. Other notable example is [2], where QCA is used to get a high-performance implementation of mathematical morphology operations (erosion and dilation) in image processing.

© Springer International Publishing AG 2017
D.I. Ignatov et al. (Eds.): AIST 2016, CCIS 661, pp. 314–325, 2017.
DOI: 10.1007/978-3-319-52920-2_29

The authors of previously mentioned works claim that QCA can provide better results than a classic cellular automata (CCA). So, complex numbers are more suitable for many image processing cases, but actually we are not limited with only these kinds of numbers. The simplest reason to use 3-D numbers ($Z \in \{\mathbb{R}, \mathbb{C}\}$) instead of 2-D ($z \in \mathbb{C}$) or 1-D ($a \in \mathbb{R}$) ones is the fact that the most images consist of three-component values (color pixels).

Three-component numbers (triplet numbers or color numbers) are the most natural means of image representation. It is very important that these will allow us to take the fact, that $R, G$ and $B$ components of all pictures are correlated (i.e. connected with each other in some manner), into consideration. This leads to the main advantage of a cellular automata with triplet numbers: we can mix all components of the neighboring pixels. Of course, a bigger amount of components for each cell will slow down the computations. That's why we will not use high-resolution cell grids in our experiments.

We should investigate the basic properties and some consistent patterns of behavior for the triplet-valued cellular automata via software modeling before proceeding to image processing itself. It is the main purpose of this work.

Firstly, we need to introduce the rules of interaction between two arbitrary triplet numbers (see [3]). Let $Z_1$ and $Z_2$ be some triplet numbers then

$$Z_1 = a_1 + b_1 \cdot \varepsilon + c_1 \cdot \varepsilon^2, \qquad Z_2 = a_2 + b_2 \cdot \varepsilon + c_2 \cdot \varepsilon^2 : \varepsilon^3 = 1. \tag{1}$$

Operation of addition is performed independently for each component of a number. Let $Z_1 + Z_2 = Z_3 = a_3 + b_3 \cdot \varepsilon + c_3 \cdot \varepsilon^2$ then

$$a_3 = a_1 + a_2; \qquad b_3 = b_1 + b_2; \qquad c_3 = c_1 + c_2. \tag{2}$$

Next we have to define the product of two triplet numbers. Let $Z_1 \cdot Z_2 = Z_4 = a_4 + b_4 \cdot \varepsilon + c_4 \cdot \varepsilon^2$ then for the components we can write

$$Z_4 = Z_1 \cdot Z_2 = (a_1 + b_1 \cdot \varepsilon + c_1 \cdot \varepsilon^2) \cdot (a_2 + b_2 \cdot \varepsilon + c_2 \cdot \varepsilon^2)$$
$$= a_1 a_2 + \varepsilon b_1 a_2 + \varepsilon^2 c_1 a_2 + c_1 b_2 + \varepsilon a_1 b_2 + \varepsilon^2 b_1 b_2 + b_1 c_2 + \varepsilon c_1 c_2 + \varepsilon^2 a_1 c_2. \tag{3}$$

As we can see, the final expressions for number's components are as shown below:

$$a_4 = a_1 a_2 + c_1 b_2 + b_1 c_2;$$
$$b_4 = b_1 a_2 + a_1 b_2 + c_1 c_2; \tag{4}$$
$$c_4 = c_1 a_2 + b_1 b_2 + a_1 c_2.$$

Also we can present, what circuit diagram would we have for one single triplet cell (see Fig. 1).

Now we can proceed to the triplet number interpretation.

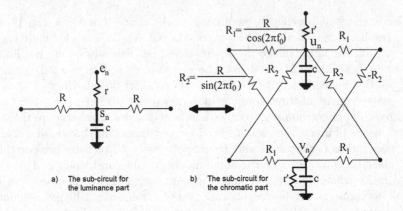

**Fig. 1.** An elementary triplet cell of two components representing the luminance and chromatic parts of a triplet number

## 2 Triplet Numbers and Color Formats

It is known that colors in digital graphics can be represented in many different formats. Colors usually have three components (for example, red, green and blue value) and it corresponds 3D numbers. We can get 3 coordinates in some space by combining 3 real numbers or by using 1 real (achromatic or luminance) and 1 complex (chromatic) number. So we can write a different expression for our arbitrary triplet number:

$$Z = \{a, b, c\} = \{A_{lum}, A_{chr}\}; \quad A_{lum} \in \mathbb{R};$$
$$A_{chr} = X_{chr} + i \cdot Y_{chr} = S \cdot e^{i \cdot H} \in \mathbb{C}. \tag{5}$$

In this expression $a, b$ and $c$ can correspond to $r, g$ and $b$ i.e. red, green and blue color components; $S$ (chromatic number's absolute value) is called saturation, and $H$ (chromatic number's phase) is called hue, $A_{lum}$ is a luminance part. We can translate color numbers from one format to other (see [4], for example):

$$\begin{bmatrix} A_{lum} \\ X_{chr} \\ Y_{chr} \end{bmatrix} = \begin{bmatrix} 1 & 1 & 1 \\ 1 & -\frac{1}{2} & -\frac{1}{2} \\ 0 & \frac{\sqrt{3}}{2} & -\frac{\sqrt{3}}{2} \end{bmatrix} \cdot \begin{bmatrix} r \\ g \\ b \end{bmatrix} = \begin{bmatrix} r + g + b \\ \frac{1}{2} \cdot (2r - g - b) \\ \frac{\sqrt{3}}{2} \cdot (g - b) \end{bmatrix}. \tag{6}$$

There is a geometrical interpretation for RGB and HSL systems relationship (see Fig. 2). RGB color space can be represented as a cube - 3-D figure with limited volume (red, green and blue components of a color also can have only limited value). If we draw a straight line from a point, that corresponds to black color (coordinates are $(0, 0, 0)$), to a point with coordinates $(255, 255, 255)$ (white color) then we will get the so-called achromatic diagonal. It is also an axis of luminance. A plane inside of our cube that is perpendicular to achromatic line is called the chromatic plane. We can construct the radial coordinate system on this plane and then get hue and saturation values of a particular color (see [5]).

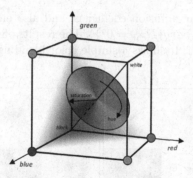

**Fig. 2.** Color space cube, an achromatic diagonal and a complex chromatic plane (Color figure online)

## 3 Lightness and Saturation of the Diffusion Coefficient

Let the diffusion coefficient $D$ from Schrodinger equation be a triplet number instead of complex number. Then we can write more detailed version of the equation:

$$\frac{d}{dt}\phi = (R_D + G_D\varepsilon + B_D\varepsilon^2)\cdot(\frac{d^2}{dx^2}\phi + \frac{d^2}{dy^2}\phi), \qquad (7)$$

because now we have

$$D = \{R_D, G_D, B_D\} = \{D_{lum}, D_{chr}\} = \{L, S \cdot e^H\}. \qquad (8)$$

Now we can adjust three different diffusion coefficient's parameters in its both representation formats. The HSL format turned out to be more useful.

We use our cellular automata with cells representing colored pixels to conduct some researches with a new $D$. Initial conditions is black field with four points of different colors on it (red, green, white and blue ones respectively). Screen is divided into four parts: left top part shows us final RGB picture, right top one - cells' lightnesses, left bottom one - cells' saturations and the last quad shows hue. Chromatic number's phase can be visualized with auxiliary gradient map (see Fig. 3 showing accordance of phases in degrees to colors). Phases for cells that have no chromatic component (shades of gray) are marked with gray color in bottom right quad. This technique is widely used (see, for example, [6]).

| 0 | 60 | 120 | 180 | 240 | 300 | 360 |

**Fig. 3.** Gradient mask for colorful phase visualization (Color figure online)

Initially let us see the metamedia's behavior for "balanced" $D$'s chromatic and achromatic parameters. It means that we should take equal luminance and saturation values for our diffusion coefficient and also eliminate the impact of chromatic phase ($Hue = arg\{D_{chr} = 0°\}$). The results of modeling for this case are represented on Fig. 4. It shows a simple color blending process.

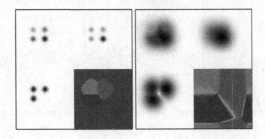

**Fig. 4.** Excitable media's state on 16th (left picture) and 210th (right picture) iterations. $D$'s luminance and saturation components are equal (Color figure online)

Figure 5 shows the process of colored spots propagation in a media with diffusion coefficient with low saturation component. Achromatic component of all pictures in this work is inverted to provide less amount of dark colors for better visual representation. So, darker shades means bigger values in top right and bottom left parts of divided images.

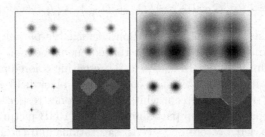

**Fig. 5.** Excitable media's state on 13th and 210th iterations. $D$'s saturation is low (Color figure online)

Note, that chromatic parts of all spots is spreading slower than achromatic ones: as you can see, the size of spots in top right quad (luminance representation) is bigger than in bottom left one (saturation representation).

## 4   Diffusion Coefficient's Chromatic Phase (Hue)

To achieve more interesting results we can adjust the $H$ value of the coefficient $D$. It is useful to change our cellular automata's initial conditions: we will use one red point in the middle instead of four ones, so the result will be analyzed easier.

For the further comparison we include the results of modeling for the case when $H = 0°$; $S = L = 0.11$ are balanced (see Fig. 6). The figure shows only final RGB pictures (top parts) and cells' chromatic phases (bottom parts).

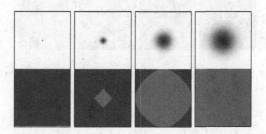

**Fig. 6.** 0th, 10th, 70th and 160th steps of red point's spot propagation for $H = 0°$ (Color figure online)

The slight increase of $D$'s chromatic phase leads to an appearance of colors, that are relatively close to a red color in hue (these are orange, yellow and green - chromatic phases $30°, 60°$ and $120°$ respectfully). New colors form thick and blurry rings in hue quad (bottom right one). Modeling results can be seen at Fig. 7.

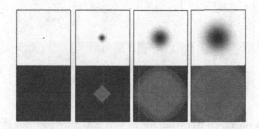

**Fig. 7.** 0th, 10th, 70th and 160th steps of red point's spot propagation for $H = 5°$ (Color figure online)

Higher chromatic number's phase gives us a very different result. A simple red point gradually transforms into a spot, that contains all colors (see Fig. 8).

Note that the color of area in the center of spreading spot doesn't change with the lapse of time, but higher $D_{chr}$'s phases will cause such effect.

## 5  $D$'s Chromatic Component and Non-Euclidean Geometry for a Chromatic Complex Plane

Even more unpredictable and complicated results will appear if we use Minkowski ($i^2 = +1$) or Galilean ($i^2 = 0$) geometry for a complex plane, that

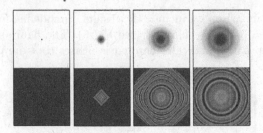

**Fig. 8.** 0th, 10th, 70th and 160th steps of red point's spot propagation for $H = 60°$ (Color figure online)

a chromatic number $D_{chr}$ belongs to. The formulas linking complex number's absolute value $(S)$ and phase $(H)$ with its real $(X_{chr})$ and imaginary $(Y_{chr})$ parts change so our excitable media will react differently. Formulas, expressions and geometric interpretations of so-called generalized complex numbers $(i^2 = k, k = -1, 0, +1)$, that were used for modeling, are represented in [7].

On Fig. 9 you can see, what the propagation process will look like for the same initial conditions, as in the previous section of our work but with Galilean geometry used for a complex chromatic plane.

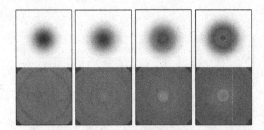

**Fig. 9.** Media states on 128th iteration for $H = 5°, 20°, 40°, 60°$ for $i^2 = 0$ (Color figure online)

Note that now we have a fixed iteration number and only $D$'s hue is changing. As we can see on Fig. 10, the further increase of diffusion coefficient's chromatic phase leads to a fast concentration and contraction of phase circle in the middle of a bottom right square. Also our red colored initial point completely turns into spot with pearl halo when $D$'s hue reaches 90°.

Also it should be mentioned that $arg\{D_{chr}\}$'s values greater than 90° don't produce any new phenomenons because of a consistent pattern, that was detected during the experiments. If we designate $arg\{D_{chr}\} = \phi_c hr$ and $\phi_c hr > 90°$ then the results of modeling will show resemblance to the output for $arg\{D_{chr}\} = \phi_{chr} - 90°$ but with inverted hue (chromatic phase). It can be seen on Fig. 11 - it is very similar to Figs. 10 and 9 taken in the reverse order.

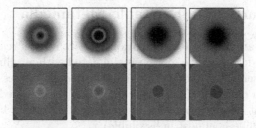

**Fig. 10.** Media states on 128th iteration for $H = 70°, 80°, 89°, 90°$ for $i^2 = 0$ (Color figure online)

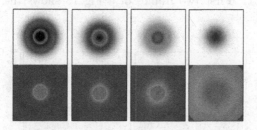

**Fig. 11.** Media states on 128th iteration for $H = 100°, 110°, 130°, 175°$ for $i^2 = 0$ (Color figure online)

The usage of Minkowski geometry ($i^2 = +1$) gives more rough results even for little $D_{chr}$'s phase values (see Fig. 12). Higher chromatic phases generate lots of cells with extremely big or, on the opposite, small absolute values in this particular geometry - these cases are quite hard to analyze.

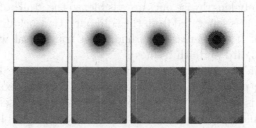

**Fig. 12.** Media states on 128th iteration for $H = 1°, 5°, 45°, 47°$ for $i^2 = +1$ (Color figure online)

It can be seen that in Minkowski geometry initial red point spawns a spot of orange colored (corresponding hue is approximately 55°) cells when $D$'s chromatic phase is being increased, whereas Galilean geometry generated a pearl spot ($hue \approx 180°$) instead. Also there is a similarity with Galilean geometry case: we can see a filled circle of equal chromatic phases around the place of our initial excited point.

# 6   Moving Particle in Excitable Metamedia

Next experiments are aimed at examination of metamedia's reaction on moving point (particle model). Particle motion is simulated with a cellular automata by continual cell excitement according to circular trajectory expression. Excited cell is being painted in red color $((R, G, B) = (255, 0, 0))$. Simulated particle's speed is adjusted so that distance between two consecutive points becomes minimal (actual trajectory line is almost indissoluble).

At first it is logical to get results for the case with an ordinary Euclidean geometry on the chromatic complex plane ($i^2 = -1$) and with relatively low hue component for $D_{chr}$. First two frames of a Fig. 13 shows the output for such metamedia conditions.

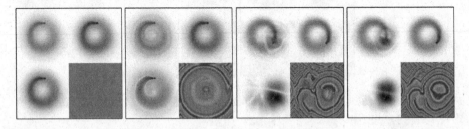

**Fig. 13.** Metamedia states for particle motion simulation on 160th iteration. $i^2 = -1$, $arg\{D_{chr}\} = H = 5°$, $H = 60°$, $H = 74°$ and $H = 85°$ from left to right (Color figure online)

There are no transversal color phase fluctuations on the first two pictures from the Fig. 13, only lengthwise ones are visible, so the "tail" of moving particle is indissoluble, but it is separated from the trail part, that belongs to the previous period of rotation. If we increase $D$'s hue, then transversal wave process will appear (see Fig. 13, 3rd and 4th frames).

Completely different results can be obtained if we replace Euclidean geometry on chromatic plane with different geometry type. For example, if Galileo geometry ($i^2 = 0$) is used then $arg\{D_{chr}\}$'s alteration will cause some interesting and even more unusual consequences, that are shown on Fig. 14. Only relatively high $D_{chr}$'s phases are represented because the changes of the output are quite smooth and there are no sharp changes while increasing hue from 0°.

It can be seen that for the case with Galilean geometry the growth of $D_{chr}$'s phase leads to increase of pearl and purple colors amount (on condition that every excited cell is initially red). Particle trail's halo on the right picture of Fig. 14 is quite bright, but there are no cells with big lightness and saturation values in the corresponding area. It is caused by irregular laws of chromatic component behavior for this geometry type.

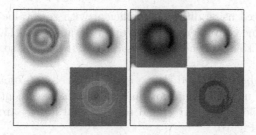

**Fig. 14.** Particle motion, 100th iteration. $i^2 = 0$, $H = 70°$ (left) and $H = 90°$ (right) (Color figure online)

## 7    The Interference in Non-Euclidean Geometries for a Chromatic Plane

The process of interference for an ordinary Euclidean geometry for complex chromatic numbers was already analyzed in the section where the luminance and saturation components of a triplet coefficient $D$ and their role were described. Now it is important to try different non-Euclidean geometries. Let the initial conditions consist of four points of different colors (just as in the first experiments). Interference simulation results for Minkowski and Galilean geometries are represented on Fig. 15.

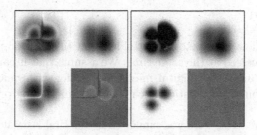

**Fig. 15.** Interference, 128th iteration, $H = 50° = const.$ $i^2 = 0$ (left) and $i^2 = 1$ (right) (Color figure online)

It can be seen that in Galilean geometry the collision of different colors produces unusual rays in the areas where occlusion happened. Minkowski and Euclidean geometry don't have such feature, but in case with $i^2 = +1$ different colors turn into the same ones after a few iterations, but these final similar colors also don't mix.

## 8    Basic Triplet Diffusion Applications

The simplest and obvious usage of triplet diffusion in digital image processing is a component-wise blurring (see Fig. 16). Chromatic or achromatic part of an image can be altered using diffusion coefficient with zero value of luminance or saturation and some convenient value of other parameter (0.11, for example).

We don't want to distort any color shades of our picture, so $arg\{D_{chr}\}$ must be equal to zero.

Also it should be mentioned that a quad with cells' chromatic phase representation provides us with an approximate image color segmentation model. All shades of a red color, that correspond to an actual flower, are separated from another colors on Fig. 16.

Initial state          Blurred chromatic component     Blurred achromatic component

**Fig. 16.** Component-wise blurring of a flower picture (Color figure online)

## 9    Conclusion

Triplet numbers allow us to take into consideration all color components of image that are being processed at the same time. We assume that it can be used in new more effective techniques of digital image processing with quantum approach. Non-Euclidean geometries drastically change the functioning process.

**Acknowledgement.** This work was supported by the Ural Federal University's Center of Excellence in "Quantum and Video Information Technologies: from Computer Vision to Video Analytics" (according to the Act 211 Government of the Russian Federation, contract 02.A03.21.0006)

## References

1. Cardenas-Barrera, J., Plataniotis, K.: QCA implementation of a multichannel filter for image processing **8**, 87–99 (2002)
2. Rosin, P., Adamatzky, A., Sun, X., Automata, C.: Cellular Automata in Image Processing and Geometry, 304 pages. Springer, Cham (2014). pp. 65–80
3. Labunets, V., Maidan, A.: Colour triplet-valued wavelets and splines. In: Image and Signal Processing and Analysis, ISpPA, p. 1. (2001)
4. Ibraheem, N.A., Hasan, M.M.: Understanding color models a review. ARPN J. Sci. Technol. **2**(3), 5–7 (2012)

5. Watts, P.: Working with RGB and HSL color coding systems in SAS software. In: Proceedings of the Twenty-Eighth SAS Users Group International Conference paper 136–28 (2003)
6. Visvanathan, A., Relchenbach, S.E.: Gradient-based value mapping for pseudocolor images. J. Electron. Imaging **16**(3), 033004 (2007)
7. Harkin, A.A., Harkin, J.B.: Geometry of generalized complex numbers. Math. Mag. **77**(2), 118–129 (2004). Mathematic association of America, p. 2

# An Efficient Algorithm for Total Variation Denoising

Artyom Makovetskii[1(✉)], Sergei Voronin[1], and Vitaly Kober[2]

[1] Chelyabinsk State University, Chelyabinsk, Russia
artemmac@mail.ru, voron@csu.ru
[2] Department of Computer Science, CICESE, Ensenada, B.C., Mexico
vkober@cicese.mx

**Abstract.** One-dimensional total variation (TV) regularization can be used for signal denoising. We consider one-dimensional signals distorted by additive white Gaussian noise. TV regularization minimizes a functional consisting of the sum of fidelity and regularization terms. We derive exact solutions to one-dimensional TV regularization problem that help us to recover signals with the proposed algorithm. The proposed approach to finding exact solutions has a clear geometrical meaning. Computer simulation results are provided to illustrate the performance of the proposed algorithm for signal denoising.

**Keywords:** Total variation · Signal denoising · Variational functional

## 1 Introduction

Total variation (TV) regularization has important applications in signal processing including denoising, deblurring, and restoration. A significant challenge in the practical use of TV regularization lies in non-differentiable convex optimization, which is difficult to solve especially for large-scale problems. The presence of noise in signals is unavoidable. To recover original signals, many noise reduction techniques have been developed to reduce or remove the noise. Noisy signals usually have high total variation. Several total variation regularization approaches have been developed to exploit the special properties of noisy signals, and they have been widely used in noise reduction in signal processing.

One of the most known techniques for denoising of noisy signals was proposed by Rudin, Osher, and Fatemi [1]. Let $J(u)$ be the following functional in the functional space $L_2$:

$$J(u) = \|u - u_0\|_{L_2}^2 + \lambda TV(u), \tag{1}$$

where $\| u - u_0 \|_{L_2}^2$ is called a fidelity term and $\lambda TV(u)$ is referred to as a regularization term. Here $u_0$ is the observed signal distorted by additive noise $n$,

© Springer International Publishing AG 2017
D.I. Ignatov et al. (Eds.): AIST 2016, CCIS 661, pp. 326–337, 2017.
DOI: 10.1007/978-3-319-52920-2_30

$$u_0 = v + n. \tag{2}$$

Consider the following variational problem:

$$u_* = \underset{u \in BV(\Omega)}{arg\,min}\, J(u), \tag{3}$$

where $u_*$ is extremal function for $J(u)$.

Numerical results have shown that TV regularization is quite useful in image restoration [2–4]. Due to non-smoothness of the TV norm, solving the large-scale TV problem is a challenging issue despite its simple form. In the past, considerable efforts have been devoted to develop efficient and scalable algorithms for TV problems. The one-dimensional total variation (1D TV), also known as the fused signal approximator, has been widely used in signal noise reduction. It has been shown to be very efficient in practice, though the convergence rate has not been established. Barabero and Sra [5] introduced a fast Newton-type method for 1D total variation regularization, and solved the 2D total variation problem using the Dykstra's method [6]. Wahlberg et al. [7] proposed the ADMM method to solve the 1D total variation problem. In this case, a linear system of equations is solved at each iteration. A dual-based approach to solve the 2D total variation problem is introduced in [8]. The 1D TV regularization can be used for signal restoration as well as image restoration in the case of anisotropic 2D TV regularization [9].

Here we consider the 1D TV regularization in framework of Strong and Chan [10, 11]. They considered the behavior of explicit solutions to the 1D TV problem when the parameter $\lambda$ in (1) is sufficiently small. In this paper, explicit solutions to (3) for any parameter $\lambda$ are analyzed. Basically, the problem is widely studied in the theory of signal and image processing. Recently, Condat [12] and Davies, Kovac [13] proposed a direct fast algorithm for 1D TV problem for discrete functions. The Condat's algorithm is very fast and has complexity of $O(n)$ for typical discrete functions. In contrast, the proposed approach to finding exact solutions has a clear geometrical meaning. This means that topological characteristics of a function of two variables [14, 15] can be considered for image denosing.

## 2 Partitions of the Bounded Variation Functions

A bounded variation (BV) function can be represented as a difference of two monotonic functions. Therefore, the points of discontinuity of the BV class are of the first kind. Let the function $u$ belongs to the BV class and has discontinuity in the point $x_d$:

$$C_{\pm} = \lim_{x \to x_d \pm 0} u(x). \tag{4}$$

If the value $u(x_d)$ is such that

$$u(x_d) \notin [C_-; C_+], \tag{5}$$

then by replacing $u(x_d)$ by either $C_-$ or $C_+$ we obtain a function with the same value of the fidelity term and with the smaller value of the regularization term in (1). Therefore, the solution to (3) in the BV space can be found in the space $BV_0$ of bounded variation functions with the condition $u(x_d) \in [C_-; C_+]$ for each discontinuity of the point $x_d$. Thus, the solution to (3) in the class BV is reduced to the solution of the problem in the class $BV_0$.

Let

$$a = r_0 < r_1 < \cdots < r_n = b \tag{6}$$

be a partition of the line segment $[a; b]$, $\Delta_i$ be the segment $[r_{i-1}; r_i]$, $i = 1, \ldots, n$. Let the partition $\delta = \{\Delta_i\}$ be a set of such segments.

The function $u$ is called piecewise constant, i.e. $u \in PC(\Omega)$, if $u(x) = u^i$ for $x \in \Delta_i$. Here $u^i$ is a constant.

Now we describe a reduction of the 1D TV problem from the $BV_0$ class to the $PC$ class. Let $PC(\delta)$ be a piecewise constant function on the segment $\Omega = [a; b]$ with the partition $\delta$.

**Theorem 1.** Suppose that $u_0 \in PC(\delta)$, then there is a function $\tilde{u} \in PC(\delta)$ such that $J(\tilde{u}) \leq J(u)$ for any $u \in BV_0$.

**Proof.** Let $\Delta_i$ be a segment of the partition $\delta$. Because belonging of the point $r_i$ to the segment $\Delta_i$ or segment $\Delta_{i+1}$ does not change the functional $J(u)$, we suppose that $\Delta_i = [r_{i-1}; r_i]$. Since $u \in BV_0$, then

$$\lim_{x \to r_{i-1}+0} u(x) = u(r_{i-1}) \text{ and } \lim_{x \to r_i-0} u(x) = u(r_i). \tag{7}$$

Suppose that the function $u(x)$ is not constant on the segment $\Delta_i$. Let us consider a function $\tilde{u}(x)$ that coincides with the function $u(x)$ outside the segment $\Delta_i$ and equal to its average value on $\Delta_i$,

$$C_i = \frac{1}{|\Delta_i|} \int_{\Delta_i} u(x) \, dx. \tag{8}$$

Then we obtain

$$
\begin{aligned}
||u - u_0||^2 - ||\tilde{u} - u_0||^2 &= \int_{\Delta_i} (u - u_0)^2 dx - \int_{\Delta_i} (\tilde{u} - u_0)^2 dx \\
&= \int_{\Delta_i} (u - u_0)^2 dx - \int_{\Delta_i} (C_i - u_0)^2 dx \\
&= \int_{\Delta_i} (u^2 - 2uu_0 + u_0^2) dx - \int_{\Delta_i} (C_i^2 - 2C_i u_0 + u_0^2) dx
\end{aligned}
$$

$$= \int_{\Delta_i} (u^2 - C_i^2 - 2uu_0 + 2C_iu_0)dx = \int_{\Delta_i} (u^2 - C_i^2 - 2u_0(u - C_i))dx$$

$$= \int_{\Delta_i} ((u - C_i)^2 - 2C_i^2 + 2uC_i - 2u_0(u - C_i))dx$$

$$= \int_{\Delta_i} ((u - C_i)^2 - 2C_i(C_i - u) - 2u_0(u - C_i))dx. \tag{9}$$

Since

$$\int_{\Delta_i} 2C_i(C_i - u)dx = 2C_i \int_{\Delta_i} (C_i - u)dx$$

$$= 2C_i(C_i|\Delta_i| - \int_{\Delta_i} u \, dx) = 2C_i(0) = 0, \tag{10}$$

and

$$\int_{\Delta_i} 2u_0 \, (u - C_i) \, dx = 0, \tag{11}$$

(9) can be rewritten in the following way:

$$\int_{\Delta_i} ((u - C_i)^2 - 2C_i(C_i - u) - 2u_0(u - C_i))dx = \int_{\Delta_i} (u - C_i)^2 dx \geq 0. \tag{12}$$

Finally, we get

$$||u - u_0||^2 \geq ||\tilde{u} - u_0||^2. \tag{13}$$

If the function $u$ in the interval $\Delta_i$ is not constant, then the inequality is strictly positive. The value of $TV(u)$ can be represented as the sum of the total variations of $u$ on the intervals $[a; r_{i-1})$, $[r_{i-1}; r_i)$ and $[r_i; b]$ (the same for $TV(\tilde{u})$). $TV(\tilde{u})$ coincides with $TV(u)$ outside the interval $[r_{i-1}; r_i)$, and $TV(u) \geq TV(\tilde{u})$ on $[r_{i-1}; r_i)$ because the function $u$ takes values greater and smaller than the average value of $C_i$.  ∎

Note that if the solution to the 1D TV problem exists, then $\arg \min J(u) \in PC(\delta)$ for $u \in BV(\Omega)$. Moreover, the existence of the minimum of the functional $J(u)$ follows from its strict convexity on the space $BV(\Omega)$ and from the general theory of convex analysis.

## 3 Formulation of TV Regularization as Finite-Dimensional Problem

By using Theorem 1, the problem in (3) in the space $BV_0$ can be reduced to the problem in the space $PC(\delta)$. Let a function $u$ belongs to $PC(\delta)$. It means that $u(x) = u^i, x \in \Delta_i, i = 1, \ldots, n$, i.e. $u(x) = \{u^1, \ldots, u^n\}$. Therefore, the functional space $PC(\delta)$ can be identified with the space $\Re^n$. For functions from $PC(\delta)$, the expression (1) takes the following form:

$$J(u) = \sum_{i=1}^{n} \left(u^i - u_0^i\right)^2 |\Delta_i| + \lambda \sum_{i=1}^{n-1} \left|u^{i+1} - u^i\right|. \tag{14}$$

The problem can be rewritten as

$$J(u) = ||u - u_0||^2 + \lambda TV(u) \to \min_{u \in PC(\delta)}, \tag{15}$$

where $||u||^2 = \sum_{i=1}^{n} (u^i)^2 |\Delta_i|$ and $TV(u) = \sum_{i=1}^{n-1} |u^{i+1} - u^i|$.

Since $||u - u_0||^2$ is strictly convex and the functional $TV(u)$ is convex, there exists a unique solution to (15).

Suppose that the function $u(x) = \{u^1, \ldots, u^n\}$ satisfies the following condition:

$$u^i \neq u^{i+1}, \ i = 1, \ldots, n - 1. \tag{16}$$

Let us define a vector $\varepsilon(u) = (\varepsilon^1, \ldots, \varepsilon^{n-1})$ for the function $u(x) = \{u^1, \ldots, u^n\}$ as follows:

$$\varepsilon^i = \begin{cases} 1, u^i < u^{i+1} \\ -1, u^i > u^{i+1} \end{cases}. \tag{17}$$

Then total variation of the function $u(x)$ can be expressed with the vector $\varepsilon(u)$,

$$TV(u) = \sum_{i=1}^{n-1} \left|u^{i+1} - u^i\right| = \sum_{i=1}^{n-1} \varepsilon^i(u^{i+1} - u^i). \tag{18}$$

One can observe that each component $u^i$ of the vector $u(x)$ is included in $TV(u)$ with the coefficient $e^i = +1 (e^i = -1)$, if $u^i$ is boundary maximum (boundary minimum); $e^i = +2 (e^i = -2)$, if $u^i$ is local maximum (local minimum) and $e^i = 0$, if $u^i$ is a "step" region of the function $u(x)$,

$$TV(u) = \sum\nolimits_{i=1}^{n} e^i u^i, \tag{19}$$

where the vector $e(u) = (e^1, \ldots, e^n)$.

Let $e_0$ be a vector of the initial function $u_0$, $u_*$ be a solution to (19), $e_*$ be a vector of the function $u_*$.

**Theorem 2.** If $e_0^i \geq 0$ ($e_0^i \leq 0$), then $e_*^i \geq 0$ ($e_*^i \leq 0$), $i = 1, \ldots, n$.

**Proof.** Suppose that $e_0^i \geq 0$ but $e_*^i < 0$. It means that $u_0^i \geq u_0^{i+1}$ and $u_*^i < u_*^{i+1}$. If $u_*^{i+1} > u_0^{i+1}$ then we reduce a little bit the value $u_*^{i+1}$. Then the fidelity term decreases and the $TV$ also decreases if $u_*^{i+2} < u_*^{i+1}$, otherwise the $TV$ does not change if $u_*^{i+2} > u_*^{i+1}$. If $u_*^{i+1} < u_0^{i+1}$ then we increase slightly the value $u_*^i$. Then the fidelity term decreases and the $TV$ also decreases if $u_*^{i-1} > u_*^i$, otherwise the $TV$ does not change if $u_*^{i-1} < u_*^i$. ∎

## 4  Reduction of TV Regularization Problem

Let $\delta_0$ be a partition of the segment $[a; b]$ generated by the function $u_0$. Let a function $u_1$ be a solution of the following variational problem:

$$\underset{u \in PC(\delta_0)}{\arg\min} \sum\nolimits_{i=1}^{n} \left(u^i - u_0^i\right)^2 |\Delta_i| + \lambda_1 \sum\nolimits_{i=1}^{n-1} \left|u^{i+1} - u^i\right|, \tag{20}$$

where $n$ is the number of segments in the partition $\delta_0$ and $\lambda_1$ is the value corresponding to the first meeting of the two neighboring segments. Let $\delta_1$ be a partition of the segment $[a; b]$ generated by the function $u_1$. Let a function $u_\varepsilon$ be a solution of the following problem:

$$\underset{u \in PC(\delta_1)}{\arg\min} \sum\nolimits_{i=1}^{k} \left(u^i - u_1^i\right)^2 |\Delta_i| + \lambda_e \sum\nolimits_{i=1}^{k-1} \left|u^{i+1} - u^i\right|. \tag{21}$$

where $k$ is the number of segments in the partition $\delta_1$ and $\lambda_\varepsilon$ is sufficiently small. We can rewrite (20) in the following form:

$$\underset{u \in PC(\delta_0)}{\arg\min} \sum\nolimits_{i=1}^{n} \left(u^i - u_0^i\right)^2 |\Delta_i| + \lambda_1 \sum\nolimits_{i=1}^{n} e_0^i u^i. \tag{22}$$

With the help of Theorem 2, (21) takes the following form:

$$\underset{u \in PC(\delta_0)}{\arg\min} \sum\nolimits_{i=1}^{n} \left(u^i - u_1^i\right)^2 |\Delta_i| + \lambda_\varepsilon \sum\nolimits_{i=1}^{n} e_1^i u^i. \tag{23}$$

Consider the following variational problem:

$$\arg\min_{u \,\in\, PC(\delta_0)} \quad \sum\nolimits_{i=1}^{n} \left(u^i - u_0^i\right)^2 |\Delta_i| + (\lambda_1 + \lambda_\varepsilon) \sum\nolimits_{i=1}^{n-1} \left|u^{i+1} - u^i\right|. \tag{24}$$

So, (24) can be written as

$$\arg\min_{u \,\in\, PC(\delta_0)} \quad \sum\nolimits_{i=1}^{n} \left(u^i - u_0^i\right)^2 |\Delta_i| + (\lambda_1 + \lambda_\varepsilon) \sum\nolimits_{i=1}^{n} e_0^i u^i. \tag{25}$$

**Theorem 3.** The solution $u_\varepsilon$ of the variational problem in Eq. (23) is also a solution in (25).

**Proof.**

$$\sum\nolimits_{i=1}^{n} \left(u^i - u_0^i\right)^2 |\Delta_i| + (\lambda_1 + \lambda_e) \sum\nolimits_{i=1}^{n} e_0^i u^i$$
$$= \sum\nolimits_{i=1}^{n} \left(u^i - u_0^i\right)^2 |\Delta_i| + \pi_1 \sum\nolimits_{i=1}^{n} e_0^i u^i + \pi_e \sum\nolimits_{i=1}^{n} e_0^i u^i$$
$$= \sum\nolimits_{i=1}^{n} \left((u^i - u_1^i) + (u_1^i - u_0^i)\right)^2 |\Delta_i| + \pi_1 \sum\nolimits_{i=1}^{n} e_0^i u^i + \pi_e \sum\nolimits_{i=1}^{n} e_0^i u^i$$
$$= \sum\nolimits_{i=1}^{n} \left(u^i - u_1^i\right)^2 |\Delta_i| + 2 \sum\nolimits_{i=1}^{n} \left(u^i - u_1^i\right)\left(u_1^i - u_0^i\right)|\Delta_i| + \sum\nolimits_{i=1}^{n} \left(u_1^i - u_0^i\right)^2 |\Delta_i| \tag{26}$$
$$+ \pi_1 \sum\nolimits_{i=1}^{n} e_0^i u^i + \pi_e \sum\nolimits_{i=1}^{n} e_0^i u^i$$
$$= \sum\nolimits_{i=1}^{n} \left(u^i - u_1^i\right)^2 |\Delta_i| + \pi_e \sum\nolimits_{i=1}^{n} e_0^i u^i + 2 \sum\nolimits_{i=1}^{n} u^i \left(u_1^i - u_0^i\right)|\Delta_i| + \pi_1 \sum\nolimits_{i=1}^{n} e_0^i u^i$$
$$+ 2 \sum\nolimits_{i=1}^{n} u_0^i \left(u_1^i - u_0^i\right)|\Delta_i| + \sum\nolimits_{i=1}^{n} \left(u_1^i - u_0^i\right)^2 |\Delta_i|.$$

From Theorem 2 follows

$$\begin{aligned}
&\sum\nolimits_{i=1}^{n} \left(u^i - u_1^i\right)^2 |\Delta_i| + \pi_e \sum\nolimits_{i=1}^{n} e_0^i u^i \\
&= \sum\nolimits_{i=1}^{n} \left(u^i - u_1^i\right)^2 |\Delta_i| + \pi_e \sum\nolimits_{i=1}^{n} e_1^i u^i.
\end{aligned} \tag{27}$$

The following terms are constant with respect to $u$:

$$2 \sum\nolimits_{i=1}^{n} u_0^i \left(u_1^i - u_0^i\right)|\Delta_i| + \sum\nolimits_{i=1}^{n} \left(u_1^i - u_0^i\right)^2 |\Delta_i|. \tag{28}$$

$$\begin{aligned}
2 \sum\nolimits_{i=1}^{n} u^i \left(u_1^i - u_0^i\right)|\Delta_i| + \pi_1 \sum\nolimits_{i=1}^{n} e_0^i u^i &= 2 \left( \sum\nolimits_{i=1}^{n} u^i \left(u_1^i - u_0^i\right)|\Delta_i| + \frac{\pi_1}{2} e_0^i u^i \right) \\
&= 2 \left( \sum\nolimits_{i=1}^{n} u^i \left((u_1^i - u_0^i)|\Delta_i| + \frac{\pi_1}{2} e_0^i \right) \right)
\end{aligned} \tag{29}$$

$$u_1^i = u_0^i - \frac{\pi_1}{2} \frac{e_0^i}{|\Delta_i|} \tag{30}$$

$$\sum\nolimits_{i=1}^{n} u^i \left(u_1^i - u_0^i\right)|\Delta_i| + \frac{\pi_1}{2} e_0^i = 0 \tag{31}$$

It means that $u_e$ is a solution to (25). ∎

# 5  Properties of Exact Solutions of 1D TV Problem and the Proposed Algorithm

Let $u_*$ be a solution to the variational problem in (15) for a given value of the parameter $\pi$. Divide the segment [a, b] into intervals with a constant sign of the function $u_0 - u_*$. We denote $S_+$ $(S_-)$ a value of the area under the plot of the function $u_0 - u_*$ in intervals where $u_0 - u_* > 0$ $(u_0 - u_* < 0)$. The properties of exact solutions of the problem follow from the proved statements:

1. For any value of the parameter $\lambda$:

$$S_+ = S_-. \tag{32}$$

2. The area under the plot of the function $u_*$ is equal to that of the function $u_0$ for any parameter $\lambda$.

Using the proved theorems, we can design a new algorithm that finds a solution of the 1D TV regularization for discrete functions. The complexity of the algorithm is $O(n^2)$.

Below we describe one of the possible implementations of the algorithm.

The input data are a discrete function $u_0$, a number $N$ of points of the function $u_0$, a real number $\pi$.

The function $u_0$ is coded by the following way:

1. Let *count* be a number of segments of $u_0$. A segment here is a few neighboring points with the same values. If a few neighbor points have a small difference of values then it is possible to unite them to the segments.
2. The array $h = [h_0, \ldots, h_{count-1}]$, where $h_i$, $i = 0, \ldots, count - 1$ is a value of the relative segment.
3. The array $r = [r_0, \ldots, r_{count-2}]$, where $r_i$, $i = 0, \ldots, count - 2$ is a coordinate of the left vertex of the segment.

The values of the parameters *count* and $\pi$ will decreasing. A code $h$, $r$ and *count* is initial data for the outside cycle of the algorithm: while (*count* > 1 and $\pi > 0$).

In the main outside cycle there is the first inside cycle: for (by $i$ from zero to *count*). In this cycle is forming an array $k = [k_0, \ldots, k_{count-1}]$, where $k_i = \pm \frac{1}{2}$ if the segment is boundary maximum or boundary minimum, $k_i = -1$ if the segment is local minimum, $k_i = 0$ if the segment is the step, $k_i = 1$ if the segment is local maximum. At the each step of the cycle is computed a value $H = |h_{i-1} - h_i|$ for every point of discontinuity. Values of the lengths $\Delta_{i-1}$ and $\Delta_i$ of the adjacent for the point of discontinuity segments are computed by the array $r$. At the each step of the cycle is computed a value of the parameter $\pi_i = \frac{H\Delta_{i-1}\Delta_i}{|k_{i-1}|\Delta_i + |k_i|\Delta_{i-1}}$ and $\pi_{min} = \min\{\pi_0, \ldots, \pi_{count-2}\}$. The output data of the first interior cycle is $\pi_{min}$ and array $k$.

In the second interior cycle are computed new values of segments $h\_new_i = h_i - k_i \frac{\pi_{min}}{\Delta_i}$, a new value *count_new*, new arrays $h\_new$ and $r\_new$. After the second cycle the old arrays are changed by new arrays: $h = h\_new$, $r = r\_new$ and *count* = *count_new*.

Then is computed $\pi := \pi - \pi_{min}$ and are checked conditions of the outside cycle.

If the outside cycle continues, then begins the first interior cycle for updated encoding data.

If the outside cycle ends, then function is restored by the encoding data. This function is a result of the algorithm.

The performance of the algorithm is illustrated in the following figures.

Figure 1 shows an initial discrete function $v(t)$.

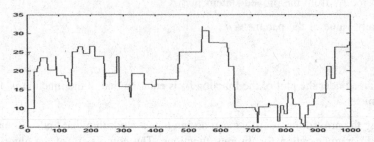

**Fig. 1.** Discrete function v(t).

Figure 2 shows a noisy function $u_0 = v + n$, where $n$ is a Gaussian noise with $\sigma = 4$.

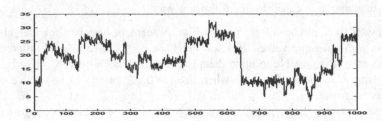

**Fig. 2.** Discrete function $u_0$.

Figure 3 shows a function $u_*$ that is the solution of the 1D TV regularization problem when $\pi = 4$.

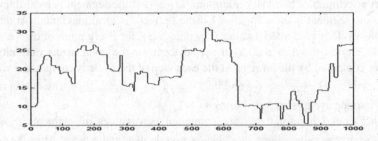

**Fig. 3.** Solution to the 1D TV regularization for $\lambda = 4$.

Figure 4 shows a function $u_*$ that is the solution to the 1D TV regularization for $\pi = 50$.

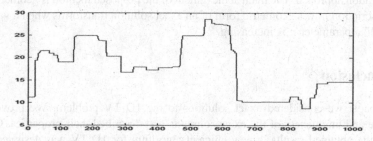

**Fig. 4.** Solution of the 1D TV regularization for $\lambda = 50$.

Figure 5 shows a function $u_*$ that is the solution of the 1D TV regularization for $\pi = 1000$.

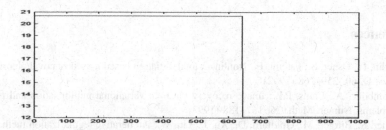

**Fig. 5.** Solution of the 1D TV regularization problem when $\lambda = 1000$.

Figure 6 shows a function $u_*$ that is the solution of the 1D TV regularization for $\pi = 5000$.

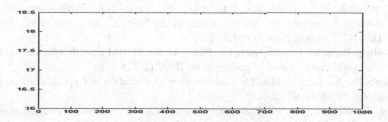

**Fig. 6.** Solution to the 1D TV regularization for $\lambda = 5000$.

The proposed mathematical approach gives an exact solution of the variational problem (15). Its computer implementation also has high precision. The Condat algorithm [12] also gives an exact solution of (15). The difference between output functions of our and Condat computer implementations is very small. For example, the difference between output functions of our and Condat computer programs for the initial function

$u_0$ on the Fig. 2 and $\pi = 4$ measured by MSE: $MSE = 5.094573910^{-13}$. Possibilities of the proposed algorithm for signal denoising are defined by general efficient of the TV regularization approach. The main achievement of the proposed method is geometrically clear description how a geometric form of an exact solution transforms when a value of the timelike parameter $\pi$ is increasing.

## 6 Conclusion

In this paper we considered exact solutions to the 1D TV problem. We proved the existence and uniqueness of the exact solutions and described their properties. On the base of the obtained results, a new efficient algorithm for 1D TV was designed. The obtained results are illustrated with the help of computer simulation.

**Acknowledgments.** The work was supported by the Ministry of Education and Science of Russian Federation (grant 2.1766.2014K).

## References

1. Rudin, L., Osher, S., Fatemi, E.: Nonlinear total variation based noised removal algorithms. Phys. D. **60**, 259–268 (1992)
2. Chambolle, A., Lions, P.L.: Image recovery via total variational minimization and related problems. Numer. Math. **76**, 167–188 (1997)
3. Osher, S., Burger, M., Goldfarb, D., Xu, J., Yin, W.: An iterative regularization method for total variation based image restoration. Multiscale Model. Simul. **4**, 460–489 (2005)
4. Chambolle, A.: An algorithm for total variation minimization and applications. J. Math. Imag. Vis. **20**, 89–97 (2004)
5. Barbero, A., Sra, S.: Fast newton-type methods for total variation regularization. In: ICML (2011)
6. Combettes, P., Pesquet, J.: Proximal splitting methods in signal processing. In: Bauschke, H. H., Burachik, R.S., Combettes, P.L., Elser, V., Luke, D.R., Wolkowicz, H. (eds.) Fixed-Point Algorithms for Inverse Problems in Science and Engineering, vol. 49, pp. 185–212. Springer, New York (2011)
7. Wahlberg, B., Boyd, S., Annergren, M., Wang, Y.: An ADMM algorithm for a class of total variation regularized estimation problems. In: IFAC (2012)
8. Chambolle, A.: An algorithm for total variation minimization and applications. J. Math. Imag. Vis. **20**(1), 89–97 (2004)
9. Yang, S., Wang, J., Fan, W., Zhang, X., Wonka, P., Ye, J.: An efficient ADMM algorithm for multidimensional anisotropic total variation regularization problems. In: Proceedings of the 19th ACM SIGKDD International Conference on Knowledge Discovery and Data Mining, pp. 641–649 (2013)
10. Strong, D.M., Chan, T.F.: Exact Solutions to Total Variation Regularization Problems, UCLA CAM Report (1996)
11. Strong, D.M., Chan, T.F.: Edge-preserving and scale-dependent properties of total variation regularization. Inverse Probl. **19**, 165–187 (2003)

12. Condat, L.: A direct algorithm for 1-D total variation denoising. IEEE Signal Process. Lett. **20**(11), 1054–1057 (2013)
13. Davies, P.L., Kovac, A.: Local extremes, runs, strings and multiresolution. Ann. Stat. **29**(1), 1–65 (2001)
14. Makovetskii, A. Kober, V.: Modified gradient descent method for image restoration. In: Proceedings of SPIE Applications of Digital Image Processing XXXVI, vol. 8856 (2013). 885608-1
15. Makovetskii, A., Kober, V.: Image restoration based on topological properties of functions of two variables. In: Proceedings of SPIE Applications of Digital Image Processing XXXV, vol. 8499 (2012). 84990A

# Classification of Dangerous Situations for Small Sample Size Problem in Maintenance Decision Support Systems

Vladimir R. Milov[1] and Andrey V. Savchenko[2(✉)]

[1] Nizhny Novgorod State Technical University n.a. R.E. Alekseev,
Nizhny Novgorod, Russia
vladimir.milov@gmail.com
[2] Laboratory of Algorithms and Technologies for Network Analysis,
National Research University Higher School of Economics,
Nizhny Novgorod, Russia
avsavchenko@hse.ru

**Abstract.** In this paper we examine the task of maintenance decision support in classification of the dangerous situations discovered by the monitoring system. This task is reduced to the contextual multi-armed bandit problem. We highlight the small sample size problem appeared in this task due to the rather rare failures. The novel algorithm based on the nearest neighbor search is proposed. An experimental study is provided for several synthetic datasets with the situations described by either simple features or grayscale images. It is shown, that our algorithm outperforms the well-known contextual multi-armed methods with the Upper Confidence Bound and softmax stochastic search strategies.

**Keywords:** Maintenance decision support systems · Classification · Small sample size problem · Nearest neighbor · Contextual multi-armed bandit

## 1 Introduction

Operations support systems perform fault analysis and monitor the state of such technical objects as gas pipes, mobile networks, etc. [1, 2]. Our experience in development and deployment of such systems in NetBoss XT and Giprogazcenter [1] has shown that the rules to discover dangerous situations are well known to the specialists in each particular area. A much more complex task is the classification of the discovered situations in order to provide a potential action, which can be used to eliminate the dangerous situation. For instance, a gas tube monitoring system can capture the video of the most vulnerable sections of the tube. The maintenance type depends on the current state of the discovered situation, i.e., the dent or the split of the tube. In this classification task it is possible to use the known failures of the observed object. The preliminarily available training data is usually absent, so online learning should be applied [3]. Moreover, the correct action in the available historic data is usually unknown, and the result of recommended action (positive or negative) is only

© Springer International Publishing AG 2017
D.I. Ignatov et al. (Eds.): AIST 2016, CCIS 661, pp. 338–345, 2017.
DOI: 10.1007/978-3-319-52920-2_31

available. Hence, this task is slightly different from the conventional classification problem, in which the labels of all instances in a training set should be given. Finally, the small sample size (SSS) problem [4] is appeared in this task, because the serious failures are quite rare events.

The purpose of our research is to develop an online classification algorithm, which can be successfully used for SSS problem. The rest of the paper is organized as follows. In Sect. 2 we recall several methods for contextual multi-armed bandit problem [5] and propose the novel algorithm based on the nearest neighbor (NN) rule [4]. In Sect. 3 experimental study results are presented for two synthetic datasets, namely, a set of grayscale images and the simple Bayesian network of features generation. Finally, concluding comments are given in Sect. 4.

## 2 Nearest Neighbor Algorithm in the Contextual Multi-armed Bandit Problem

Let the dangerous situation discovered by the maintenance system [2] be characterized by the feature vector $\mathbf{x}$ with dimensionality $M$. It is required to assign the input observation $\mathbf{x}$ to one of $A \geq 2$ potential repair operations (or *actions*). We assume that all possible actions are independent, and the training set is available $X_N = \{(\mathbf{x}_n, a_n, r_n)\}, n = \overline{1, N}$, where $r_n = 1$ if the recommended action $a_n \in \{1, \ldots, A\}$ successfully resolved the $n$-th dangerous situation described with the feature vector $\mathbf{x}_n$, and $r_n = 0$ otherwise. The preliminarily set of $N_0 \geq 0$ training instances with the known correct actions should be provided by a decision maker. Other instances are added to the training set of an online learning system while discovering the new dangerous situation $\mathbf{x}$. Namely, the decision maker estimates the result $r \in \{0, 1\}$ of an action $a$ recommended by the system, and the tuple $(\mathbf{x}, a, r)$ is added to the historic data as the $(N + 1)$-th instance [3]. This task is rather different from the traditional classification problem [4], in which the correct class label of each reference instance is known. The presence of the negative samples ($r_n = 0$) with unknown correct labels makes the task more complicated. One of the most reasonable ways to solve this task is to use the known contextual multi-armed bandit methods [5] with the *context* $\mathbf{x}$ and two-stage iterative "exploitation-exploration" procedure [6]. At first, *exploitation*, stage an optimal decision (an *arm* of the bandit) is obtained for available historic data with the principle of maximum expected utility [6]:

$$a^* = \underset{a \in \{1, \ldots, A\}}{argmax} \; \bar{r}(a, \mathbf{x}), \tag{1}$$

where $\bar{r}(a, \mathbf{x}) = M\{r|a, \mathbf{x}\}$ is an expected *reward* of the action $a$ for the situation $\mathbf{x}$. This expectation is equal to the probability $p(r = 1|a, \mathbf{x})$ for the binary reward. This probability can be estimated with the probabilistic neural network (PNN) [7]

$$\hat{p}(r = 1|a, \mathbf{x}) = \frac{\sum\limits_{\mathbf{x}_n \in X_N^+(a)} K(\mathbf{x}, \mathbf{x}_n)}{\sum\limits_{\mathbf{x}_n \in X_N^+(a)} K(\mathbf{x}, \mathbf{x}_n) + \sum\limits_{\mathbf{x}_n \in X_N^-(a)} K(\mathbf{x}, \mathbf{x}_n)}, \tag{2}$$

where $X_N^+(a) = \{\mathbf{x}_n|(\mathbf{x}_n, a, 1) \in X\}$ is a set of positive examples for the action $a$, $X_N^-(a) = \{\mathbf{x}_n|(\mathbf{x}_n, a, 0) \in X\}$ is the training set of negative instances, and $K(\mathbf{x}, \mathbf{x}_n)$ is the Gaussian Parzen kernel computed with Euclidean distance $\rho(\mathbf{x}, \mathbf{x}_n)$. At the second step the solution space is *explored* with the stochastic strategies. One of the known exploration strategies is the softmax rule, in which the action-selection probabilities are determined by ranking the expected reward (1), (2) using a Boltzmann distribution [6]. Another popular UCB (upper confidence bound) method optimistically chooses the action with the highest upper bound of an expected reward [5]. The confidence bound can be estimated by using the known distribution of the nonparametric Nadaraya-Watson kernel regression [7]. This regression is equivalent to the PNN (1), (2) in the case of the binary reward $r$.

Unfortunately, such an approach was designed for the samples, which size is large enough to estimate the features probability density (2) [6]. This method is not expected to provide the best accuracy for the SSS problem [4]. In such a case, we can use other pattern recognition methods suitable for small samples, e.g., the NN search. For instance, the following straightforward implementation of the k-NN method can be used. The sets $X_N^-(a)$ of negative examples of each action form the separate class, and the action cannot be recommended if $k$ negative instances from $X_N^-(a)$ are closer to the feature vector $\mathbf{x}$, than $k$ positive samples from $X_N^+(a)$.

However, such an approach is not appropriate, if the positive and negative training samples are imbalanced. Hence, we propose in this paper slightly different solution (Algorithm 1). Here the minimum distances $\rho(\mathbf{x}, \mathbf{x}_n)$ from the input observation $\mathbf{x}$ and sets $X_N^+(a)$ and $X_N^-(a)$ are made ordered, and we recommend the first action, for which the minimum distance $\rho^+(a)$ to positive set $X_N^+(a)$ is less than the minimum distance $\rho^-(a)$ to negative set $X_N^-(a)$.

---

**Algorithm 1.** The proposed NN-based algorithm for classification of dangerous situations.

---

**Data**: The dangerous situation $\mathbf{x}$, initial training set $X_N$

**Result**: Recommended action $a^*$, modified training set $X_{N+1}$

1.  Assign $a^* := 0$
2.  For each action $a \in \{1,...,A\}$

    2.1.  Assign $\rho^+(a) := \min\limits_{\mathbf{x}_n \in X_N^+(a)} \rho(\mathbf{x},\mathbf{x}_n)$, $\rho^-(a) := \min\limits_{\mathbf{x}_n \in X_N^-(a)} \rho(\mathbf{x},\mathbf{x}_n)$

3.  Obtain the sequence of actions $\{a_1,...,a_A\}$ for which $\rho^+(a_1) \leq ... \leq \rho^+(a_A)$.
4.  For each action $a$ in the sequence $\{a_1,...,a_A\}$

    4.1.  If $\rho^+(a) < \rho^-(a)$, then

        4.1.1.  Assign $a^* := a$
        4.1.2.  Break

5.  If $a^* = 0$, then

    5.1.  Assign $a^* := \mathrm{argmax}\limits_{a \in \{1,...,A\}} \rho^-(a)$

6.  Recommend an action $a^*$ and request its correctness $r$.
7.  Add new tuple $(\mathbf{x}, a^*, r)$ to the training set $X_N$

---

# 3 Experimental Results

In this section the proposed Algorithm 1 is experimentally compared with: (1) the k-NN described earlier; (2) the PNN (1), (2); and (3) the contextual multi-armed methods (UCB and softmax exploration strategies for the density (2)). The number of actions $A = 3$, the prior probability of every action is identical. The total size of the testing set $N$ is equal to 100 instances. Every observation from the test set is added to the training set (step 7 in the Algorithm 1) right after classification. Thus, the accuracy of every method is *increased* with an increase of the test set size $N$. It is expected that each situation is rather rare event. Hence, we varied the size of the initial set $N_0$ from 0 to 20, i.e., it is a very small training sample. The increase of the size $N_0$ of the training set should theoretically lead to the lower error rate.

In the first experiment we simulated the dangerous situations discovered at the gas tube cross sections (Fig. 1) [1]. The cross section in each state is represented as the ellipse with the nominal size 30 of both axes. A uniformly generated noise from the range $[-2; 2]$ is added independently to the size of each ax. The first defect (Fig. 1, left) is simulated with the small arc, which axes are uniformly generated from the range [13; 17]. The difference between starting and ending angles is equal to $180°$, and the uniformly generated noise from the range $[-10°; 10°]$ is added. The second defect (Fig. 1, center) is created similarly, but the dent is $180°$ rotated. The exfoliation in the third defect (Fig. 1, right) is represented as a line with the 30-pixel length. In the last

**Fig. 1.** Sample schematic images of the gas tube cross sections

case, the additive noise from the range $[-5; 5]$ is presented in each instance of this type of defect.

In the SSS case the modern deep neural networks trained with external dataset (e.g., ImageNet) [8] are useless in this task, because the differences in gas tube failures (Fig. 1) will be processed as a noise and ignored by such feature extractor. Moreover, the variability of the real gas tube images is usually to high to apply texture features. Thus, the features are extracted with the noise-resistant morphological algorithm [9] by using the following image processing stages [10]: adaptive thresholding, contour detection, combining the closed contours, search for the convex hull, flood filling of all contours, binary erosion with $5 \times 5$ circle, and Canny edge detection. The following features are extracted in the resulted shape: its area, perimeter, x and y coordinates of the gravity center, the convexity of this shape (Boolean flag), and an area of the result of morphological opening with the $28 \times 28$ circle. Hence, the feature vector **x** contains $M = 6$ features.

An experiment with the random generation of the training and test sets is repeated 1000 times, and the average error rate is estimated. The average error rates are presented in Fig. 2.

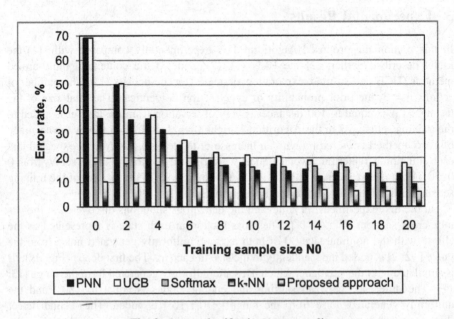

**Fig. 2.** Image classification error rate, %

Here the softmax exploration strategy allows achieving the best accuracy in comparison with other multi-armed bandit methods, if the training sample is relatively small ($N_0 \leq 6$). However, the simple maximum expected utility principle implemented with the PNN (1), (2) is more preferable in the case of $N_0 > 12$, as such size of the training sample is enough to estimate the class distributions quite accurately. The k-NN implementation outperforms contextual multi-armed bandit methods when $N_0 \geq 10$, but it is worth, than the softmax strategy for the very-small samples. Nevertheless, though the proposed Algorithm 1 is also based on the NN method, it is the best choice in all the cases. The gain in accuracy is rather high: the accuracy of our algorithm is 3–8% higher, than the accuracy of other best methods (k-NN and softmax strategy for the PNN (1), (2)).

In the second experiment the vector $\mathbf{x}$ contains one binary feature $x_1$ and one real feature, i.e. $x_2$. In the first two states these features are independent: $p(x_1 = 0) = 0.2$ and $x_2 \sim N(-0.5; 0.04)$ for the first action, and $p(x_1 = 0) = 0.8$ and $x_2 \sim N(1.5; 0.01)$ for the second action. For the third action the first feature is uniformly distributed, and if $x_1 = 1$, then $x_2 \sim N(0; 0.02)$, otherwise $x_2 \sim N(1; 0.04)$. An artificially generated noise with distribution $N(0; 0.005)$ was added to $x_2$, and the first feature is changed with the probability 0.2. The experimental results are shown in Fig. 3.

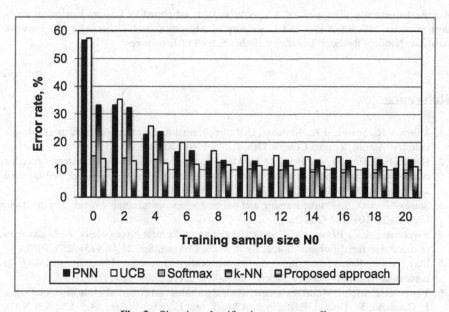

**Fig. 3.** Situation classification error rate, %

Here the proposed algorithm is 1–5% more accurate, than the softmax and the k-NN for the small training set ($N_0 \leq 8$). However, if $N_0$ is increased, the multi-armed bandit methods outperform the NN-based approaches in 1–2%, as the decision space in this experiment is quite simple and can be efficiently explored by the known exploration methods [6].

## 4   Conclusion and Future Work

In this paper we formulate the problem of classification of dangerous situations in terms of the contextual multi-armed bandit task. We proposed the novel adaptive algorithm on the basis of the NN method. It was experimentally demonstrated, that this algorithm is rather accurate even in the case of very small sample size [4], when the historic data of the monitoring system do not contain enough information about the serious failures. However, the known exploitation-exploration methods with softmax stochastic strategy [5, 6] outperform the NN-based approach in a long-term perspective with the increase of the size of available training set.

Thus, the main direction for further research is the creation of the hybrid techniques, which perform fusion of the NN search (Algorithm 1) and the conventional multi-armed bandit procedure. These techniques should depend on the size $N$, so that the NN search is preferred in the case of the small samples, but the multi-armed bandit algorithm is used when $N$ becomes large enough. Another important research direction is the application of our procedure to the real data gathered, for instance, by the recently deployed gas tube monitoring system.

**Acknowledgements.** The work of A.V. Savchenko is supported by Russian Federation President grant no. MD-306.2017.9 and Laboratory of Algorithms and Technologies for Network Analysis, National Research University Higher School of Economics.

## References

1. Milov, V.R., Suslov, B.A., Kryukov, O.V.: Intellectual management decision support in gas industry. Autom. Remote Control **72**(5), 1095–1101 (2011)
2. Huynh, K.T., Barros, A., Berenguer, C.: Maintenance decision-making for systems operating under indirect condition monitoring: value of online information and impact of measurement uncertainty. IEEE Trans. Reliab. **61**(2), 410–425 (2012)
3. Shalev-Shwartz, S.: Online learning and online convex optimization. Found. Trends Mach. Learn. **4**(2), 107–194 (2011)
4. Savchenko, A.V., Belova, N.S.: Statistical testing of segment homogeneity in classification of piecewise–regular objects. Int. J. Appl. Math. Comput. Sci. **25**(4), 915–925 (2015)
5. May, B.C., Korda, N., Lee, A., Leslie, D.S.: Optimistic Bayesian sampling in contextual-bandit problems. J. Mach. Learn. Res. **13**(1), 2069–2106 (2012)
6. Vermorel, J., Mohri, M.: Multi-armed bandit algorithms and empirical evaluation. In: Gama, J., Camacho, R., Brazdil, P.B., Jorge, A.M., Torgo, L. (eds.) ECML 2005. LNCS (LNAI), vol. 3720, pp. 437–448. Springer, Heidelberg (2005). doi:10.1007/11564096_42
7. Li, Q., Racine, J.S.: Nonparametric Econometrics: Theory and Practice. Princeton University Press, Princeton (2007)
8. Szegedy, C., Liu, W., Jia, Y., Sermanet, P., Reed, S., Anguelov, D., Erhan, D., Vanhoucke, V., Rabinovich, A.: Going deeper with convolutions. In: Proceedings of the IEEE Conference on Computer Vision and Pattern Recognition (CVPR), pp. 1–9 (2015)

9. Chernousov, V.O., Savchenko, A.V.: A fast mathematical morphological algorithm of video-based moving forklift truck detection in noisy environment. In: Ignatov, D.I., Khachay, M.Y., Panchenko, A., Konstantinova, N., Yavorskiy, Rostislav, E. (eds.) AIST 2014. CCIS, vol. 436, pp. 57–65. Springer, Heidelberg (2014). doi:10.1007/978-3-319-12580-0_5
10. Sonka, M., Hlavac, V., Boyle, R.: Image Processing, Analysis, and Machine Vision, 4th edn. Cengage Learning, Stamford (2014)

# Ortho-Unitary Transforms, Wavelets and Splines

Ekaterina Ostheimer[1(✉)], Valery Labunets[2], and Ivan Artemov[2]

[1] Capricat LLC, Pompano Beach, FL, USA
katya@capricat.com
[2] Ural Federal University, Yekaterinburg, Russia
vlabunets05@yahoo.com

**Abstract.** Here we present a new theoretical framework for multidimensional image processing using hypercomplex commutative algebras that codes color, multicolor and hypercolor. In this paper a family of discrete color–valued and multicolor–valued 2–D Fourier–like, wavelet–like transforms and splines has been presented (in the context of hypercomplex analysis). These transforms can be used in color, multicolor, and hyperspectral image processing. In our approach, each multichannel pixel is considered not as an K–D vector, but as an K–D hypercomplex number, where K is the number of different optical channels. Orthounitary transforms and splines are specific combination (Centaurus) of orthogonal and unitary transforms. We present several examples of possible Centuaruses (ortho–unitary transforms): Fourier+Walsh, Complex Walsh+Ordinary Walsh and so on. We collect basis functions of these transforms in the form of iconostas. These transforms are applicable to multichannel images with several components and are different from the classical Fourier transform in that they mix the channel components of the image. New multichannel transforms and splines generalize real–valued and complex–valued ones. They can be used for multichannel images compression, interpolation and edge detection from the point of view of hypercomplex commutative algebras. The main goal of the work is to show that hypercomplex algebras can be used to solve problems of multichannel (color, multicolor, and hyperspectral) image processing in a natural and effective manner.

**Keywords:** Ortho–unitary transforms · Wavelets · Splines · Image processing · Multichannel image

## 1 Introduction

We develop a conceptual framework and design methodologies for multichannel image processing systems with assessment capability. The term multichannel image is used for an image with more than one component. They are composed of series of images $f_{\lambda_0}(x,y), f_{\lambda_1}(x,y), ..., f_{\lambda_{K-1}}(x,y)$ in different optical bands at wavelengths $\lambda_0, \lambda_1, ..., \lambda_{K-1}$, called spectral channels, where $K$ is the number of different optical channels.

© Springer International Publishing AG 2017
D.I. Ignatov et al. (Eds.): AIST 2016, CCIS 661, pp. 346–356, 2017.
DOI: 10.1007/978-3-319-52920-2_32

For processing and recognition of 2-D, 3-D and $n$-D retinal images we turn the perceptual spaces into corresponding hypercomplex algebras (and call them perceptual algebras). We give algebraic models for two general levels (retina and VC) of visual systems using different hypercomplex and Clifford algebras. In the algebraic-geometrical approach, each multichannel pixel is considered not as a $K$-D vector, but as a $K$-D hypercomplex number. We will interpret multichannel retinal images as multiplet-valued signals

$$\mathbf{f}(\mathbf{x}) = (f_0(\mathbf{x}), f_1(\mathbf{x}), ..., f_{K-1}(\mathbf{x})) = f_0(\mathbf{x})1 + f_1(\mathbf{x})\epsilon^1 + ... + f_{K-1}(\mathbf{x})\epsilon^{K-1} \quad (1)$$

which take values in the so called multiplet (visual, perceptual) algebra $Alg_k^{Vis} = Alg_k^{Vis}\left(\mathbf{R} \mid 1, \varepsilon^1, \varepsilon^2, ..., \varepsilon^{k-1}\right)$ where $\varepsilon^k = -1, 0, +1$ and $1, \varepsilon^1, ..., \varepsilon^{K-1}$ are hyper-imaginary (multicolor) units with the commutative multiplication rule $\varepsilon^s \cdot \varepsilon^r = \varepsilon^r \cdot \varepsilon^s = \varepsilon^{r\oplus s(modK)}$. In this context, the full machinery of ordinary grey-level signal processing theory can be transposed in multichannel image processing theory.

## 2    Orthounitary Transforms for Color Images

Classical Fourier analysis based on orthogonal and unitary transforms plays an important role in digital image processing. Transforms, notable the classical Discrete Fourier Transform (DFT), are extensively used in digital image processing (filtering, power spectrum estimation and so on). A natural question that arises in our approach is the definition of color and multicolor transforms that can be used efficiently in color and multicolor image processing. We propose a wide library of so-called *orthounitary (color-valued or multicolor-valued) Fourier transforms* for using in image compression, processing and pattern recognition applications.

2-D discrete color $(N \times N)$-image $\mathbf{f} := [\mathbf{f}(i,j)]_{i,j=1}^{N}$ can be defined as a 2-D $(N \times N)$-array in the (R,G,B) or (LC) formats

$$\mathbf{f}(i,j) = f_R(i,j)1 + f_G(i,j)\varepsilon + f_B(i,j)\varepsilon^2 : \mathbf{Z}_N^2 \to Alg_k^{Vis}\left(\mathbf{R} \mid 1, \varepsilon^1, \varepsilon^2, \right),$$
$$\mathbf{f}(i,j) = f_{lu}(i,j)\mathbf{e}_{lu} + \mathbf{f}_{ch}(i,j)\mathbf{E}_{ch} = (f_{lu}(i,j), \mathbf{f}_{ch}(i,j)) : \mathbf{Z}_N^2 \to \mathbf{R} \oplus \mathbf{C} \quad (2)$$

where $\mathbf{Z}_N^2$ is 2-D $(N \times N)$-array. Here, every color pixel $\mathbf{f}(i,j)$ at position $(i,j)$ is a triplet number in (R,G,B)- or in LC-formats, respectively. All images $\mathbf{f} := [\mathbf{f}(i,j)]_{i,j=1}^{N}$ from $N^2$-dimension space over the triplet algebra: $\left(Alg_3^{Vis}\right)^{N^2}$. We say the operator $L_{2D} : \left(Alg_3^{Vis}\right)^{N^2} \to \left(Alg_3^{Vis}\right)^{N^2}$ (or $L_{2D}[\mathbf{f}_{col}] = \mathbf{F}_{col}$) is ortho-unitary if it conserves the norm of color images. It should be noted that orthogonal transforms keep the norm of real-valued (gray-level) images, unitary transforms keep the norm of complex-valued (bichromatic) images. For this reason, ortho-unitary transforms are a generalization of orthogonal and unitary transforms for color images.

In LC format ortho-unitary transforms can be constructed with help an orthogonal $O_{2D}$ and unitary $U_{2D}$ transforms $L_{2D} = O_{2D}\mathbf{e}_{lu} + U_{2D}\mathbf{E}_{ch} = (O_{2D}, U_{2D})$.

The simplest form of ortho-unitary transform for image processing is a separable 2-D transform formed from two 1-D transforms by tensor product $L_{2D} = (O_1 \otimes O_2)\mathbf{e}_{lu} + (U_1 \otimes U_2)\mathbf{E}_{ch}$, where $\otimes$ is the symbol of tensor product. Using separable transforms reduces the problem of designing efficient ortho-unitary 2-D transforms to a one-dimensional problem. It is possible to use one pair of orthogonal and unitary transforms, when $O_1 = O_2 = O$ and $U_1 = U_2 = U$. Every pair $(O, U)$ of an orthogonal $O$ an unitary $U$ transforms generates ortho-unitary (triplet-valued) 1D $L_{1D} = O\mathbf{e}_{lu} + U\mathbf{E}_{ch}$ and 2-D ortho-unitary transforms: $L_{2D} = (O \otimes O)\mathbf{e}_{lu} + (U \otimes U)\mathbf{E}_{ch}$, where $\otimes$. They are *Centaurus* of orthogonal and unitary transforms. It is known that *Centaurus* is a combination of half-men and half-horse. By this reason we can called an ortho-unitary (color) transform as a *Centaurus* of orthogonal and unitary transforms. For example, the Table 1 shows some Centuaruses:

**Table 1.** *"Centauruses"* of orthogonal and unitary transforms

| | $F$ | $\dot{W}$ | $\dot{H}d$ | $\dot{W}v$ |
|---|---|---|---|---|
| $W$ | $W \cdot \mathbf{e}_{lu} + F \cdot \mathbf{E}_{ch}$ | $W \cdot \mathbf{e}_{lu} + \dot{W} \cdot \mathbf{E}_{ch}$ | $W \cdot \mathbf{e}_{lu} + \dot{H}d \cdot \mathbf{E}_{ch}$ | $W \cdot \mathbf{e}_{lu} + \dot{W}v \cdot \mathbf{E}_{ch}$ |
| $Hd$ | $Hd \cdot \mathbf{e}_{lu} + F \cdot \mathbf{E}_{ch}$ | $Hd \cdot \mathbf{e}_{lu} + \dot{W} \cdot \mathbf{E}_{ch}$ | $Hd \cdot \mathbf{e}_{lu} + \dot{H}d \cdot \mathbf{E}_{ch}$ | $Hd \cdot \mathbf{e}_{lu} + \dot{W}v \cdot \mathbf{E}_{ch}$ |
| $Ht$ | $Ht \cdot \mathbf{e}_{lu} + F \cdot \mathbf{E}_{ch}$ | $Ht \cdot \mathbf{e}_{lu} + \dot{W} \cdot \mathbf{E}_{ch}$ | $Ht \cdot \mathbf{e}_{lu} + \dot{H}d \cdot \mathbf{E}_{ch}$ | $Ht \cdot \mathbf{e}_{lu} + \dot{W}v \cdot \mathbf{E}_{ch}$ |
| $Hr$ | $Hr \cdot \mathbf{e}_{lu} + F \cdot \mathbf{E}_{ch}$ | $Hr \cdot \mathbf{e}_{lu} + \dot{W} \cdot \mathbf{E}_{ch}$ | $Hr \cdot \mathbf{e}_{lu} + \dot{H}d \cdot \mathbf{E}_{ch}$ | $Hr \cdot \mathbf{e}_{lu} + \dot{W}v \cdot \mathbf{E}_{ch}$ |
| $Wv$ | $Wv \cdot \mathbf{e}_{lu} + F \cdot \mathbf{E}_{ch}$ | $Wv \cdot \mathbf{e}_{lu} + \dot{W} \cdot \mathbf{E}_{ch}$ | $Wv \cdot \mathbf{e}_{lu} + \dot{H}d \cdot \mathbf{E}_{ch}$ | $Wv \cdot \mathbf{e}_{lu} + \dot{W}v \cdot \mathbf{E}_{ch}$ |

Here $W, Hd, Ht, Hr, Wv$ are Walsh, Hadamard, Hartley, Haar, and Wavelet orthogonal transforms, respectively, and $F, \dot{W}, \dot{H}d, \dot{W}v$ are Fourier, complex Walsh, complex Hadamard and complex wavelet transforms, respectively.

If $O = [\varphi_k(n)]_{k,n=0}^{N-1}$ and $U = [\psi_k(n)]_{k,n=0}^{N-1}$ are an orthogonal and unitary transforms with real-valued and complex-valued basis functions $\{\varphi_k(n)\}_{k,n=0}^{N-1}$ and $\{\psi_k(n)\}_{k,n=0}^{N-1}$, respectively, then

$$(O \otimes O)\mathbf{e}_{lu} + (U \otimes U)\mathbf{E}_{ch}$$
$$= ([\varphi_{k_1}(n_1)] \otimes [\varphi_{k_2}(n_2)])\,\mathbf{e}_{lu} + ([\psi_{k_1}(n_1)] \otimes [\psi_{k_2}(n_2)])\,\mathbf{E}_{ch}$$
$$= [\varphi_{k_1}(n_1)\varphi_{k_2}(n_2)]\,\mathbf{e}_{lu} + [\psi_{k_1}(n_1)\psi_{k_2}(n_2)]\,\mathbf{E}_{ch}, \tag{3}$$

is orthounitary (color) transform, where $\{\varphi_{k_1}(n_1)\varphi_{k_2}(n_2)\}_{k_1,k_2=0\,n_1,n_2=0}^{N-1\quad N-1}$ and $\{\psi_{k_1}(n_1)\psi_{k_2}(n_2)\}_{k_1,k_2=0\,n_1,n_2=0}^{N-1\quad N-1}$ are $N^2$ orthogonal and unitary basis $(N \times N)$-pictures. We are going to collect them in the form of *iconostas* (in a Russian orthodox church, the "Iconostas" is literally the "Stand of Icons" that rise up at the front of the Sanctuary). For example, Iconostasis of Walsh, Cosine and Haar transforms shown in Fig. 1.

Some examples of Iconostasis of Centaurus transforms shown in Fig. 2.

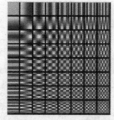

| (a)  Walsh basis of 2-D 8x8-functions | (b) Cosine basis 2-D 8x8-functions | (c) Haarh basis of 2-D 8x8-functions |

**Fig. 1.** *Iconostasis* of 2D basis functions of orthogonal transforms

## 3    Orthounitary Wavelets

Let $\psi^R(x)$ a real–valued mother wavelet and $\psi^R_{s,\tau}(x)$ its scaled and shifted versions

$$\psi^R_{s,\tau}(x) = \frac{1}{\sqrt{|s|}}\psi^R\left(\frac{x-\tau}{s}\right), \; s,\tau \in R, \; s \neq 0 \tag{4}$$

form the orthogonal basis of the space $L_2(\mathbf{R})$. We define the complex–valued (chromatic) wavelet as an analytic signal:

$$\begin{aligned}
\psi^{Ch}_{s,\tau}(x) &= \psi^{Re}_{s,\tau}(x) + jH\left\{\psi^{Re}_{s,\tau}(x)\right\} = \psi^{Re}_{s,\tau}(x) + i\psi^{Im}_{s,\tau}(x) \\
&= \frac{1}{\sqrt{|s|}}\psi^{Re}\left(\frac{x-\tau}{s}\right) + i\frac{1}{\sqrt{|s|}}\psi^{Im}\left(\frac{x-\tau}{s}\right),
\end{aligned} \tag{5}$$

| (a) Color Walsh-Complex Haar basis of 2-D 2x2-functions | (b) Color Walsh-Complex Walsh basis 2-D 2x2-functions | (c) Color Walsh-Complex Walsh basis 2-D 2x2-functions |

**Fig. 2.** *Iconostasis* of 2-D basis functions of ortho-unitary transforms

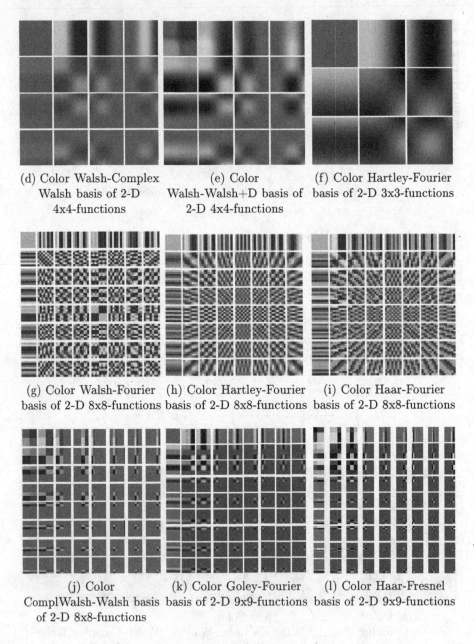

(d) Color Walsh-Complex Walsh basis of 2-D 4x4-functions    (e) Color Walsh-Walsh+D basis of 2-D 4x4-functions    (f) Color Hartley-Fourier basis of 2-D 3x3-functions

(g) Color Walsh-Fourier basis of 2-D 8x8-functions    (h) Color Hartley-Fourier basis of 2-D 8x8-functions    (i) Color Haar-Fourier basis of 2-D 8x8-functions

(j) Color ComplWalsh-Walsh basis of 2-D 8x8-functions    (k) Color Goley-Fourier basis of 2-D 9x9-functions    (l) Color Haar-Fresnel basis of 2-D 9x9-functions

**Fig. 2.** (*continued*)

where $\psi_{s,\tau}^{Im}(x) := H\left\{\psi_{s,\tau}^{Re}(x)\right\}$ is the Hilbert transform of the real–valued mother wavelet (4). Now we construct triplet–valued (color) wavelet basis by

$$\psi_{s,\tau}^{Col}(x) = \varphi_{s,\tau}^{lu}(x) \cdot \mathbf{e}_{lu} + \psi_{s,\tau}^{Ch}(x) \cdot \mathbf{E}_{ch}$$
$$= \varphi_{s,\tau}^{lu}(x) \cdot \mathbf{e}_{lu} + \left[\psi_{s,\tau}^{Re}(x) + jH\left\{\psi_{s,\tau}^{Re}(x)\right\}\right] \cdot \mathbf{E}_{ch}, \qquad (6)$$

where $\varphi_{s,\tau}^{lu}(x)$ is a real–valued wavelet basis for luminance terms and $\psi_{s,\tau}^{Ch}(x)$ is a complex–valued wavelet basis for chromatic term. We can take $\varphi_{s,\tau}^{lu}(x) \equiv \psi_{s,\tau}^{Re}(x)$ In this case we obtain a color wavelet generated by a single real-valued wavelet $\psi_{s,\tau}^{Re}(x)$:

$$\begin{aligned}\psi_{s,\tau}^{Col}(x) &= \psi_{s,\tau}^{Re}(x) \cdot \left[\mathbf{e}_{lu} + jH\left\{\psi_{s,\tau}^{Re}(x)\right\}\right] \cdot \mathbf{E}_{ch} \\ &= \psi_{s,\tau}^{Re}(x) \cdot \left[\mathbf{e}_{lu} + \mathbf{E}_{ch}\right] + jH\left\{\psi_{s,\tau}^{Re}(x)\right\} \cdot \mathbf{E}_{ch}.\end{aligned} \qquad (7)$$

We define 2D direct triplet–valued (color) *continuous orthounitary wavelet transform* (COUT) by

$$F_{COUT}^{col}(s_1,\tau_1,s_2,\tau_2) = \frac{1}{\sqrt{|s_1||s_2|}} \int\limits_{-\infty}^{+\infty}\int\limits_{-\infty}^{+\infty} \mathbf{f}_{col}(x,y)\psi_{s,\tau}^{Col}(x)\psi_{s,\tau}^{Col}(y)dxdy \qquad (8)$$

Examples of chromatic and triplet–valued wavelets are shown in Fig. 3. They are *Centaurus* of orthogonal and unitary wavelets.

## 4   Orthounitary Splines

Similarly to color wavelets, it is possible to construct color splines. Let $Spl(x)$ be a real–valued spline. We define the complex–valued (chromatic) spline as an analytic signal:

$$Spl^{Ch}(x) = Spl^{Re}(x) + jH\left\{Spl^{Re}(x)\right\} = Spl^{Re}(x) + iSpl^{Im}(x), \qquad (9)$$

where $Spl^{Im}(x) = H\left\{Spl^{Re}(x)\right\}$ is the Hilbert transform of the real–valued spline. Now we construct triplet–valued (color) wavelet basis by

$$\begin{aligned}Spl^{Col}(x) &= Spl^{lu}(x) \cdot \mathbf{e}_{lu} + Spl^{Ch} \cdot \mathbf{E}_{ch} \\ &= Spl_{\tau}^{lu}(x) + \left[Spl^{Re}(x) + jH\left\{Spl^{Re}(x)\right\}\right] \cdot \mathbf{E}_{ch}\end{aligned} \qquad (10)$$

where $Spl^{lu}(x)$ is a real–valued wavelet basis for luminance terms and $Spl^{Ch}(x)$ is a complex–valued wavelet basis for chromatic term. We can take $Spl^{lu}(x) \equiv Spl^{Re}(x)$. In this case we have

$$\begin{aligned}Spl^{Col}(x) &= Spl^{Re}(x) \cdot \mathbf{e}_{lu} + \left[Spl^{Re}(x) + jH\left\{Spl^{Re}(x)\right\}\right] \cdot \mathbf{E}_{ch} \\ &= Spl^{Re}(x) \cdot \left[\mathbf{e}_{lu} + \mathbf{E}_{ch}\right] + jH\left\{Spl^{Re}(x)\right\} \cdot \mathbf{E}_{ch}\end{aligned} \qquad (11)$$

Let $BSpl(x)$ be, for example, a B–spline. B–splines are symmetrical, bell shaped functions constructed from a rectangular pulse:

$$BSpl_0(x) = \begin{cases} 1, & -1/2 < x < 1/2, \\ 1/2, & |x| = 1/2, \\ 0, & \text{otherwise} \end{cases} \qquad (12)$$

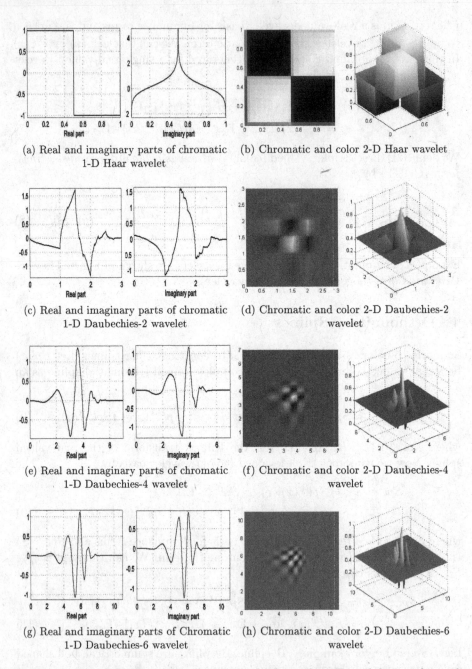

(a) Real and imaginary parts of chromatic 1-D Haar wavelet

(b) Chromatic and color 2-D Haar wavelet

(c) Real and imaginary parts of chromatic 1-D Daubechies-2 wavelet

(d) Chromatic and color 2-D Daubechies-2 wavelet

(e) Real and imaginary parts of chromatic 1-D Daubechies-4 wavelet

(f) Chromatic and color 2-D Daubechies-4 wavelet

(g) Real and imaginary parts of Chromatic 1-D Daubechies-6 wavelet

(h) Chromatic and color 2-D Daubechies-6 wavelet

**Fig. 3.** Examples of chromatic and triplet–valued wavelets. From left to right: real part of chromatic wavelet, imaginary part of chromatic wavelet, chromatic wavelet as a function of two spatial coordinates, triplet-valued wavelet as a function of two spatial coordinates (third axis is intensity, surface is coloured according to chromatic component values)

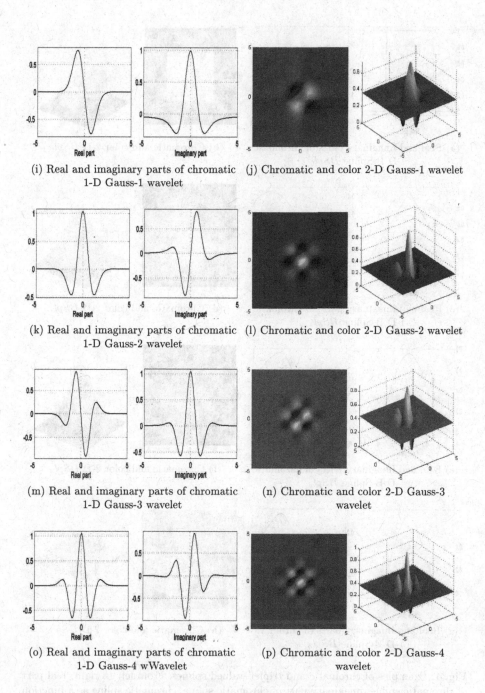

(i) Real and imaginary parts of chromatic 1-D Gauss-1 wavelet

(j) Chromatic and color 2-D Gauss-1 wavelet

(k) Real and imaginary parts of chromatic 1-D Gauss-2 wavelet

(l) Chromatic and color 2-D Gauss-2 wavelet

(m) Real and imaginary parts of chromatic 1-D Gauss-3 wavelet

(n) Chromatic and color 2-D Gauss-3 wavelet

(o) Real and imaginary parts of chromatic 1-D Gauss-4 wWavelet

(p) Chromatic and color 2-D Gauss-4 wavelet

**Fig. 3.** (*continued*)

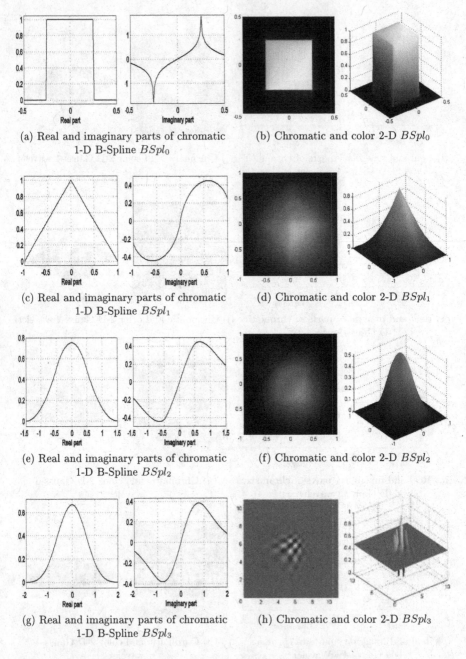

(a) Real and imaginary parts of chromatic 1-D B-Spline $BSpl_0$

(b) Chromatic and color 2-D $BSpl_0$

(c) Real and imaginary parts of chromatic 1-D B-Spline $BSpl_1$

(d) Chromatic and color 2-D $BSpl_1$

(e) Real and imaginary parts of chromatic 1-D B-Spline $BSpl_2$

(f) Chromatic and color 2-D $BSpl_2$

(g) Real and imaginary parts of chromatic 1-D B-Spline $BSpl_3$

(h) Chromatic and color 2-D $BSpl_3$

**Fig. 4.** Examples of chromatic and triplet–valued splines. From left to right: real part of chromatic spline, imaginary part of chromatic spline, chromatic spline as a function of two spatial coordinates, triplet-valued spline as a function of two spatial coordinates (third axis is intensity, surface is coloured according to chromatic component values)

by $BSpl_n(x) = (Bspl_{n-1} * Bspl_0)(x)$, where $*$ is the symbol of convolution. According to (12) the color B–spline has the following form:

$$BSpl_n^{Col}(x)$$
$$= BSpl_n^{Re}(x) \cdot \mathbf{e}_{lu} + \left[BSpl_n^{Re}(x) + jH\left\{BSpl_n^{Re}(x)\right\}\right] \cdot \mathbf{E}_{ch}$$
$$= BSpl_n^{Re}(x) \cdot \left[\mathbf{e}_{lu} + \mathbf{E}_{ch}\right] + jH\left\{BSpl_n^{Re}(x)\right\} \cdot \mathbf{E}_{ch} \tag{13}$$

Examples of chromatic and triplet–valued splines are shown in Fig. 4. They are *Centauruses* of real-valued and complex-valued splines.

## 5   Edge Detection

One of the primary applications of this work could be in edge detection and color image compression. For the edge detection, we convolve the color $(3 \times 3)$ masks $\mathbf{M}_{col}(i, j)$ with color image $\mathbf{f}_{col}(i, j)$ We use *color Prewitt's-like* masks for detection of horizontal, vertical, and diagonal edges. As entries instead of real numbers these masks have triplet numbers:

$$\mathbf{M}_{col}^H = \begin{bmatrix} 1 & \varepsilon & \varepsilon^2 \\ 0 & 0 & 0 \\ -1 & -\varepsilon & -\varepsilon^2 \end{bmatrix}, \mathbf{M}_{col}^V = \begin{bmatrix} 1 & 0 & -1 \\ \varepsilon & 0 & -\varepsilon \\ \varepsilon^2 & 0 & -\varepsilon^2 \end{bmatrix},$$

$$\mathbf{M}_{col}^{LD} = \begin{bmatrix} \varepsilon & \varepsilon^2 & 0 \\ 1 & 0 & -\varepsilon^2 \\ 0 & -1 & -\varepsilon \end{bmatrix}, \mathbf{M}_{col}^{RD} = \begin{bmatrix} 0 & 1 & \varepsilon \\ -1 & 0 & \varepsilon^2 \\ -\varepsilon & -\varepsilon^2 & 0 \end{bmatrix}.$$

We see that triplet color detector is realized without multiplications. Figure 5 shows result of color edge detecting

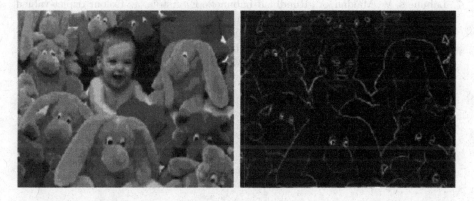

**Fig. 5.** Color edge detector. Left: original image, right:detected edges.

# 6    Conclusion

We developed a novel algebraic approach based on hypercomplex algebras to algebraic models of color, multicolor and hyperspectral images. It is our aim to show that the use of hypercomplex algebras fits more naturally to the tasks of recognition of multicolor patterns than does the use of color vector spaces [1], [2], [3], [4, Chap. 25], [5], [6], [7], [8].

# References

1. Greaves, C.: On algebraic triplets. Proc. Irisn Acad. **3**, 51–54 (1847). 57–64, 80–84, 105–108
2. Ekaterina, L.-R., Nikitin, I., Labunets, V.: Unified approach to fourier-clifford-prometheus sequences, transforms and filter banks. In: Byrnes, J. (ed.) Computational Noncommutative Algebra and Applications. NAII, vol. 136, pp. 389–400. Springer, Dordrecht (2003)
3. Ekaterina, L.-R., Maidan, A., Novak, P.: Color wavelet-haar-prometheus transforms for image processing. In: Computational Noncommutative Algebra and Applications (2003)
4. Labunets, V., Rundblad, E., Astola, J.: Is the brain a 'clifford algebra quantum computer'? In: Dorst, L., Doran, C., Lasenby, J. (eds.) Applied Geometrical Algebras in Computer Science and Engineering, pp. 486–495. Birkhäuser Boston, New York (2002)
5. Labunets-Rundblad, E., Labunets, V., Astola, J.: Is the visual cortex a "fast clifford algebra quantum computer"? In: Brackx, F., Chisholm, J.S.R., Souĉek, V. (eds.) Clifford Analysis and Its Applications, NATO Science Series II: Mathematics, Physics and Chemistry, vol. 25, pp. 173–183. Springer, Dordrecht (2001)
6. Labunets, V., Maidan, A., Rundblad-Labunets, E. Astola, J.: Colour triplet-valued wavelets and splines. In: Image and Signal Processing ana Analysis, ISPA 2001, pp. 535–541 (2001)
7. Labunets, V., Maidan, A., Rundblad-Labunets, E., Astola, J.: Colour triplet-valued wavelets, splines and median filters. In: Spectral Methods and Multirate Signal Processing, SMMSP 2001, pp. 61–70 (2001)
8. Labunets-Rundblad, E.: Fast fourier-clifford transforms design and application in invariant recognition. In: Spectral Methods and Multirate Signal Processing, SMMSP 2001, p. 265 (2000)

# A Real-Time Algorithm for Mobile Robot Mapping Based on Rotation-Invariant Descriptors and Iterative Close Point Algorithm

A. Vokhmintcev[1](✉) and K. Yakovlev[2]

[1] Research Laboratory, Chelyabinsk State University, Chelyabinsk, Russia
vav@csu.ru
[2] Computer Science and Control of Russian Academy of Sciences,
National Research University Higher School of Economics, Moscow, Russia
yakovlev@isa.ru

**Abstract.** Nowadays many algorithms for mobile robot mapping in indoor environments have been created. In this work we use a Kinect 2.0 camera, a visible range cameras Beward B2720 and an infrared camera Flir Tau 2 for building 3D dense maps of indoor environments. We present the RGB-D Mapping and a new fusion algorithm combining visual features and depth information for matching images, aligning of 3D point clouds, a "loop-closure" detection, pose graph optimization to build global consistent 3D maps. Such 3D maps of environments have various applications in robot navigation, real-time tracking, non-cooperative remote surveillance, face recognition, semantic mapping. The performance and computational complexity of the proposed RGB-D Mapping algorithm in real indoor environments is presented and discussed.

**Keywords:** Fusion · Simultaneous location and mapping · Iterative closest point algorithm · Matching algorithm · Histograms of oriented gradients · Depth map

## 1 Introduction

Over the last decade numerous methods of SLAM (Simultaneous Location and Mapping) have been suggested [1, 2]. There are many algorithms for simultaneous navigation and mapping which are based on the following approaches: Fast SLAM; Mono-SLAM, Toro-SLAM, Extended Kalman filter, Graph-SLAM, Visual SLAM. The existing projects SLAM provide research-based software platforms, among which are the following successful commercial realizations: HOG-Man, ORB-SLAM, TORO-SLAM, RGBD-Slam, EKF-Mono-SLAM. Here are the basic research projects in the field of SLAM:

- Stanley (DARPA, Stanford Artificial Intelligence Lab, the Stanford Racing Team, Volkswagen Electronics Research Laboratory);
- Robots Podcast: 3D SLAM (Department of Defense Research and Development Laboratory, Massachusetts Institute of Technology (MIT) Lincoln Laboratory);

© Springer International Publishing AG 2017
D.I. Ignatov et al. (Eds.): AIST 2016, CCIS 661, pp. 357–369, 2017.
DOI: 10.1007/978-3-319-52920-2_33

- Autonomous and Perceptive Systems research (University of Groningen);
- Active 3-D Visual SLAM (General Electric Company);
- Recognition of Non-cooperative Individuals based on SLAM technology (University of Southern California);
- SLAM using vision (Imperial College London);
- DARPA Virtual Robot Challenge (TRACLabs);
- Adaptive Visual SLAM (University of Notre Dame).

Since multisensory cameras, such as a visible range camera, an infrared camera, a RGB-D camera, provide visual and depth information, a data fusion from two type sources are useful. The most successful systems for mobile robot mapping in large maps environments are based on the metric-topological and visual approach (Visual SLAM). The robotic navigation problem in this case can be presented by the following steps. First, the local mapping using metrics to assess the trajectory; second, recognition of places to solve the problem "loop closure detection"; finally, global optimization and pose graph optimization. Some successful real-time systems for 3D dense map building have been proposed, using monocular cameras [3] and multi-view stereo approach [4]. In order to solve the data association problem more robustly, sparse features are extracted from RGB images [5–7]. Projective geometry is used to define the spatial relationship between features [8, 9]. A scheme of the proposed fusion algorithm is represented on Fig. 1.

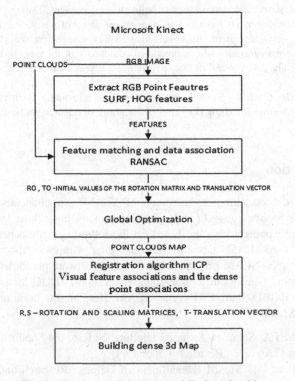

**Fig. 1.** A scheme of the proposed Fusion algorithm.

The RGB-D Mapping algorithm consists of three components: spatial alignment of consecutive data frames, detection of loop-closures, and globally consistent alignment of the data frame sequence. The problem of consistent aligning of 3D point data is known as registration task. In order to solve registration task Iterative Closest Point (ICP) algorithm is often used [10, 11]. In this paper we propose to use visual features to significantly improve the initial point of the ICP algorithm, and the alignment between frames is computed by joint optimization of appearance and shape matches [12], also visual features descriptors, such as SIFT (Scale Invariant Feature Transform) [13] and SURF (Speeded-Up Ro-bust Features) [14] could be useful for loop-closure detection and global localization.

The paper is organized as follows. In Sect. 2, the proposed fast iterative close point and fusion algorithms are presented. In Sect. 3, the loop-closure detection algorithm is considered. Computer simulation results are provided in Sect. 4. Section 5 summarizes our conclusions.

## 2 Affine Transformations and Fusion Algorithm

In this topic the Fusion algorithm based on recursive calculation of oriented gradient histograms over several circular sliding windows and Iterative Close Points is present-ed. The proposed fusion algorithm can be described by the next main steps:

Step 1. Down-sampling the images and down-sampling dense 3D point clouds into voxel grid.

Step 2. Feature extraction in 2D from visible images and thermal images based on rotation descriptors in 2D.

Step 3. Feature matching: nearest neighbors on the feature space, using kd-trees;

Step 4. Alignment of consecutive 3D data frames, calculated with RANSAC [12].

Step 5. Pose graph optimization based on the g2o framework, and detecting loop closures.

Step 6. Point clouds registration and building the global 3D map of the scene.

We compute the histograms of oriented gradients (HOGs) in several circular windows to extract visual features. In paper [15, 16] rotation descriptor was presented in details. The gradient magnitudes $\{Mag_i (x, y): (x, y) \in W_i\}$ and orientation values $\{\varphi i (x, y): (x, y) \in W_i\}$, quantized for Q levels, the histogram of oriented gradients can be computed as follows:

$$HOG_i(\propto) = \begin{cases} \sum_{(x,y) \in W_i} \delta(\alpha - \varphi_i(x,y)), & if \, Mag_i(x,y) \geq Med \\ 0, otherwise \end{cases}, \qquad (1)$$

where $\alpha = \{0, \ldots, Q - 1\}$ are histogram values (bins), Med is the median value inside of the circular window, and $\delta (z) = \{1, if \, z = 0; 0$, otherwise is the Kronecker delta function. The correlation output for the i-th circular window at position k can be computed with the help of the fast Inverse Fourier Transform as follows:

$$C_i^k(\alpha) = IFT\left[\frac{HS_i^k(\omega)HR_i^*(\omega)}{\sqrt{Q\sum_{q=0}^{Q-1}(HOG_i^k(q))^2 - (HS_i^k(0))^2}}\right],$$    (2)

where $HS_i^k(\omega)$ is the Fourier Transform of the histogram of oriented gradients inside of the i-th circular window over the frame fragment, and HRi $(\omega)$ is the Fourier Transform of $\overline{HOG_i^R(\alpha)}$. The correlation peak is a measure of similarity of the two histograms, which can be obtained as follows: $P_i^k = \max_\alpha\{C_i^k(\alpha)\}$. A scheme of the matching algorithm is presented on Fig. 2.

Let $X = \{x_1,\ldots,x_n\}$ be a first point cloud and $Y = \{y,\ldots,y_m\}$ be a second (target) point cloud in $\mathbb{R}^3$. In many works [10, 11] the ICP algorithm is considered as a geometrical transformation for rigid objects mapping $X$ to $Y$:

$$RSx_i + T,$$    (3)

where $R$ is a rotation matrix, $T$ is a translation vector, $i = 1,\ldots,n$, $S$ is a scaling matrix.

Let us consider ICP variational problem for the case of an arbitrary affine transformation. Let $J(A,T)$ be the following function [17]:

$$J(A,T) = \sum_{i=1}^n \|Ax_i + T - y_i\|^2.$$    (4)

where

$$A = \begin{pmatrix} a_{11} & a_{12} & a_{13} \\ a_{21} & a_{22} & a_{23} \\ a_{31} & a_{32} & a_{33} \end{pmatrix}, \ T = \begin{pmatrix} t_1 \\ t_2 \\ t_3 \end{pmatrix}, \ x_i = \begin{pmatrix} x_{1i} \\ x_{2i} \\ x_{3i} \end{pmatrix}, \ y_i = \begin{pmatrix} y_{1i} \\ y_{2i} \\ y_{3i} \end{pmatrix}.$$    (5)

Then ICP variational problem can be seen that

$$\begin{aligned} J(A,t) = \sum_{i=1}^n &(a_{11}x_{1i} + a_{12}x_{2i} + a_{13}x_{3i} + t_1 - y_{1i})^2 \\ &+ (a_{21}x_{1i} + a_{22}x_{2i} + a_{23}x_{3i} + t_2 - y_{2i})^2 \\ &+ (_{31}x_{1i} + a_{32}x_{2i} + a_{33}x_{3i} + t_3 - y_{3i})^2. \end{aligned}$$    (6)

Let new coordinates $x_{ki}, y_{ki}$, be expressed through old coordinates as follows:

$$x_{ki} = x_{ki} - \frac{1}{n}\sum_{j=1}^n x_{kj}, \ y_{ki} = y_{ki} - \frac{1}{n}\sum_{j=1}^n y_{kj}, k = 1,\ldots,3, i = 1,\ldots,n.$$    (7)

The elements of the first row of the matrix $A$ are computed as

$$a_{11} = \frac{\sum_{i=1}^n (y_{1i} - a_{12}x_{2i} - a_{13}x_{3i})x_{1i}}{\sum_{i=1}^n x_{1i}^2},$$    (8)

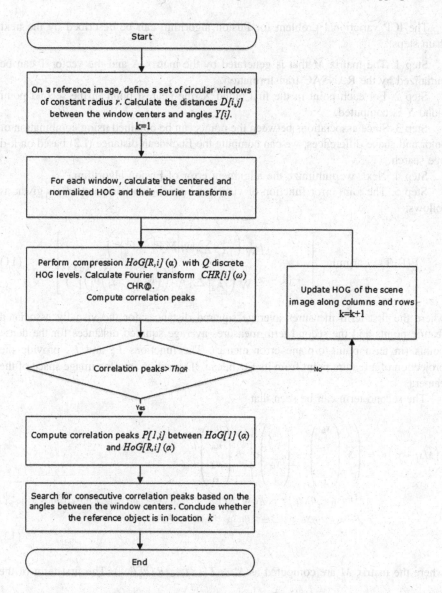

**Fig. 2.** A scheme of the fast matching algorithm.

$$a_{12} = \frac{\sum_{j=1}^{n} \gamma_j \, \alpha_j - a_{13} \sum_{j=1}^{n} \beta_j \, \alpha_j}{\sum_{j=1}^{n} \alpha_j^2},$$    (9)

$$a_{13} = \frac{\sum_{k=1}^{n} \varphi_k \, \psi_k}{\sum_{k=1}^{n} \varphi_k^2}.$$    (10)

The ICP variational problem for Fusion algorithm can be described by the next main steps:

Step 1. The matrix $\widehat{M}$ that is generated by the matrix A and the vector T can be initialized by the RANSAC transformation.

Step 2. For each point in the first cloud X, the nearest point in the second point cloud Y is computed.

Step 3. Since associations between the points can be obtained using combination of color and shape differences, we can compute the Euclidean distance (L2) based on k-d tree search.

Step 4. Next, we minimize the alignment error of Fusion algorithm.

Step 5. The joint error function of visual and point cloud associations is given as follows:

$$
E(R, T) = \underset{\widehat{M}}{argmin} \left[ \begin{array}{c} \alpha \frac{1}{W} \left( \frac{1}{|A_f|} \sum_i w_i \left| \widehat{M} F_s^i - F_d^i \right|^2 \right) \\ + (1 - \alpha) \frac{1}{W} \left( \frac{1}{|A_d|} \sum_j w_j \left| \left( \left( \widehat{M} x_j - y_j \right) n_j \right) \right|^2 \right) \end{array} \right], \tag{11}
$$

where the first term measures average squared distances for the visually associated feature points and the second term measures average squared distances for the dense points (in term point-to-plane error metric). The functions $F_s^i$ and $F_d^i$ provide the projection of a feature point from its Euclidean 3D position into the image space of the camera.

The second term can be seen that

$$
\left( \widehat{M} x_i - y_i \right) n_i = \left( \widehat{M} \begin{pmatrix} x_{ix} \\ y_{iy} \\ y_{iz} \\ 1 \end{pmatrix} - \begin{pmatrix} y_{ix} \\ y_{iy} \\ y_{iz} \\ 1 \end{pmatrix} \right) * \begin{pmatrix} n_{ix} \\ n_{iy} \\ n_{iz} \\ 0 \end{pmatrix}
$$

$$
= \left[ \left( n_{iz} x_{iy} - n_{iy} x_{iz} \right) \alpha + \left( n_{ix} x_{iz} - n_{iz} s_{ix} \right) \beta + \left( n_{iy} x_{ix} - n_{ix} x_{iy} \right) \gamma + n_{ix} t_x + n_{iy} t_y + n_{iz} t_z \right]
$$
$$
- \left[ n_{ix} y_{ix} + n_{iy} y_{iy} + n_{iz} y_{iz} - n_{ix} x_{ix} - n_{iy} x_{iy} - n_{iz} x_{iz} \right]
$$

$$\tag{12}$$

where the matrix $\widehat{M}$ are computed as $\widehat{M} = T \left( t_x, t_y, z \right) \widehat{R} \left( \alpha, \beta, \gamma \right)$. The first term of the variational problem $\alpha \frac{1}{W} \left( \frac{1}{|A_f|} \sum_i w_i \left| \widehat{M} F_s^i - F_d^i \right|^2 \right)$ can be obtained in a similar manner. The projection of a feature point is computed through triangulation. We compute $F_s^i$ and $F_d^i$ using relation between pixel coordinates and global coordinates of pair point clouds. The two terms in Eq. (11) are weighted using a coefficient $\alpha$, which is chosen experimentally based on observation conditions in indoor environment [17, 18].

# 3   Loop-Closure Detection

First, the loop-closure detection is needed to be recognized when a camera has returned to a previously visited location. Second, the map must be corrected to merge duplicate regions. We developed a fast algorithm for computing of the dynamic position and orientation of the robot. For each step, we can determine the displacement of the robot and we can predict the new robot position. We can identify the location of special features on the new frame, and calculate the new position of the robot relating to the special features. We updated the weight of the features and movement trajectory of the robot based on difference between these two robot position estimates. To solve this problem would be an approach based on a vocabulary tree and a "bag of words" [19]. The general algorithm of the loop-closure detection consists the of next steps:

Step 1. Initialing the "loop-closure candidate" database.

Step 2. Searching the most similar previous images based on a vocabulary tree.

Step 3. Prevention of already detected loop when the robot is still at the same location.

Step 4. Determining the similarities of the current image with each resulting candidate based on their BoW vectors.

Step 5. Disregard of images whose similarity is below a specified threshold.

Step 6. Returning up to images with the highest similarity as loop-closure candidates.

Step 7. The current image is added to the database images, so it can later be found as a "loop-closure candidate".

Step 8. Updating the history of camera positions and movements of the robot.

Step 9. The model of the scene is saved as a graph, whose vertices correspond to features in time.

Step 10. Determining the displacement of the robot.

# 4   Computer Simulation

In this section, we presented and discussed the results of computer simulation. We evaluated a RGB-D Mapping in indoor environments: hall of Chelyabinsk State University, a room campus and our research laboratory. Let's define the key constraints for the indoor environments and objects in the scene. All objects in the scene have a rigid form. The mobile robot has a 6D movement (degree of freedom). All objects and details of the discovered scene are obtained using the following sensors: infrared camera Flir Tau 2, Kinect 360 XBox 2.0 and the two visible range of the camera Beward B2720. We haven't an environment plan of the university, and coordinates of the movement of the robotic platform are given in geographic coordinates. We didn't use odometer information for SLAM simulation. All objects in the scene are static. The accuracy of the calculation of the trajectory is determined by the type of sensors and time constraints. We evaluated mobile robot Mapping in the following experiments. The first experiment: the visible range camera was carried by a person in the direction of travel. In this experiment 23 RGB-D key-frames were captured. Then the group of hall images were down-sampled into 320 * 240 resolution. Then point cloud of the hall

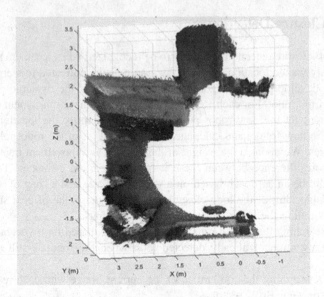

**Fig. 3.** The visualization fusion algorithm: 2D maps. Top view of the office.

of our university was down-samlped 0.1 voxel grid size. The Fig. 3 represented 2D maps built for large loops in environments. The loop consists of 906 frames over a length of 84 m. To assess the consistency of these maps we overlaid our 3D maps onto 2D layouts generated by different means. For clarity, we removed most floor, stairs and ceiling points from the discharged 3D maps. The second experiment: the Kinect 2.0 camera was mounted on a mobile robotic platform (Odyssey 6 Wheel Drive Robot). This platform is a flexible tools-kit which can be mounted from a number of different hardware and software configurations, such as a wheel version, gear-motors, chassis and SLAM frameworks. Then the sequence of RGB-D frames was captured in the room in our campus. In this case the loop consists of 3854 frames over a length of 453 m. The discharged 3D maps were overlaid onto 2D layouts-maps to assess the consistency of these maps. The scene volume of approximately $0.8 \times 0.8 \times 0.6$ was captured with a resolution of $586 \times 586 \times 256$ voxels. The processing time is approximately 48 s real-time operation. The third experiment: the Flir Tau 2.0 camera was mounted on a mobile robotic platform (Odyssey 6). Then the sequence of thermal frames was captured in the research laboratory. In this case the loop consists of 245 frames over a length of 76 m and 34 key-thermal frames. After that thermal frames were merging with 2D layouts-maps to assess the consistency of these maps. The processing time is approximately 11 s real-time operation.

The performance of rotation descriptor was compared with the Scale Invariant Feature Transform, the Speeded-Up Robust Features, the Oriented Fast and Rotated BRIEF algorithm [20]. The rotation descriptors were tested in the following conditions: small scaling, translation in-plane in the scene, small out-plane/in-plane rotations in the scene. The performance was estimated in terms of the number of correct matches and the processing time. Tables 1, 2, 3, 4, 5 and 6 summarizes the experiment

**Table 1.** Precision of matching (in %) of various algorithms (in-plane rotation) for the experiment with visual range of the camera

| Matching algorithm | Angle of rotation (in degress) | | | | | |
|---|---|---|---|---|---|---|
| | 45° | 90° | 135° | 180° | 225° | 270° |
| Scale Invariant Feature Transform | 100 | 98 | 99 | 98 | 97 | 95 |
| Speeded-Up Robust Features | 74 | 69 | 74 | 69 | 74 | 69 |
| Oriented FAST and Rotated BRIEF | 87 | 85 | 83 | 86 | 85 | 87 |
| Fusion algorithm | 100 | 98 | 96 | 98 | 97 | 95 |

**Table 2.** Precision of matching (in %) of various algorithms (small out-plane rotation) for the experiment with visual range of the camera

| Matching algorithm | Angle of rotation (in degress) | | | | | |
|---|---|---|---|---|---|---|
| | 5° | 10° | 15° | 20° | 25° | 30° |
| Scale Invariant Feature Transform | 98 | 91 | 78 | 64 | 58 | 47 |
| Speeded-Up Robust Features | 82 | 77 | 64 | 55 | 38 | 29 |
| Oriented FAST and Rotated BRIEF | 96 | 84 | 77 | 61 | 58 | 54 |
| Fusion algorithm | 83 | 78 | 74 | 72 | 76 | 72 |

**Table 3.** Precision of matching (in %) of various algorithms (small scaling) for the experiment with visual range of the camera

| Matching algorithm | Scaling | | | | |
|---|---|---|---|---|---|
| | 0.8X | 0.9X | 1.0 X | 1.1X | 1.2 X |
| Scale Invariant Feature Transform | 92 | 95 | 100 | 98 | 91 |
| Speeded-Up Robust Features | 79 | 90 | 99 | 97 | 92 |
| Oriented FAST and Rotated BRIEF | 78 | 79 | 90 | 83 | 89 |
| Fusion algorithm | 84 | 94 | 100 | 99 | 91 |

**Table 4.** Precision of matching (in %) of various algorithms (in-plane rotation) for the experiment with infrared camera Flir Tau 2

| Matching algorithm | Angle of rotation (in degress) | | | | | |
|---|---|---|---|---|---|---|
| | 45° | 90° | 135° | 180° | 225° | 270° |
| Scale Invariant Feature Transform | 100 | 100 | 99 | 98 | 97 | 97 |
| Oriented FAST and Rotated BRIEF | 91 | 85 | 83 | 87 | 83 | 89 |
| Fusion algorithm | 100 | 100 | 98 | 98 | 97 | 97 |

results obtained for visual range of the camera Beward B2720 and the infrared camera Flir 2 Tau2. The Fusion algorithm, namely rotation descriptor, gives more stable matching performance than other algorithms for small out-of-plane rotations (Tables 2 and 5) and small scaling (Tables 3 and 6).

**Table 5.** Precision of matching (in %) of various algorithms (small out-plane rotation) for the experiment with infrared camera Flir Tau 2

| Matching algorithm | Angle of rotation (in degress) | | | | | |
|---|---|---|---|---|---|---|
| | 5° | 10° | 15° | 20° | 25° | 30° |
| Scale Invariant Feature Transform | 90 | 89 | 83 | 68 | 61 | 54 |
| Oriented FAST and Rotated BRIEF | 93 | 87 | 79 | 64 | 62 | 58 |
| Fusion algorithm | 89 | 84 | 78 | 76 | 82 | 79 |

**Table 6.** Precision of matching (in %) of various algorithms (small scaling) for the experiment with infrared camera Flir Tau 2

| Matching algorithm | Scaling | | | | |
|---|---|---|---|---|---|
| | 0.8X | 0.9X | 1.0 X | 1.1X | 1.2 X |
| Scale Invariant Feature Transform | 94 | 93 | 100 | 99 | 95 |
| Oriented FAST and Rotated BRIEF | 82 | 86 | 91 | 85 | 95 |
| Fusion algorithm | 99 | 98 | 100 | 99 | 98 |

**Table 7.** Matching precision (%) and processing time (sec) for the experiment with Kinect camera.

| Registration algorithm | Rotation angle (degrees) | | | | |
|---|---|---|---|---|---|
| | 5 | 10 | 15 | 20 | 25 |
| Fast Iterative Close Point | 91/0,06 | 80/0,1 | 73/0,1 | 66/0,13 | 56/0,15 |
| Iterative Close Point (point-to-plane metric) | 97/1,56 | 95/1,86 | 91/2,3 | 86/3,8 | 81/4,17 |
| Fusion algorithm | 98/1,45 | 99/1,78 | 98/2,4 | 96/2,95 | 91/3,2 |

The processing time of the rotation descriptor algorithm and Oriented FAST/Rotated BRIEF is about 1.7 s. The processing time of the SIFT algorithms is 9.82 s. The processing time of the SURF is 0.95 s. Then, in this section the proposed algorithm in Eq. (11) is compared with the Fast ICP [10] and ICP [11] algorithms. The performance of the Fusion algorithm is shown in Table 7 comparing with that of the tested algorithms. Each row in Table 7 has two values: the first value is matching precision (%), the second value is processing time in sec. Figure 4 presents the performance of the tested algorithms in terms of mean square error. One can observe that the Fusion algorithm has the best performance.

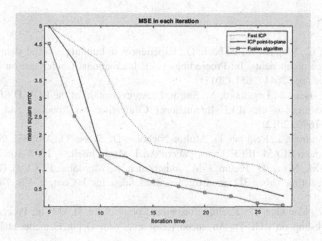

**Fig. 4.** The performance of the algorithms in terms of mean square error.

## 5    Conclusions

In this research the following tasks were solved. That is, first, the new fusion method for building dynamic three-dimensional maps in indoor environments based on multisensory information and algorithms for dynamic image matching and point cloud registration was suggested; second, robust navigation methods for mobile devices (robots) based on dynamic three-dimensional environment maps were developed; finally, the effective real-time Visual SLAM algorithm for Mobile Robot Mapping was proposed for personal computers with graphics processors and using parallel programming technologies. The obtained experimental results were represented with the help of computer and real simulation. The suggested Visual SLAM algorithm has been tested in indoor environments with a limited number of obstacles and without distortion (uneven lighting scenes, geometric matrix receivers noise, impulse noise cluster, linear spatially inhomogeneous interference). The proposed Visual SLAM algorithm has the better performance in terms of accuracy than the known Visual SLAM algorithms. The following tasks will be solved in the future in the framework of the project: the development of a new robust method of comparing rigid objects ESM (Efficient Second-order Minimization); the algorithm for computing the dynamic position and orientation of the robot based on the Kalman filtering options (EKF- Extended Kalman Filter, UKF - Unscented Kalman Filter, EIF - Extended Information Filter).

**Acknolwledgments.** The work was supported by the RFBR, project no 16-08-00342 and the Ministry of Education and Science of Russian Federation, grant no.2.1766.2014.

# References

1. Hertzberg, C., Wagner, R., Birbach, O.: Experiences in building a visual slam system from open source components. In: Proceedings IEEE International Conference on Robotics and Automation, pp. 2644–2651 (2011)
2. Endres, F., Hess, J., Engelhard, N., Sturm, J.: An evaluation of the RGB-D SLAM system. In: Proceedings of the IEEE International Conference on Robotics and Automation, pp. 1691–1696 (2012)
3. Davison Andrew, J., Reid Ian, D., Molton Nicholas, D., Stasse, O.: MonoSLAM Real-Time Single Camera SLAM. IEEE Trans. Pattern Anal. Mach. Intell. **7**, 1052–1067 (2007)
4. Pollefeys, M., Nister, D., Frahm, J.-M., Akbarzadeh, A., Mordohai, P., Clipp, B., Engels, C.: Detailed real-time Urban 3D reconstruction from video. Int. J. Comput. Vis. **78**(2), 143–167 (2008)
5. Fioraio, N., Konolige, K.: Realtime visual and point cloud SLAM. In: Proceedings of the RSS Workshop on RGB-D: Advanced Reasoning with Depth Cameras at Robotics, no. 27 (2011)
6. Konolige, K., Agrawal, M., Sola, J.: Large scale visual odometry for rough terrain. In: Proceedings of the International Symposium on Robotics Research, 201–212 (2010)
7. Konolige, K., Agrawal, M., Bolles, R.C., Cowan, C., Fischler, M., Gerkey, B.: Outdoor mapping and navigation using stereo vision. In: Proceedings of the International Symposium on Experimental Robotics, pp. 179–190 (2006)
8. Nister, D.: An efficient solution to the five-point relative pose problem. IEEE Trans. Pattern Anal. Mach. Intell. **26**, 756–777 (2004)
9. Snavely, N., Seitz, S., Szeliski, R.: Photo tourism: exploring photo collections in 3D. Proc. ACM Trans. Graphics **25**(3), 835–846 (2006)
10. Besl, P., McKay, N.: A method for registration of 3-D shapes trans. IEEE Trans. Pattern Anal. Mach. Intell. **14**(2), 239–256 (1992)
11. Chen, Y., Medioni, G.: Object modeling by registration of multiple range images. J. Image Vis. Comput. **10**(3), 145–155 (1992). Elsevier
12. Fischler, M., Bolles, R.: Random sample consensus: a paradigm for model fitting with applications to image analysis and automated cartography. In: Graphics and Image Processing, pp. 381–395 (1981)
13. Lowe, D.G.: Object recognition from local scale invariant features. In: Proceedings of the 7th International conference on Computer Vision, vol. 2, pp. 1150–1157 (1999)
14. Bay, H., Ess, A., Tuytelaars, T., Van Gool, L.: SURF: speeded up robust features. Comput. Vis. Image Underst. **110**, 346–359 (2008)
15. Vokhmintcev, A.V., Sochenkov, I.V., Kuznetsov, V.V., Tikhonkikh, D.V.: Face recognition based on matching algorithm with recursive calculation of local oriented gradient histogram. Dokl. Math. **466**(3), 261–266 (2016)
16. Miramontes-Jaramillo, D., Kober, V., Diaz-Ramirez, V.H., Karnaukhov, V.: A novel image matching algorithm based on sliding histograms of oriented gradients. J. Commun. Technol. Electron. **59**(12), 1446–1450 (2014)
17. Vokhmintsev, A., Makovetskii, A., Kober, V., Sochenkov, I., Kuznetsov, V.: A fusion algorithm for building three-dimensional maps. In: Proceedings. SPIE's Annual Meeting: Applications of Digital Image Processing XXXVIII, vol. 8452, p. 9599-81 (2015)

18. Henry, P., Krainin, M., Herbst, E.: RGB-D mapping: Using depth cameras for dense 3D modeling of indoor environments. In: Proceedings of the 12th International Symposium on Experimental Robotics, pp. 477–491 (2014)
19. Josef, S.: Efficient visual search of videos cast as text retrieval. IEEE Trans. Pattern Anal. Mach. Intell. **31**(4), 591–605 (2009)
20. Rublee, E., Rabaud, V., Konolige, K., Bradski, G.: ORB: an efficient alternative to SIFT or SURF. In: IEEE International Conference Computer Vision (2011)

# Author Index

Printed in the United States
By Bookmasters